Frontiers in Civil Engineering

(Volume 6)

Occupant Behaviour in Buildings: Advances and Challenges

Edited by

Enedir Ghisi
Federal University of Santa Catarina
Department of Civil Engineering
Laboratory of Energy Efficiency in Buildings
Florianópolis-SC
Brazil

Ricardo Forgiarini Rupp
Technical University of Denmark
Department of Civil Engineering
International Centre for Indoor Environment and Energy
Kongens Lyngby
Denmark

&

Pedro Fernandes Pereira
University of Porto
Department of Civil Engineering
Building Physics Laboratory
Portugal

Frontiers in Civil Engineering

Volume # 6.

Occupant Behaviour in Buildings: Advances and Challenges

Editors: Enedir Ghisi, Ricardo Forgiarini Rupp and Pedro Fernandes Pereira

ISSN (Online): 2468-4708

ISSN (Print): 2468-4694

ISBN (Online): 978-1-68108-832-7

ISBN (Print): 978-1-68108-833-4

ISBN (Paperback): 978-1-68108-834-1

Published by Bentham Science Publishers – Sharjah, UAE. All Rights Reserved.

need for a court order if at any point you breach any terms of this License Agreement. In no event will any delay or failure by Bentham Science Publishers in enforcing your compliance with this License Agreement constitute a waiver of any of its rights.

3. You acknowledge that you have read this License Agreement, and agree to be bound by its terms and conditions. To the extent that any other terms and conditions presented on any website of Bentham Science Publishers conflict with, or are inconsistent with, the terms and conditions set out in this License Agreement, you acknowledge that the terms and conditions set out in this License Agreement shall prevail.

Bentham Science Publishers Ltd.
Executive Suite Y - 2
PO Box 7917, Saif Zone
Sharjah, U.A.E.
Email: subscriptions@benthamscience.net

BENTHAM SCIENCE

CONTENTS

PREFACE ii

LIST OF CONTRIBUTORS v

CHAPTER 1 EXPLORING THE POTENTIAL OF COMBINING TECHNOLOGICAL
INNOVATIONS WITH QUALITATIVE METHODS IN OCCUPANT BEHAVIOUR
RESEARCH 1
Mateus V. Bavaresco, Ricardo F. Rupp and *Enedir Ghisi*
    INTRODUCTION 2
    INDOOR ENVIRONMENT, CLIMATE AND OCCUPANT BEHAVIOUR 4
    POTENTIALLY APPLICABLE TECHNOLOGICAL INNOVATIONS 6
        Behavioural Sensing 7
        Wearable Devices 10
        Internet-of-Things 11
        Virtual Reality 12
    QUALITATIVE METHODS TO STUDY OCCUPANT BEHAVIOUR 13
        Questionnaires 14
        Interviews 16
        Personal Diaries 18
        Post-Occupancy Evaluation 19
        Comparison of Qualitative Methods 20
    COMBINATION OF TECHNOLOGICAL INNOVATIONS AND QUALITATIVE
    METHODS 21
    CONCLUDING REMARKS 26
    CONSENT FOR PUBLICATION 26
    CONFLICT OF INTEREST 26
    ACKNOWLEDGEMENTS 26
    REFERENCES 27

CHAPTER 2 MONITORING OCCUPANT WINDOW OPENING BEHAVIOUR IN
BUILDINGS: A CRITICAL REVIEW 38
Shen Wei, Yan Ding and *Wei Yu*
    INTRODUCTION 39
    MONITORING OCCUPANT WINDOW OPENING BEHAVIOUR 40
        Self-Recording Window Usage By Building Occupants 42
        Recording Window Usage by Electronic Measuring Devices 43
        Recording Window Usage by Surveyor Observations 44
        Recording Window Usage by Self-Estimation from Building Occupants 45
        Camera-Based Estimation 46
        Method Comparison 46
            Popularity 46
        Sample Size 53
        Measurement Interval 54
        Measurement Duration 54
        Whether Measuring Window State or Window Opening Angle 55
        Whether Occupants' Window Opening/Closing Action has been Analysed ... 56
    CAPTURING IMPORTANT INFLUENTIAL FACTORS 56
        Outdoor Environmental Factors 57
        Indoor Environmental Factors 58
        Building- and System-Related Factors and Occupant-Related Factors 60
        Time-Dependent Factors and Other Factors 61

CONCLUSION 61
CONSENT FOR PUBLICATION 63
CONFLICT OF INTEREST 63
ACKNOWLEDGEMENT 63
REFERENCES 63

CHAPTER 3 SUPPORTING THE DECISION-MAKING PROCESS OF BUILDING USERS IN THE SELECTION OF ENERGY-EFFICIENT HEATING SOLUTIONS BY IDENTIFYING AND EVALUATING CO-BENEFITS 72
Ricardo Barbosa and *Manuela Almeida*
INTRODUCTION 72
CO-BENEFITS: ADDING VALUE TO ENERGY EFFICIENCY 75
ADDRESSING CO-BENEFITS THROUGH A QUALITATIVE SURVEY 79
DEGREE OF RELEVANCE AND WILLINGNESS TO PAY FOR CO-BENEFITS 83
COMPARING THE PERSPECTIVES OF NATIONAL CONTEXTS ON CO-BENEFITS 85
    Portugal 85
    Spain 86
    Italy 87
    France 89
    Germany 90
CO-BENEFITS AS A SUPPORT FOR DECISION-MAKING 92
CONCLUSION 96
CONSENT FOR PUBLICATION 97
CONFLICT OF INTEREST 97
ACKNOWLEDGEMENTS 97
REFERENCES 98

CHAPTER 4 THE IMPACT OF OCCUPANTS IN THERMAL COMFORT AND ENERGY EFFICIENCY IN BUILDINGS 101
António Ruano, Karol Bot and *Maria da Graça Ruano*
INTRODUCTION 101
THERMAL COMFORT AND HVAC CONTROL 102
    Thermal Comfort Metrics 102
    PMV Computation 105
    Occupancy Estimation and Detection, and Its Impact on Thermal Comfort 109
    Eletric Consumption and Energy Efficiency 114
        Data Acquisition System 115
    Electric Consumption and Occupation 116
    Energy Consumption Forecasting Models and Occupancy 124
CONCLUDING REMARKS 131
CONSENT FOR PUBLICATION 133
CONFLICT OF INTEREST 133
ACKNOWLEDGEMENT 133
REFERENCES 133

CHAPTER 5 DETECTING OCCUPANT ACTIONS IN BUILDINGS AND THE DRIVERS OF THEIR BEHAVIOUR 138
Pedro F. Pereira and *Nuno M. M. Ramos*
INTRODUCTION 139
    Scope 139
    Occupant Behaviour 139
    State-of-the-Art 141

Occupant Behaviour Data-acquisition 141
Intelligent Buildings 141
Summarising 143
Digital Twins 144
Machine Learning Techniques Used in Occupant Behaviour 147
Gap and Objectives 156
MATERIALS AND METHODS 156
.... 157
Case Study 157
Data Acquisition Strategy 158
Monitoring System 158
Surveys 159
Software Used 160
Detection of Occupant Actions 160
Change Point Analysis (CPA) 160
New Technique Based on the Outlier Detection 163
Decision Tree 163
Comparison Between Methodologies of Occupant Actions Detection 164
Detection of the Drivers of Occupant Behaviour 164
Spearman's Rank Correlation Coefficient 164
Logistic Regression 165
Comparison Between Methodologies of Occupant Actions Drivers Detection 168
CONCLUSION 169
CONSENT FOR PUBLICATION 170
CONFLICT OF INTEREST 170
ACKNOWLEDGEMENT 170
REFERENCES 170

CHAPTER 6 THE (NOT SO) CLOSE RELATIONSHIP BETWEEN OCCUPANCY AND WINDOWS OPERATION 178
Aline Schaefer, João Vitor Eccel and *Enedir Ghisi*
INTRODUCTION 178
Studies on Occupant Behaviour 181
Explanatory Data Analysis 183
METHOD 188
Obtaining the Data 188
Descriptive Data Analysis 190
Cluster Analysis Application 192
RESULTS 195
Descriptive Analysis 195
Cluster Analysis 205
DISCUSSION 206
CONCLUDING REMARKS 209
CONSENT FOR PUBLICATION 209
CONFLICT OF INTEREST 209
ACKNOWLEDGEMENTS 209
REFERENCES 210

CHAPTER 7 INVESTIGATING THE UNCERTAINTIES OF OCCUPANT BEHAVIOUR IN BUILDING PERFORMANCE SIMULATION: A CASE STUDY IN DWELLINGS IN BRAZIL 213
Arthur Santos Silva
INTRODUCTION 214

LITERATURE REVIEW 216
DEVELOPING OPERATIONAL SCHEDULES 219
    Data Acquisition 221
    Measurements of the Average Power of Equipment and Lighting 221
    Questionnaires on the Operational Schedules 222
    Data Treatment 223
    Developing Schedules for Occupancy and Openings Operation 223
    Developing Schedules for Equipment and Lighting Usage 225
    Representation of the Schedules 226
    Schedules of Occupancy 226
    Schedules of the Operation of Windows and Doors 227
    Equipment and Lighting Schedules 229
    Internal Loads with Equipment and Lighting 231
BUILDING PERFORMANCE SIMULATION 231
    Building Simulation Programme 231
    Building Model 233
CONSTRUCTION AND MATERIALS 234
    Climate Consideration 236
    Simulation Experiments 237
    Natural Ventilation Mode 239
    Hybrid Ventilation Mode 243
    Output Variables 246
    Statistical Methods and Data Treatment 248
RESULTS AND DISCUSSION 249
    Natural Ventilation Mode 249
    Global Sensitivity Analysis 250
    Uncertainty Analysis 252
    Hybrid Ventilation Mode 257
    Global Sensitivity Analysis 258
    Uncertainty Analysis 262
DISCUSSION 267
CONCLUSION 269
CONSENT FOR PUBLICATION 270
CONFLICT OF INTEREST 270
ACKNOWLEDGEMENTS 270
REFERENCES 270

CHAPTER 8 INDOOR CLIMATE MANAGEMENT OF MUSEUMS: THE IMPACT OF
VENTILATION ON CONSERVATION, HUMAN HEALTH AND COMFORT 275
    Hugo Entradas Silva and *Fernando M. A. Henriques*
INTRODUCTION 276
STATE-OF-THE-ART 277
    General Considerations 277
    IAQ Based on Comfort Perception 278
        Traditional Approach 278
    The Adaptation Concept 280
    International Guidelines and Standards 281
        EN 16798-1 282
        ASHRAE 62.1 284
        CIBSE Guide A 285
        EN 15759-2 286

    IAQ Based on Human Health 286

   IAQ in Museums 288

METHODOLOGY 291

   General Considerations 291

   Case Study – Geometry, Envelope and Internal Gains 291

      Building Geometry 291

   Envelope 293

   Internal Loads 295

      Visitors Flux 295

   Sensible and Latent Heat 296

   CO2 Generation Rate 299

   Lighting 300

   Simulation Study 303

   Weather File 305

   Data Analysis 306

RESULTS 307

CONCLUSION 316

CONSENT FOR PUBLICATION 317

CONFLICT OF INTEREST 317

ACKNOWLEDGEMENT 317

REFERENCES 317

SUBJECT INDEX 324

PREFACE

Occupant behaviour in buildings has been a matter of concern all over the world. Buildings are responsible for a significant portion of energy consumption; therefore, improving the thermal and energy performance of such buildings requires knowledge about the variables that influence them. However, to increase the potential for improving thermal and energy performance of buildings, studies must also consider the occupant's interactions with the built environment. The occupant behaviour influences the conditions of the internal environment through the occupation of the spaces and through the interaction with building elements, such as air-conditioning, lighting, blinds and windows. Thus, the objective of this e-book is to put together some of these aspects, presenting advances and challenges, by means of eight chapters written by renowned researchers.

Due to recent technological innovations related to Information and Communication Technologies (ICTs), buildings are undergoing some evolutions and incorporating technologies that endow them with intelligence. However, the requirement of building intelligence to be related to the response of the occupants' needs leads to the consideration of the buildings as Cyber-Physical-Social Systems (CPSS). Combining technical and social dimensions in this new generation of buildings, occupants' satisfaction and energy use can be improved. Mateus V. Bavaresco, Ricardo F. Rupp and Enedir Ghisi concluded that by enriching data collection and presentation, more professionals can access previous outcomes and adapt their practices towards achieving comfortable and energy-efficient buildings.

People's behaviour can significantly impact both the energy consumption and the indoor thermal environment of the buildings, and of particular interest is their window opening behaviour. A better understanding of why, when and how occupants open windows is, therefore, essential in the quest to achieve low-carbon buildings. Shen Wei provided systematic criteria for selecting a suitable monitoring method for their specific research objectives. Additionally, the author demonstrates the need for a standard method for monitoring relevant influential factors, as these varied considerably between existing studies with respect to the accuracy, interval and location. Such variation clearly has the potential to influence the ability to perform cross-study comparison.

Changing and improving the heating systems have been systematically associated with a wide range of effects, such as thermal comfort and improved air quality, which are often termed as co-benefits or ancillary benefits. Literature shows that co-benefits can be decisive when users choose a heating solution. Ricardo Barbosa and Manuela Almeida used international qualitative surveys to identify, quantify and evaluate the co-benefits associated with heating solutions, to clarify the relevance of the co-benefits in the decision-making process of building users. The results suggest that both the degree of relevance and the willingness to pay for co-benefits vary significantly amongst different national contexts.

Occupancy is a paramount factor to achieve energy efficiency. The authors António Ruano, Karol Bot and Maria Ruano, proposed a new methodology to estimate the occupancy and analysed the impacts of occupants on thermal comfort and energy efficiency in buildings from two distinct sectors: residential and educational.

The knowledge of occupant actions and needs is determinant for the proper function of an intelligent building. Therefore, the building management systems (BMS) must be supplied with data from the occupants. However the data by itself does not ensure the knowledge of occupants' needs and the ability to predict their behaviours. To do that, BMSs must be gifted with artificial intelligence (AI) and machine learning (ML) techniques to data mine the information provided by the monitoring systems. Pedro F. Pereira and Nuno M. M. Ramos compared methodologies used to detect occupant actions and occupants' needs in the same case study. The compared methodologies have the ability of self-learning and, therefore, can the used in multiple circumstances.

The variability of human behaviour is not taken into account in many thermal and energy performance studies, causing inconsistencies between simulation results and reality. One of the reasons for these inconsistencies also relies on adopting an opening availability schedule which is strictly limited to the occupancy schedule of a room, especially in residential buildings. Aline Schaefer, João Vitor Eccel and Enedir Ghisi studied the dependency relationship between the room's occupancy schedule and the operation of openings in low-income houses in Florianópolis, southern Brazil. The main result has shown that the opening operation schedule often does not depend on whether the room is occupied or not and seems to rely more accordingly to a daily routine, such as the time one wakes up or goes to sleep, or leaving and coming back home.

The gap between the estimate and actual thermal and energy performance is directly and indirectly attributed to occupants. To address such issue, Arthur Santos Silva investigated the uncertainties of occupant behaviour in building performance simulation through a probabilistic approach. The author showed that the number of occupants, the schedules of occupancy of the bedrooms, the setpoint temperatures for operating the openings, the cooling setpoint of the Heating, Ventilation and Air-Conditioning system (HVAC) and the limits for operative temperatures of the rooms were the most influent variables for the thermal and energy performance, especially in the heating period. The uncertainty was up to 65.6% for estimating the degree-hours for heating (in the natural ventilation mode) and up to 59.3% for estimating the total electricity consumption with HVAC (in the hybrid ventilation mode), indicating that these operational uncertainties had a great impact on the simulation results.

Cultural heritage plays an important role in society, not only in cultural terms but also due to its touristic interest. However, it is necessary to ensure that conservation and comfort conditions are not affected, since the human body releases heat, moisture, CO_2 and odours. Hugo Entradas Silva and Fernando M. A. Henriques analysed the impact of the binomial ventilation *vs.* occupancy, simulating various combinations of ventilation and air recirculation on the indoor air quality, conservation and energy consumption in museums. Since the visits to major national museums take usually long periods, the concept of adaptation was analysed to reduce the airflow of fresh air per visitor.

Chapter 1, written by Mateus V. Bavaresco (of the Federal University of Santa Catarina, Brazil), Ricardo F. Rupp (of Technical University of Denmark) and Enedir Ghisi (of the Federal University of Santa Catarina), explores the potentials of combining objective information gathered from technological innovations with subjective inputs obtained through qualitative methods in occupant behaviour research.

Chapter 2, written by Shen Wei, of the University College London, UK, introduces existing methods that have been used to monitor occupant window opening behaviour in buildings based on a comprehensive literature review. The author also points out relevant influential factors and discusses the advantages and disadvantages of each method.

Chapter 3 was written by Ricardo Barbosa and Manuela Almeida of the Department of Civil Engineering of the University of Minho, Portugal. The authors support the decision-making process of building users in the selection of energy-efficient heating solutions by identifying and evaluating co-benefits.

Chapter 4, written by António Ruano, Karol Bot, Maria da Graça Ruano of University of Algarve, Portugal, studied the impact of occupants in thermal comfort and energy efficiency.

Chapter 5, written by Pedro F. Pereira and Nuno M. M. Ramos of the Faculty Engineering of the University of Porto, Portugal, compared different machine learning techniques used for the detection of occupant actions in buildings and the drivers of their behaviour.

Chapter 6 was written by Aline Schaefer, João V. Eccel and Enedir Ghisi, of the Federal University of Santa Catarina, Brazil. The authors investigate the dependency relationship between the room's occupancy schedule and the operation of openings in low-income residential buildings.

Chapter 7, written by Arthur S. Silva, of the Federal University of Mato Grosso do Sul, Brazil, investigates the uncertainties of occupant behaviour in building performance simulation through a probabilistic approach.

Hugo Entradas Silva and Fernando M. A. Henriques, of the Department of Civil Engineering, Faculty of Science and Technology, Portugal, wrote Chapter 8. The authors analyse the impact of ventilation on conservation, human health and comfort in museums.

We would like to thank all the authors who have contributed to this e-book, and the editorial team for their valuable work and completion of this e-book.

The views and opinions expressed in each chapter of this e-book are those of the authors.

Enedir Ghisi
Federal University of Santa Catarina
Department of Civil Engineering
Laboratory of Energy Efficiency in Buildings
Florianópolis-SC
Brazil

Ricardo Forgiarini Rupp
Technical University of Denmark
Department of Civil Engineering
International Centre for Indoor Environment and Energy
Kongens Lyngby
Denmark

&

Pedro Fernandes Pereira
University of Porto
Department of Civil Engineering
Building Physics Laboratory
Portugal

List of Contributors

Aline Schaefer	Federal University of Santa Catarina, Florianópolis, Brazil
António Ruano	Universidade do Algarve, Faro, Portugal IDMEC, Instituto Superior Técnico, Universidade de Lisboa, Lisboa, Portugal
Arthur Santos Silva	Laboratory of Analysis and Development of Buildings, Campo Grande, Mato Grosso do Sul, Brazil
Enedir Ghisi	Laboratory of Energy Efficiency in Buildings, Department of Civil Engineering, Federal University of Santa Catarina, Florianópolis, Brazil
Fernando M. A. Henriques	Departamento de Engenharia Civil Faculdade de Ciências e Tecnologia, FCT, Universidade NOVA de Lisboa, Caparica, Portugal
Hugo Entradas Silva	Departamento de Engenharia Civil Faculdade de Ciências e Tecnologia, FCT, Universidade NOVA de Lisboa, Caparica, Portugal
João Vitor Eccel	Federal University of Santa Catarina, Florianópolis, Brazil
Karol Bot	Universidade do Algarve, Faro, Portugal
Manuela Almeida	ISISE, Department of Civil Engineering, University of Minho, Guimarães, Portugal
Maria da Graça Ruano	Universidade do Algarve, Faro, Portugal CISUC, University of Coimbra, Coimbra, Portugal
Mateus V. Bavaresco	Laboratory of Energy Efficiency in Buildings, Department of Civil Engineering, Federal University of Santa Catarina, Florianópolis, Brazil
Nuno M. M. Ramos	CONSTRUCT – LFC Faculty of Engineering (FEUP), University of Porto, Porto, Portugal
Pedro F. Pereira	CONSTRUCT – LFC Faculty of Engineering (FEUP), University of Porto, Porto, Portugal
Ricardo Barbosa	ISISE, Department of Civil Engineering, University of Minho, Guimarães, Portugal
Ricardo F. Rupp	International Centre for Indoor Environment and Energy, Department of Civil Engineering, Technical University of Denmark, Kongens Lyngby, Denmark
Shen Wei	The Bartlett School of Sustainable Construction, University College London, London, United Kingdom
Wei Yu	Joint International Research Laboratory of Green Buildings and Built Environments (Ministry of Education), Chongqing University, Chongqing-400045, China National Centre for International Research of Low-carbon and Green Buildings (Ministry of Science and Technology), Chongqing University, Chongqing-400045, China
Yan Ding	School of Environmental Science and Engineering, Tianjin Key Laboratory of Indoor Air Environmental Quality Control Tianjin University, Tianjin - 072, China

CHAPTER 1

Exploring The Potential of Combining Technological Innovations with Qualitative Methods in Occupant Behaviour Research

Mateus V. Bavaresco[1,*], Ricardo F. Rupp[2] and Enedir Ghisi[1]

[1] *Laboratory of Energy Efficiency in Buildings, Department of Civil Engineering, Federal University of Santa Catarina, Florianópolis, Brazil*

[2] *International Centre for Indoor Environment and Energy, Department of Civil Engineering, Technical University of Denmark, Kongens Lyngby, Denmark*

Abstract: The literature emphasises the important role that occupants play regarding the energy performance of buildings. Scholars have applied several methods to assess occupants' preferences and practices in their field studies. Technological innovations such as Internet-of-Things (IoT) may capture valuable objective information that can be translated into mathematical models. Such models are vital in Building Performance Simulation (BPS) practices as they are expected to reduce performance gaps between expected and real energy use in buildings during operational phase. However, data-driven models strictly related to physical parameters exclude essential subjective information like occupant preferences and needs. There is enough evidence showing that individual differences impact on thermal preferences and levels of comfort indoors, which must also be considered in occupant behaviour studies. Aside from individual preferences, there is also social influence when occupants share spaces and the control of building systems. Several methods commonly used in social science studies are expected to incorporate the needed subjective information in this field if properly used. Therefore, this chapter explores the potentials of combining objective information gathered from technological innovations with subjective inputs obtained through qualitative methods.

Keywords: Behavioural sensing, Building control, Building operation, Comfort, Energy, Energy efficiency in buildings, Indoor environment, Internet of things, Occupant behaviour, Social science, Technology.

---

* **Corresponding author Mateus V. Bavaresco:** Laboratory of Energy Efficiency in Buildings, Federal University of Santa Catarina, Florianópolis, Santa Catarina, Brazil; Tel: +55 48 3721-5184; Fax: +55 48 3721-5191; E-mail: bavarescomateus@gmail.com

INTRODUCTION

Buildings are commonly related to a high share of energy use worldwide. According to the last report by the International Energy Agency (IEA), the buildings and construction sector were responsible for 36% of the final energy use and 39% of energy and process-related CO_2 emissions [1]. Therefore, huge opportunities for energy savings and reduction of CO_2 emissions may be achieved by improving buildings. The energy use in buildings was the object of study of a group of 100 researchers from 15 countries, who gathered together and conducted strong research under the IEA and Energy in Buildings and Communities (EBC) Programme, in the IEA-EBC Annex 53 "Total energy use in buildings - Analysis and evaluation methods" [2]. Researchers concluded that there are six main factors that impact the energy use of buildings, *i.e.* climate, building envelope, building equipment, operation and maintenance, occupant behaviour, and indoor environmental conditions. As they argued, the three first are technical and physical factors, while the three last ones are human-influenced factors. Technical and physical characteristics cannot be considered as the only aspects when optimisation of energy use in buildings is intended. Indeed, technological and envelope-based interventions may reduce energy use in buildings; however, it is important to consider that they cannot guarantee this outcome alone [3]. When it comes to building operation, the literature also highlights that people in modern societies tend to spend about 85% of their time indoors [4]. It is then clear that the way occupants interact with buildings largely impact their total energy use.

Along these lines, recent research has evolved regarding the evaluation of human-related factors that impact building energy use. Considering the success of Annex 53, a different group of collaborative research was made to work on many unanswered questions. This new group, IEA-EBC Annex 66 "Definition and simulation of occupant behaviour in buildings", was then established based on the main takeaways from Annex 53. IEA-EBC Annex 66 main objectives relied on enhancing occupant behaviour research in terms of data collection, model representation and evaluation, and integrating such models in building performance simulation practices [5]. This field presented huge improvements with the completion of Annex 66, and a large amount of work was conducted throughout the world. Then, a follow-up research group was established following the conclusion of Annex 66 since there is a need for implementing advanced occupant modelling in practical activities. IEA-EBC Annex 79 "Occupant-Centric Building Design and Operation" is developing new knowledge about occupant behaviour, focusing on applying and transferring knowledge to practitioners [6]. This new research group involves a multidisciplinary team with expertise in engineering, architecture, computer science, psychology, and sociology. Its scope encompasses the conception of guidelines, recommendations for codes and standards, the establishment of data-driven methods, as well as the creation of new occupant models and simulation tools.

A common practice in occupant behaviour research is relying on sensor-based information to objectively assess indoor conditions as well as occupant presence and actions. For instance, environmental parameters may be used to infer as well as explain occupant behaviour or presence through statistical analyses or machine learning algorithms. By inferring, we mean using such environmental parameters to deduce certain actions, *e.g.*, by evaluating carbon dioxide concentration indoors, one may assume that a space is occupied or not [7]. When the actual occupant behaviour is also monitored, environmental parameters may be used to explain and determine boundaries for building adjustments. For instance, one may evaluate typical temperature thresholds that drive air-conditioning use [8]. In this way, the literature shows that several environmental parameters may be linked to the adjustment of building systems. Aside from occupancy, CO_2 concentration was also related to window control [9 - 11]. Indoor [12 - 14] and outdoor air temperatures [12,15,16] have been related to window, blind/shade, and HVAC (Heating, Ventilation, and Air Conditioning) control, as well as adaptive actions like drinking a cold drink. Indoor humidity has been associated with thermostat adjustments [17]. Specific choices like the degrees of opening in residential windows were also related to indoor and outdoor air temperatures [18]. Solar radiation and indoor/outdoor air temperature [12,19,20] were also related to the adjustments of blinds or shades. Although several environmental parameters were already linked with occupant behaviour in buildings, there is evidence that subjective aspects also play an important role in this field.

It is evident that occupant-behaviour related studies are increasing fast in the last few years; however, more work is still necessary to properly evaluate how occupants use different building systems, as several aspects influence this role [21]. More specifically, the literature supports that multi-domain physical variables, contextual and personal factors affect both occupants' perceptions and behaviours in buildings [22]. A huge body of research has focused on the influence of physical variables on occupant behaviour. However, there are still uncertainties related to the impact of contextual and personal factors in this field. Therefore, behavioural theories also significantly contribute to understanding personal factors and their relation with occupant behaviour, and a literature review synthesising the most commonly used theories was presented [23]. Authors acknowledged 27 approaches used in the literature, and they come mainly from the fields of psychology, sociology, and economics. Specifically, psychological theories were the most common, and the Theory of Planned Behaviour was the most frequent. Relying on the potential of applying qualitative knowledge on occupant-related research, another literature review presented methods commonly used in social sciences that are feasible to assess the human dimension of use in buildings [24]. Authors argued that broader use of qualitative methods is expected to provide building stakeholders with practical knowledge that may be helpful to achieve user-centric design and operation of buildings.

In this panorama, it is evident that both quantitative and qualitative data are vital to understand and model occupant behaviour patterns in a better way. On the one hand, technological innovations are key to collect a huge amount of quantitative data regarding building operation and objective aspects associated with the adjustments performed. On the other hand, personal and contextual variables may be missed if evaluations are solely based on objective aspects. Expertise from social sciences is necessary for this field and several methods are available. Therefore, the objective of this chapter is to assess the possibility of combining technological innovations with qualitative methods aiming to enhance occupant behaviour research practices.

INDOOR ENVIRONMENT, CLIMATE AND OCCUPANT BEHAVIOUR

The indoor environment, the climate and the occupant behaviour have an intrinsic relationship. For example, when occupants are feeling a warmer sensation, they could choose to open a window when it is cooler outside, which will decrease the indoor temperature. They could also opt to change their clothes or drink a cold beverage. Occupants could choose to turn on a fan or the cooling system, if such systems are available. The choice for any of those adaptive opportunities may vary between occupants due to their individual preferences and needs. This way, predictions of building energy consumption considering a poor representation of occupant behaviour would result in an unrealistic estimate of actual energy consumption, *i.e.* the so-called "performance gap" between predictions and actual energy use [25,26]. Even when certified buildings are considered, the literature supports that expected and measured energy consumptions are different [27,28]. Besides uncertainties related to climatic conditions and simulation programmes, it is evident that a better representation of occupant behaviour in simulation practices may mitigate such performance gap [29,30].

The influence of occupant behaviour on building energy use should be considered during both building design and operation. During building design, proper representation of occupant behaviour is expected to enrich the development of user-centric buildings. Additionally, computer simulation models can use a reliable representation of human-related aspects to turn actual simulation approaches more dependable. During the operation phase, understanding occupants' preferences and behaviours are also key to maintain indoor conditions comfortable while energy is efficiently used.

As previously shown, occupant behaviour research has increased in number and importance recently. A primary aspect on this field comprises data collection; indeed, this part is not trivial, and researchers must plan their approaches carefully, considering that costs, occupant privacy, and socioeconomic factors influence it [5]. A large body of research is available on innovative sensing technologies and approaches to incorporate them into buildings to explore occupants' behaviours, as highlighted by a literature review [31]. In fact, by collecting and processing data on this field, better modelling strategies may be achieved, as well as improvements on the indoor environmental quality of spaces. Physical monitoring relies mostly on adaptive behaviour monitoring, which includes both occupants' actions and environmental parameters related to them [32]. Innovative approaches to collect data on occupant behaviour include, but are not limited to Building Automation Systems (BAS). Indeed, BAS may collect real-time data and provide it to building stakeholders to enhance the control algorithms of automated systems. Such an approach relies mostly on quantitative data like the occupancy of different spaces and environmental parameters that affect adjustments of building systems through their interfaces. The literature shows that some studies relied on improving occupant representation aiming to adapt the algorithms of BAS [33 - 35]. Other strategies are also available, as researchers should not necessarily rely on data from BAS. Individual data loggers also play an important role in data collection. Regarding residences, a literature review showed that indoor air temperature, relative humidity, and air velocity are the parameters most frequently measured in such studies [36].

In addition to those commonly measured parameters, several other aspects may also be gathered to have a deeper understanding of triggers for occupant behaviour. For instance, one of the outcomes from IEA-EBC Annex 66 is the proposition of the DNAS (Drivers, Needs, Actions, and Systems) framework to improve occupant behaviour research considering that human cognition covers a complex combination of people "inside world" (*i.e.* Drivers and Needs) and "outside world" (*i.e.* Actions and Systems) [3]. Thus, Internet-of-Things (IoT) based monitoring may benefit from several passive and active sensors available to characterise human-related influence on building operation [37]. A considerable benefit of using innovative approaches is the possibility of collecting a huge amount of data, which can be translated into boundaries for occupant-centric building design and operation.

Further improvements were reached by combining the DNAS framework with social-psychology theories to encompass subjective aspects of occupant behaviour [38]. An interdisciplinary survey was then achieved through the integration of the DNAS framework with constructs from Social Cognitive Theory (SCT) and Theory of Planned Behaviour (TPB). SCT was explained by Bandura [39]; it states that personal and environmental factors influence human behaviours, *i.e.-* people's perceptions, beliefs, and acts affect their behaviours. TPB was introduced by Ajzen [40], and it evidences the impact of individual intention to behave on the behaviour itself; also, the theory supports that one's intention to behave is influenced by attitudes, subjective norms, and perceived control towards the exercised behaviour. This interdisciplinary framework was already implemented in several office settings worldwide, and interesting conclusions were achieved [41 - 42].

For instance, it enabled the assessment of human-building interactions through the lenses of the Five-Factor Model (FFM) to associate occupants' personality traits with common behavioural patterns [41]. Differences on adaptive actions undertaken by occupants to restore their thermal comfort under hot and cold discomfort, as well as conformity to social norms towards sharing the control of building systems, were shown as an indicator of energy use in offices [43]. A proposition of a broad theoretical framework to evaluate the link between indoor environmental quality and the perceived productivity of office occupants was also presented [44]. Finally, results from specific countries also added valuable information to the literature. From the Brazilian case study, the authors used the framework to conduct a theoretical-driven Structural Equation Modelling and determine the primary subjective aspects that influence occupants' adaptive actions [45]. Results support that interventions based on social-psychology theories play an important role to boost occupants' adaptive opportunities. The Hungarian case study added information about the importance of knowledge to control building systems – especially when complicated controls are used – which supports that training programmes may be conducted throughout the country [42]. Finally, the Italian case study synthesised all the surveyed factors in different regions of the country (north, centre, and south) to illustrate why and how knowledge from social sciences may provide valuable information to building stakeholders [42].

POTENTIALLY APPLICABLE TECHNOLOGICAL INNOVATIONS

Modern buildings can be understood as Cyber-Physical Systems (CPS) considering the advance of diverse technological innovations, which provides the opportunity to link some characteristics of the physical environment with occupants' behaviours or preferences to reach user-centred services [46]. Cyber-Physical Systems can be understood as systems in which the physical world is combined with cyber components, and information can be exchanged between them [37]. Building Automation Systems (BAS) may benefit from these advances and can provide user-centred controls aiming to reach comfortable and energy-efficient targets for building operation. Similarly, Energy Management Systems

(EMS) may become more reliable as the role of occupants is continually evaluated under these conditions; as a consequence, user-related uncertainties regarding building energy use may become less challenging.

Indeed, optimal operation is a key aspect to guarantee energy efficiency in buildings without compromising indoor environmental quality conditions. This aspect emphasises the importance of including knowledge from occupants' preferences in building maintenance, especially considering that current building controls are mainly focused on energy-savings rather than occupants' preferences [47]. However, the literature already supports that intelligent and autonomous controls for buildings can connect occupants with such systems by including users preferences in decision-making processes [48]. It synthesises the potential of turning the current building stock into Cyber-Physical Systems. By capturing occupants' preferences and needs, they can be included in the loop of building control to increase indoor environmental quality while high energy efficiency levels are reached. As a consequence, building stakeholders must be aware of several opportunities provided by technological innovations in this field. By combining human preferences with up-to-date technologies, meaningful improvements may be reached through the twofold relation created. Intelligent building systems may inform occupants about the best options to control a building, *e.g.* opening internal blinds or shades to increase daylight penetrations and reduce artificial lighting need. However, occupants' actions may also provide valuable knowledge that may improve the algorithms of an intelligent system, *e.g.* thresholds for indoor conditions may be updated when occupants adjust thermostats. Therefore, this subsection presents information about up-to-date technologies that may be used to evaluate occupants' preferences and behaviours as well as to control building systems based on the sensed information.

Behavioural Sensing

Several up-to-date behavioural sensing technologies are available to assess the physical aspects related to buildings. Those sensors can help to understand indoor conditions, *e.g.* air temperature and humidity, indoor air quality, noise levels, illuminance, *etc.* Additionally, occupant presence and actions (OPA) can be assessed throughout them, *e.g.* occupancy, window opening and closing, HVAC usage, *etc.* The literature highlights two main groups of sensors that may be used in buildings: passive and active sensors. This topic shows the differences between these sensors and some potentials related to their use in the building sector.

Passive sensors can be characterised by their low energy use compared to active ones [49]. This is expected because passive sensors do not emit any energy to probe the space since they rely on others' body energy. Such devices are widely used to track localisations, movements, as well as behaviours performed by building occupants. The most common passive sensor is the Passive Infrared (PIR) sensor, which is highly used to detect occupancy indoors [50]. Therefore, automated systems may be controlled according to the occupant's presence and absence; for instance, artificial lighting or HVAC systems may be turned off when no occupancy is detected in a given space. This alternative is important concerning the reduction of energy wasting during building operation. However, the actions undertaken by an automated system should be reliable to avoid bothering building occupants with the unexpected shutdown of the systems. This trend is evidenced by the literature when the control is based solely on passive sensors' data. As these devices rely on others' body energy, "false-off" is commonly observed in such spaces because the sensors tend to fail in detecting stationary bodies [51]. In this manner, some solutions may be considered to minimise the "false-off" issue, and the literature supports that passive sensors may be combined with other technologies to increase reliability. Indeed, capacitive sensors were presented as a solution to detect long-term stationary occupancy indoors as a promising tool to control HVAC systems [52]. A prototype using passive infrared array sensors was also created to anonymously collect occupancy data [53]. Authors concluded that such a device could detect stationary occupants, especially when lower occupancy level is detected. Passive sensors may also be combined with active ones, and the outcomes reached minimise fails in stationary detections. For instance, PIR sensors (to detect occupancy) were combined with Hall effect sensors (to detect when a door is opened or closed), and authors proposed machine-learning-based strategies to enhance occupant presence detection and activity recognition [54]. Similarly, PIR sensors were combined with plug-load meters in offices to detect energy consumption at the desk level [55]. Authors achieved high accuracies from the predictions of both presence and absence during the work (up to 99% and 96%, respectively). Therefore, it is important to highlight that even with some hindrances, low-cost approaches may be applied in buildings to improve occupant detection practices. As a consequence, reliable systems for building control may be reached throughout the building operation phase.

Passive sensors are not limited to occupancy detection in buildings, and some solutions were created to probe occupant behaviours as well. A passive wireless prototype that combines PIR with other sensors like accelerometer and environmental parameters sensors was proposed as a way to recognise activities of daily life [56]. Authors concluded that room-level resolution of activities recognition might be achieved with this non-intrusive system. Similarly, PIR sensors were combined with a piezoresistive accelerometer to develop wearable sensors and track humans' location while also estimating their behavioural state [57]. Specific behaviours like real-time bed-egress were proposed with passive radio-frequency identification integrated with an accelerometer [58]. Passive and active sensors were combined in order to monitor specific behaviours of elderlies, and the created device was able to detect eating behaviours by estimating when items have been removed from the refrigerator and when plates have been placed in a table [59]. Some of these alternatives are highly important in households or health centres with elderly that need assistance throughout the day. However, it is important to highlight that they are not limited or exclusively valid for those situations, and all these opportunities found in the literature may be used in smart building contexts as well as in occupant behaviour studies. Indeed, the inclusion of passive sensors in low-cost solutions for the built environment may be a way to improve user-centred practices in this field. Advanced statistics, as well as machine learning approaches, are also expected to boost the usage of data from those solutions as many learning classification algorithms are currently popular.

While passive sensors rely on others' body energy (*e.g.* an infrared emitting source), active sensors need internal power to operate. The literature supports that active sensors can use self-generated signals to evaluate a space as well as rely on motion to probe the intended variables [60]. On this topic, occupancy may also be detected through active sensing technologies. Occupancy estimation was proposed by evaluating indoor acoustic properties with ultrasonic chirps [61]. Such an alternative can transmit ultrasonic chirps and then assess how these signals dissipate over time to estimate indoor occupancy with algorithms. Self-generated signals and self-motion can also be combined to boost sensing capability. An example of this case can be the combination of laser range sensor and pan-tilt camera with the ability to move and detect humans even when they are positioned in blind spots – like behind other occupants [62]. Self-controlled servo-motor was also presented as an alternative to improve the detection of stationary subjects when combined with pyroelectric infrared sensors [63]. Detection of indoor localisations was proposed with stickers enabled by Bluetooth Low Energy (BLE) signals and beacons under points with the availability of Wi-Fi [64]. As argued by the authors, indoor tracking is important to improve the control of systems as well as the reliability of assistive-living services.

Another remarkable aspect of this field is that active sensors can convert one form of energy into another. Energy harvesting from environmental sources is a well-known technique, and it enables a set of green solutions like harvesting wind power through turbines. However, alternatives for doing so in micrometric scale are also available, and nanostructured piezoelectric transducers were shown as a viable solution [65]. Authors argued that this technology enables the conversion of slow fluids like human breaths into energy. Additionally to biomechanical energy harvesting, this innovation was also used to detect gait cycles [66]. Although all these alternatives still seem to fit monitoring for health care purposes, they are feasible regarding personalised monitoring, which can increase knowledge about human behaviour indoors. A positive outcome would be enhancing smart building control algorithms as well as the proposition of personalised recommendations that are expected to satisfy occupants at the same time that energy-efficient targets are maintained.

Wearable Devices

Gathering data in buildings is important for further improvements in the control algorithms or on the understanding of how the operation phase of the building impacts their energy use. Therefore, relying on the wide availability of technologies, and the improvement of sensing technologies, a set of wearable devices are also currently available. The concept of active sensors that can act like nanogenerators and harvest energy was applied to create wearable devices. The literature supports the use of triboelectric nanogenerator to create fabrics that can be used in smart clothing applications [67]. Advances in this field of intelligent clothing were presented, and the outcomes show that smart clothing can be independent of external power sources and, even so, can be integrated with other technology-driven controls [68]. Such a prototype may be used to control devices as well as monitor occupants in health centres. Another application to understand humans in health centres was based on wearable sensors to qualitatively assess arm movements in stroke survivors that need assistance [69]. Regarding the control of systems, wearable sensors with the ability to recognise human voices were also proposed [70]. Authors showed that voiceprint recognition enabled by the system was able to assess the password and the speaker, which can be used to control devices and building systems.

On the one hand, some wearable devices seem to be still in their infancy stage, as well as have theoretical applications and their use in the building sector is not common yet. On the other hand, there are pieces of evidence supporting that building stakeholders may already apply some wearable technologies to understand occupant behaviour in built environments better. For instance, wearable devices were used to understand human perceptions of an indoor environment with a view on personal attributes [71]. Such wearable devices were used to collect electrocardiogram, electrodermal, and electroencephalogram signals as physiological aspects that may be linked to occupant perception of environmental parameters, which directly influences their behaviour. Wearable sensors were also combined with stationary ones to learning occupants interactions with their workplace using machine learning algorithms [72]. An APP for Fitbit smartwatch was created to enhance the collection and labelling of comfort-related data provided by occupants [73]. This alternative is remarkable since authors enable the free download of the APP aiming a broad application throughout the world. A positive outcome reached with this approach is that instead of giving thermal comfort votes in fixed times, this wearable technology enables collection of data whenever participants want. This alternative may capture best the moments of peak discomfort indoors to tailor environmental conditions to meet occupants' requirements. Advances within this study were recently published, and other dimensions of indoor environmental quality were included (visual and acoustic) [74]. Authors argue that humans can act as sensors within these conditions, and important improvements in this field may be reached.

Internet-of-Things

The literature shows that by monitoring the operation of buildings as much as possible, the comprehension of building performance based on data-driven outcomes will improve [37]. Authors argued that data-driven decision-making may guide the proper discovery of both user-related adversity and system malfunctions. Solutions for those problems may be set by building stakeholders with the wider availability of information. Relying on the availability of several sensing technologies, a relevant fact to turn current buildings into Cyber-Physical Systems, as previously mentioned, is with the concept of Internet-of-Things (IoT). In fact, IoT-based solutions can connect several objects (physical) with cyber components that provide opportunities to better understand their relationship. Such an alternative is meaningful for the building sector because it can be embedded in automation systems to enable smart control of devices (*e.g.* smart lighting [75]). It is worth mentioning that those alternatives may represent high costs, which is an evident hinder for its application, especially in developing countries. However, the concept of IoT may be reached in several creative manners. The literature supports the use of everyday objects like smartphones to evaluate the patterns of HVAC usage in households [76]. Smartphones were also used to sense the magnetic field inside buildings, which was used as a proxy to determine occupant localisation indoors [77]. Such approaches are feasible considering that smartphones present several built-in sensors, which can probe intended variables and be included in the loop of an IoT-based system. In a broader perspective, a recent literature review highlighted that the advance of IoT and wireless networks empowered a quick increase in occupant-centric urban data [78].

Components of an IoT system also provide valuable information that can be used in the building-control loop. For instance, device-free occupancy detection and counting were proposed by evaluating the channel state information (CSI) observed in IoT systems [79]. Authors used the propagation signals of Wi-Fi as a proxy to identify whether there are occupants, as well as count them, in a built environment. Similarly, based on channel state information and advanced machine learning algorithms, it is possible to infer occupant activity inside buildings [80]. Such approaches highlight the wide possibilities provided by IoT-based systems. Indeed, the propagation of Wi-Fi signals from transmitters to receivers' components encapsulates meaningful information, as physical components like walls, doors, furniture, and occupants impact the way such signals are propagated. By recognising the impact of fixed elements on the signals, machine-learning-based approaches can infer the extent that occupants' presence and actions influence this characteristic. Besides that, IoT-based systems are also useful for real-time monitoring of indoor conditions, which is expected to increase the knowledge about healthy and comfortable built environments. Indoor air quality monitoring was proposed with a low-cost IoT-based tool that relies on measurements of environmental parameters (air quality, temperature, and humidity) coupled with a Raspberry Pi microprocessor and cloud storage [81]. Additionally, a literature review regarding technologies and practices regarding occupant-centric thermal comfort in buildings highlighted that the advances in IoT enable the vast collection of occupants' responses, which is promising to include occupants' feedback into building control [82]. IoT frameworks are handy in this case and have been used to improve the communication between users and HVAC systems to enable user-centric control [83].

In addition to occupant votes being used to tailor HVAC operation in a user-centric manner, such information is also important as a way to understand occupant preferences and behaviours. The role of feedback is presented in the literature as an important alternative in IoT-based systems [84]. Concerning this matter, it is important to understand that in addition to occupant preferences being a source of knowledge to adapt systems control, the existing system may provide some feedback to occupants. By means of integrated platforms, occupants may adjust their actions in order to save energy without compromising their preferences regarding indoor environmental quality (IEQ). With this broad applicability, the literature emphasised a clear path to deliver buildings with high IEQ levels and low energy use. For instance, with real-time monitoring of indoor conditions, feedback and behavioural-based consumption change may be achieved, which is a key to enhance energy management systems in buildings continually [37].

Virtual Reality

Virtual reality and immersive environments are emerging as promising tools to improve occupant behaviour research. These technologies enable longitudinal studies regarding occupants' preferences and behaviours in short-term experiments [85]. A wide variability of conditions can be tested under virtual representations instead of real-world experiences. This alternative may reduce the costs of a given experiment as smaller time frames are enough to drive conclusions [86]. The literature emphasised that virtual environments were used to collect occupant-related information like lighting preferences [87], internal blinds' adjustments [86], as well as people movements [88]. Although promising, this field still needs more evidence to support that outcomes reached within virtual environments actually correspond to real-world sensations [89]. Additionally, the literature supports that some people may face cyber or motion sickness when experience some virtual environment [89,90]. Researchers must be aware of this issue and guarantee some practices that are expected to reduce discomfort of those participants. For instance, the literature supports the use of a Simulator Sickness Questionnaire to detect the sickness tendency of different individuals previously to the realisation of the experiment [90]. Occupants with different tendencies of cyber sickness can be therefore selected for pilot studies of the designed experiment to adjust the practice before a higher sample is reached. Additionally, limiting the time of each experiment is also a potential alternative to reduce cyber sickness, as long exposition to virtual environments may be disliked by participants.

QUALITATIVE METHODS TO STUDY OCCUPANT BEHAVIOUR

Driving factors for occupant behaviour in buildings are largely discussed in the literature, but standardised methods for assessing them are still lacking [91]. On the one hand, this might be interpreted as a weakness of this study area because results obtained with different experiments may be hardly compared. On the other hand, such a lack of standardised approaches provides practitioners with several opportunities for tailoring occupant behaviour evaluations to current needs as well as available resources. The literature already supports that understanding subjective aspects that lead to energy use in buildings is crucial, as focusing specifically on physical parameters may not be enough [92]. In fact, occupant behaviour research needs multidisciplinary efforts [30], highlighting the importance of adding qualitative data to those studies. Although some subjective or contextual factors may not enhance the mathematical representation of building performance, they are expected to provide practical advice through case studies for improvements in building design and operation [93]. As emphasised by Sovacool [94], energy studies need social science approaches, and this combination can make energy research practices more socially oriented, interdisciplinary and heterogeneous. As a consequence of mixing qualitative and quantitative data, analytic excellence and social impact are more likely to be achieved.

Questionnaires and interviews are the most commonly used approaches when subjective evaluations are included in occupant behaviour research, as shown in a book on occupant behaviour studies [31]. The book presented a chapter about the qualitative methods mentioned (questionnaires and interviews) and provided important information about the state-of-the-art and future steps. Literature reviews focusing specifically on the use of questionnaires in this field were also published [95,96]. With advances in this area, energy efficiency studies are becoming more user-centric and several underlying effects that may be explained by behavioural theories are being incorporated in research practices as highlighted by a recent literature review [23]. Therefore, considering the importance of assessing subjective aspects in occupant behaviour research, several potential methods used in social science studies are available [24,97]. In this topic, some qualitative methods are presented as a promising way to include subjective data on occupant behaviour evaluation.

Questionnaires

Questionnaires are highly used in energy-related research, and previous literature reviews have presented an in-depth evaluation of approaches and challenges related to them. Considering residential contexts, Carpino *et al.* [95] concluded that, although the use of this method has increased recently, the use of non-standardised nomenclature may represent a source of uncertainty. Authors then provided some guidance for future research, considering the need to present detailed information about the sample evaluated, as well as the homogenisation of the nomenclature used. Another literature review focused on the use of cross-sectional questionnaires in occupant behaviour research [96]. In this case, the authors concluded that the projects reviewed were mostly focused on environmental and engineering factors. Therefore, they suggested that future research should encompass multidisciplinary approaches to gather more representative knowledge from field studies. Another literature review provided comprehensive information about different approaches and specific features when questionnaires are applied [24]. Authors reported common types and scales used, as well as the types of questions frequently employed.

Regarding the types of questionnaires, right-here-right-now, cross-sectional, and longitudinal ones are commonly used in the field [24]. Right-here-right-now questionnaires comprise the approach of asking subjects for their right-in-time opinions, perceptions or behaviours. As the questions are focused on the current moment, retrospective biases are more likely avoided when compared to approaches in which the respondent must provide information about past events. Right-here-right-now questionnaires are a standard procedure to collect occupants' perceptions about the thermal environment, and many researchers rely on the ASHRAE 55 method [98]. Besides its frequent use in thermal comfort studies, such an approach may be applied for other purposes as well. For instance, different dimensions of indoor environmental quality (*e.g.* lighting, air quality, and acoustics) may be assessed through right-here-right-now questionnaires. If combined with concurrent objective measurement, they can provide meaningful information to understand multi-modal comfort aspects better. Cross-sectional questionnaires comprise approaches in which larger timeframes are evaluated [99], and they enable understanding tendencies on the topic of interest. As previously shown, this approach is common in occupant behaviour research [96]. Different from right-in-time questions, this approach may cause some confusion in the respondents when a large timeframe is comprised. It is recommended to limit the scope of the questions to the current season or month instead of asking about whole-year experiences of occupants [24]. The literature supports the use of this approach to assess patterns of occupant behaviours [100 - 102], as well as constructs related to it [103 - 105]. Finally, longitudinal questionnaires represent the scenario in which participants' opinions are asked more than once in a given period. Both right-here-right-now and cross-sectional methods can be combined in a large-scale longitudinal evaluation [24]. This approach can provide practitioners with valuable information about differences in occupants' perceptions between seasons [106] or after interventions on the building [107]. An important aspect of this kind of research is the frequency of questionnaire applications since intensive data collection may bother participants and do not result in high-quality outcomes.

The type of questions used in survey-based studies is another important aspect. Commonly, both close-ended and open-ended questions are applied in energy research. Close-ended questions are those in which pre-defined options are presented to participants, and they are asked to either choose one (mutually exclusive) or to select all options that apply (collectively exhaustive). In practical terms, mutually exclusive questions can facilitate the comparison among trends reported; however, only one option must apply to avoid frustration during the participation [97]. A common approach to guarantee this is the use of Likert questions, which relies on asking participants to what extent they agree or disagree with a given statement. Similarly, Likert-like questions are highly used in this field to ask varied opinions of participants regarding the topic of study – not necessarily if they agree or not. For instance, the literature emphasises the use of Likert-like questions to assess occupants satisfaction with varied aspects of indoor environmental quality [108 - 110]. One key aspect when using this approach is guaranteeing symmetry on the options regarding agreement and disagreement [97]. Collectively exhaustive questions – also known as check-all-that-apply questions – provide to participants the option to select as many options as they want. Importantly, both literature review and pilot studies are needed to create those close-ended questions, because options given must be related to the participants' experiences; otherwise, inconclusive responses may be obtained.

Finally, open-ended questions enable the participants to provide their opinion. It requires more cognitive efforts from them, but important and unexpected information may be obtained in the evaluation. Besides the possibility of one question (or even the whole questionnaire) be open-ended in a study, an open-ended feature can be included in a close-ended question by providing the option "Other" and allowing participants to write an explanation about this difference [24].

Interviews

Interviews are also highly used in occupant-behaviour-related research, and both individual or in-group data collection may be considered [31]. Besides the different approaches used, interviews can be compared to questionnaires as a set of questions are asked to participants; therefore, data collection can also be structured or open-ended. Fully-structured interviews are those in which all the participants respond to a set of pre-defined questions. This approach may be helpful when specific aspects need to be evaluated [111]. Semi-structured interviews are not completely pre-defined, but a previous structure is created as well. This method allows for adding topics that emerged during the discussions in the evaluation. Finally, open-ended interviews allow participants to explain their opinions about a few pre-established points of interest. Fully-structured interviews enable the collection of structured data, which may facilitate the comparison of all the responses. On the other hand, open-ended questions may facilitate the discussion between the interviewer and the participants [111], as well as enable the collection of powerful stories [112]. Collecting stories is shown as a valuable way to inform building stakeholders about malfunctions as well as opportunities to improve building performance, especially considering that stories can be more easily remembered comparing to objective outcomes like numbers [112].

Individual interviews can be conducted either face-to-face or remotely. Both approaches are valid, and stakeholders should determine which method suits best their research interest. For instance, face-to-face interviews are expected to be more time- and resource-consuming, especially when conducted *in situ*, compared to an interview *via* telephone or video. However, the literature highlights that *in situ* interviews may provide some underlying information or aspects that can be observed instead of asked [112]. In some cases, the interviews must be performed *in situ*. For instance, asking questions to specific occupants like children in kindergartens requires an *in situ* interaction between the child and the interviewer in order to allow the child to become familiar with the researcher before conducting the interview [113,114]. Additionally, interviewers can assess directly with occupants some daily practices regarding building control, *e.g.* dwellers can describe and re-enact common practices that lead to energy use in their households [115]. Such an approach is characterised as a situated and embodied "telling" technique that improves users participation and communication of their practices [116]. In monitoring studies, such *in situ* interviews can be conducted when researchers install the sensors or equipment [117], as well as when they remove all the devices. Less chance of bothering participants can be achieved when both needs are combined in one visit. The concept of follow-up interviews is also presented in the literature, and this approach can be used after an initial study (for instance, a questionnaire application) to deepen the understanding of the responses previously provided [118]. As a broad of valuable information can be collected from this method, researchers must be careful about the way they conduct the studies. A key aspect is the span of an interview-based evaluation, since participants may be bothered when long interventions are made. The literature reports a high variability of the duration of interviews, and time intervals ranging from 30 to 150 minutes were found [92,119].

Differently from individual interviews, focus groups comprise collective interviews in which informal conversations and dynamic interactions are encouraged [24]. There are no specific guidelines to conduct a focus group and, similarly to individual interviews, several aspects can be assessed through this method. Satisfaction levels about rented apartments were evaluated using a focus group approach, and this method was important to gather further information as previous questionnaires and individual interviews were also used [120]. This method was also valid to understand the intention of house owners regarding refurbishment activities of recently built housing [121]. Besides understanding the occupant perspectives, focus groups were also used to evaluate opinions and approaches used by professionals of the building sector regarding building energy modelling [122]. Such an alternative can drive changes on the design process since key shortcomings on the process can be assessed, especially when a group of professionals within an organisation is considered. Along these lines, this method is an important alternative to gather knowledge from experts in a field. The literature supports the use of focus groups in workshops to understand specific aspects of building design [123], but this trend can be extended to other aspects of the building sector. For instance, workshops or conferences can provide valuable opportunities to conduct focus groups with experts in the field of occupant behaviour in buildings and user-centric design to discuss advances in these areas. Therefore, direct adaptation in daily practices could be implemented in occupant behaviour research. Similarly to individual interviews, the duration of a focus group must be carefully decided as long data collection may bother participants. As there are no specific rules, varied durations like 45 minutes up to

180 minutes were found [124,125]. Besides time, one further aspect that must be considered is the sample used in a focus group; commonly, six to eight people are recommended [97].

Personal Diaries

Diaries can be understood as a self-administered questionnaire, and this method allows gathering contemporaneous and frequent information if participants are engaged with the experiment. Besides the important role in data collection, personal diaries may also have an indirect effect on occupants' awareness of their impact on the energy use or performance of a building [126]. Authors state that personal diaries offer a chance to occupants reflect upon daily habits as well as explain them, which provides valuable information regarding building control. Additionally, data collection can be both paper-based or electronic, and the latter is less time-consuming for the participants and researchers [127]. Electronic data collection also enables the inclusion of photos or videos to contextualise the actions undertaken. On the one hand, this method allows gathering data in natural settings, which may reduce uncertainties like retrospective bias. On the other hand, it requires commitment from the participants, and low response rates or even blank diaries may be achieved at the end of the experiment [97].

Many studies relied on personal diaries or similar approaches to drive meaningful conclusions about occupant-related energy use in buildings. For instance, Time Use Surveys (TUS) are largely adopted throughout the world, and they consist of requesting participants to record their presence and activities in small intervals during a few days. A case study using TUS method was conducted in Denmark, and the outcomes enabled to profile occupant behaviour in dwellings to enhance occupant representation in building performance simulations (BPS) [128]. Authors relied on a large-scale data collection from 4679 households. However, an important aspect is that each participant provided their daily habits only for one weekday and one day of the weekend – both determined by the researchers. By doing so, more commitment is expected from the participants compared to long-term experiments. With some variations in the details, a case study with TUS was also conducted in the U.S.A [129]. In this case, there were 17 pre-established activities, and participants should choose among them for the intervals they reported in the diary. Therefore, activity types were achieved by grouping the responses in categories like "away from home", "sleeping", "cooking", "dishwashing", and so forth. Such a combination can also provide valuable contributions to modelling occupant presence and actions for BPS. Many other case studies were also conducted using data from TUS, and the literature supports that advanced data mining or machine learning techniques can extract reliable models of occupant behaviour [129 - 131].

Other kinds of experiments may also include diaries. For instance, the literature supports that this method helps to understand the satisfaction levels of occupants, and it can be combined with objective measurements to have a deeper comprehension of occupant preferences [132]. Such an approach can be adopted to evaluate the comfort thresholds of occupants. Therefore, instead of asking them to provide information about a whole day, diaries can be used to register specific moments in which participants felt discomfort indoors. Besides dissatisfaction, such in-time registration may provide ground truth when sensors are being developed to capture energy-related events in buildings [133]. It is evident that several approaches are valid in occupant behaviour research, and experiments can achieve good results by relying on available methods. Then, it is up to the researchers to choose specific details of the experiment undertaken, as well as selecting a representative sample. Regarding personal diaries, as previously mentioned, small timeframes for data collection is expected to reduce bothering levels and, as a consequence, increase the adherence to the research. However, it is worth mentioning that long-term data collection with personal diaries was also conducted [134]. The use of long-term evaluations, meaningful information about seasonal variations within the environment, as well as the strategies to deal with perceived problems, are expected to enrich studies in this field.

Post-Occupancy Evaluation

Post-occupancy evaluations (POE) comprise a process of methodological data collection about building-related aspects in facilities that have been previously occupied [135]. Such an approach can drive meaningful conclusions about misfunctioning parts of a building, as well as provide insights to solve current problems and enhance indoor quality by understanding occupants' needs and preferences. However, the literature supports that POE is not commonly conducted, and it is especially common when the budget allows or when a specific problem is observed [112]. Besides that, POE provides valuable opportunities for building designers to continually improve their practices by understanding what is and what is not working as intended in a facility. A recent POE-based study also highlighted the valuable insights provided by qualitative information reported by occupants [136]. In this case, authors showed that allowing occupants to submit photographs about their workplaces provide contextual information about building interfaces, which is handy even for improvements on the interface design. However, the authors also raised questions about privacy issues since the inclusion of photos may be sensitive and hinder anonymisation procedures. Thus, future research must carefully consider such ethics and privacy if photograph-based evaluations are intended. POE is also linked with advances in green building certificates, since one may understand if a certified building is performing as expected. Concerning this matter, a comprehensive literature review was presented, and aspects like building energy consumption, indoor environmental quality performance, and occupant satisfaction in green buildings were evaluated [137]. Although the authors concluded that green buildings tend to present some advantages on operating performance, there is still a huge performance gap between operation and design phases. Therefore, it is important to popularise POE to continually evaluate field data with a view of building occupants to improve other green buildings' design and certifications.

To have a comprehensive overview of the building during its operation, POE can use several approaches. For instance, the literature shows that three main categories of data collection can compose a POE: perception (*e.g.* questionnaires and interviews), monitoring (*e.g.* objective measurements and benchmarking), and observation (*e.g.* walkthroughs and historical records) [138]. Authors argued that a given POE should not, necessarily, apply varied techniques from all the categories; however, by combining them, a comprehensive overview of the building may be reached. By associating surveys to understand occupants' satisfaction levels with indoor monitoring and walkthroughs, researchers can dive deep on specific aspects that either boost or hinder occupant wellbeing indoors. As an indirect consequence of the variability of methods and their combinations allowed in occupant behaviour research, the literature supports that POE can rely on the assessment of diverse aspects. In other words, POE can be case-specific in many situations, and results tend to be hardly generalisable. It should not be a discouragement for further research, as studies with similar approaches provided fruitful contributions to understand occupant-related aspects [112]. Finally, as a consequence of the diversity of methods applied, POE can present various performance indicators for the building community as well. Among them, the literature highlights design quality, indoor environmental quality, and quality of services [139].

Comparison of Qualitative Methods

As many qualitative methods may be applied in this field, researchers should pay attention to the associated pros and cons of each method available [24]. Therefore, Table **1** synthesises some advantages and disadvantages of each method presented in this chapter. The purpose of this analysis is not disclosing one method as better than others; rather, it is aimed to facilitate further experiments' design by showing the convenience of adopting each method.

COMBINATION OF TECHNOLOGICAL INNOVATIONS AND QUALITATIVE METHODS

This chapter provided literature-based evidence that both technological innovations and qualitative methods can significantly improve occupant behaviour research. The former is expected to provide practitioners with objective measures, while the latter is vital to contextualise and inform subjective aspects related to buildings' operation and performance. As presented throughout this chapter, modern buildings can be characterised as Cyber-Physical Systems (CPS) with the advance of technological innovations that can combine physical aspects with cyber components [37,46]. Such a combination is fruitful as information may be exchanged between them; indeed, automation systems and building occupants alike may learn and improve practices with a so-called CPS. In this way, the literature also supports that modern buildings are emerging as Cyber-Physica--Social Systems (CPSS) by including the social dimension into a CPS framework [24]. Qualitative knowledge is therefore a vital aspect within this field, and methods used in social sciences are expected to enhance the representation and understanding of the social dimension in a given CPSS context. Therefore, multidisciplinary approaches are expected to keep cyber, physical, and social components working together in the loop of building performance. By combining up-to-date technologies with qualitative data collection, plenty of information to understand the extent that social components influence a physical feature is expected.

Table 1. Pros and cons of each method presented in this chapter.

Method	Advantages	Disadvantages
Questionnaire	- A variety of types of questions as well as questionnaire applications have been used in this field, which facilitates data collection; - Possibility to collect data online, which reduces the cost of the survey and enables achieving a high number of responses; - Questionnaires can be anonymous, respecting privacy concerns.	- Participants can feel bored and give up before completing their answers; - If printed questionnaires are adopted, data curation and analysis can be tricky; - Although enabling the collection of many answers, researchers must be careful when inviting participants to have a representative sample – especially if the results are expected to be generalised at a population level.

(Table 1) cont.....

Method	Advantages	Disadvantages
Interview	- Interviews can be used as a follow-up method, *e.g.*, after participating in a project, some subjects may be invited on interviews to clarify trends; - The conversational approach provides flexibility, and the interviewer can adjust the questions accordingly; - If *in situ* interviews are used, researchers may observe complementary aspects.	- Comparing to questionnaires, data analyses are harder because interviews must be recorded and transcript; - Also, this method is more time-consuming since each response demand the full-time involvement of an interviewer; - If *in situ* interviews are used, both time and resources for travelling should be considered.
Personal diary	- Right-in-time data collection is enabled in multiple spaces even more than once a day; - Rigorous data collection can provide ground truth information for measurements in the field; - Diaries can provide some consciousness regarding building operation as occupants become more aware of their practices.	- Data collection depends completely on the participants' willingness to fill the diary. Thus, even blank instruments may be returned at the end of the monitoring period; - Highly demanding designs (*e.g.*, participants reporting their opinions more than once a day) may result in high boring levels. - Training sessions may be needed to guarantee proper data collection.
Post-occupancy evaluation (POE)	- POE is a good way to include professionals in the loop of building performance, aiming to evaluate if design or retrofit decisions are performing well; - As researchers are expected to visit the building, they can observe underlying aspects of systems operation and control; - Several methods may be combined in a traditional POE, enabling triangulation of data, and higher reliability may be reached.	- Although it enables triangulation of data, the lack of standard protocols to conduct POEs may lead to hard-to-compare results as well as exhausting data analysis practices; - More than one researcher may be required during a POE practice, which may increase the cost of applying this method; - Although different actors may be included in this practice, there is still a lack of consensus about who should be responsible for conducting such evaluations.

When it comes to technological innovations, this chapter provided insights on several behavioural sensing that apply to this field. Indeed, passive and active sensors, as well as wearable devices, are expected to probe specific aspects related to occupants' presence and actions indoors. It is up to building stakeholders to target their needs along with their facilities. For instance, passive sensors are reported as less expensive than the other ones, and they are largely used to detect occupant presence in buildings. Active sensors, differently from passive ones, rely on internal power to operate instead of using other bodies' energy. Therefore, they are expected to be more reliable when it comes to detecting stationary subjects inside a building. Additionally, active sensors can convert energy from one form to another, which benefits the creation of nanogenerators that can be used in smart clothing or other smart devices. Indeed, such wearable devices can drive important changes in the understanding of occupants' physiological aspects, daily routines and their relation to building adjustments. Wearable devices like Fitbit smartwatches were provided with an APP to assess occupants' preferences regarding indoor conditions [73]. It can be characterised as a big improvement in the field, as humans can then act as sensors for buildings [74]. Data security should be ensured and dealt with properly in order to avoid any risk to occupants. By putting humans in a central position in the loop of building assessment and control, a greater acceptance of indoor conditions can be achieved.

In this manner, Internet-of-Things (IoT) technology is a key concept when it comes to turning actual buildings into Cyber-Physical Systems. Indeed, by combining all the needed sources of information (*e.g.* by using diverse sensing technologies), IoT-based frameworks can gather necessary knowledge to understand trends on occupant behaviours and preferences. In fact, a huge body of data can be collected when those devices are installed in a building, which can provide professionals with guidelines for designing user-centric buildings. Additionally, another consequence of evaluating these aspects is the creation of better algorithms to control building systems – *i.e.* occupant-centric control (OCC) [140]. Finally, virtual reality and immersive environments can be used to represent real-world scenarios to reach responses from occupants. This practice can reduce the time of longitudinal studies as several variations can be performed within the graphical representation.

However, although varied technological innovations are likely to improve occupant behaviour understanding and representation, focusing solely on objective measurement may not be enough. Additionally, it is discussed in the literature that occupant acceptance of such devices is worthy of consideration. Both high costs and privacy concerns may reduce the acceptance of new technologies in the building sector. Trust is a fundamental aspect when it comes to adopting these features in buildings, and a model based on the theory of Technology Acceptance Model was proposed [141]. Occupant trust and acceptance are also necessary for smart buildings with innovative automation systems. For instance, when systems exclude people's preferences during building operation, little control is perceived by occupants, and poor acceptance may be a consequence [142]. Such a trend shows the need for multidisciplinary approaches to understand occupants' preferences inside buildings, as well as to propose trusted and accepted innovative models for building control.

On this topic, meaningful improvements are expected by involving different stakeholders with varied expertise to boost building design and operation. Indeed, the literature supports that qualitative methods like surveys, interviews and post-occupancy evaluations (POE) can provide stories about building performance, and designers can learn from occupants to guarantee that future projects will not present similar issues [112]. Importantly, the authors argued that stories and qualitative data might be memorable to occupants, owners and designers compared to quantitative data like graphs and advanced statistics [112]. Therefore, this chapter presented updated information about qualitative methods commonly used in social science studies that suit energy research, especially considering occupant-related aspects. As shown, questionnaires are highly used in this field and can provide contextual information to enhance current practices [24,95,96]. Specific formats can be applied according to the needs of a given project – *e.g.* both cross-sectional and longitudinal survey-based evaluations may be used. Cross-sectional evaluations can give an overview of a target topic and present tendencies reported by the participants considering a given time interval. Longitudinal questionnaires, on the other hand, rely on at least two scenarios (*e.g.* summer *versus* winter or pre- *versus* post-retrofit) to drive conclusions on occupant-related aspects. As no standardised methods are available to assess driving aspects for occupant behaviours in buildings [91], each project may benefit from local features that enrich evaluations. Additionally, considering questionnaire-based evaluations, differences in the questions asked are also highly denoted, and this chapter presented varied structures that may apply. It is, therefore, up to professionals involved in such studies to determine the best practices considering local needs.

Although questionnaires are commonly used, the scope of qualitative data collection should not be limited to this method. This chapter also provided some insights regarding the use of interviews, personal diaries and post-occupancy evaluations. Interviews are indeed great ways to approach building occupants either individually or in-group. Their scope can vary from open-ended to fully structured, depending on whether all the participants respond strictly to the same questions or more conversational parts are allowed. Professionals can apply interviews to gather new knowledge on a given topic, as well as to deepen the understanding of specific aspects after a questionnaire application [118]. When conducted *in situ*, interviewers can directly observe facts that are not likely to be reported, as the presence of mould or smoke smell that indicate poor ventilation rates [143]. Additionally, participants can re-enact their common practices so that researchers can truly understand the personal influence on building control, as well as detailed manners that different people deal with building interfaces [115].

Personal diaries are also promising in this field since occupants can be placed in a fundamental position of data collection. Such an aspect result in a twofold issue, because if occupants are not engaged enough in the experiment, very low response rates will be achieved, and conclusions will be hindered [97]. Even so, international efforts have relied on personal diaries data to enhance occupant behaviour research. It is the case of applying Time Use Surveys (TUS) in local contexts to assess patterns of occupant presence and actions [128 - 131]. As shown throughout this chapter, TUS provide a great opportunity to study occupant behaviour as each participant is expected to report his/her actions during a few days. If a large sample is reached, a meaningful representation of the typology of study is expected. Besides that, large data collections are also possible using personal diaries. Although being conditional upon participants' willingness to report their actions or sensations, the literature supports that previous research that relied on this method also reached fruitful outcomes [134]. Finally, this chapter provided some insights on post-occupancy evaluations (POE), which represent a series of methodological data collections after a period of occupancy in a building [135]. POE is widely used by the scientific community. Although promising to evaluate how the design is performing after construction, the literature supports that it is not a common practice among stakeholders, and it is mostly conducted when specific problems are observed or when the budget allows [112]. However, understanding how a facility is performing during its operation is key to enhance other designs as well as present solutions to improve misfunctioning aspects in the current one.

It is clear that both technological innovations and qualitative methods are likely to provide important pieces of information to understand and represent occupant behaviour in buildings. Specifically, when it comes to technologies, they can also control building features within an automation system. However, occupants and their preferences must be included in the loop of building control to guarantee trustiness and acceptance of such systems. Therefore, objective and subjective methods alike are needed in this field of research. Even by including up-to-date behavioural sensing in an IoT-based building automation system, the social dimension of a so-called Cyber-Physical-Social System needs to be carefully and continually evaluated. Careful assessments are necessary to understand as many relations between humans and buildings as possible. For instance, how occupants respond to different indoor environmental conditions as well as non-physical parameters [144]; to what extent the internal layout may affect their satisfaction and productivity [145]; to what extent occupants perceive that they are in control of building systems [146]; or even how they deal with different building interfaces [21]. Continuous assessments are also important, considering that varied climatic conditions – as extreme events – are likely to happen more frequently due to climate change and affect building performance [147]. Additionally, it can guarantee that indoor conditions are satisfactory even if spaces are converted to different uses or occupants [148]. Therefore, combining innovative technologies with qualitative data collection can deepen the understanding and representation of occupant presence and actions indoors. As no standardised method is available, building stakeholders should rely on accessible resources and start collecting information and stories with building occupants.

CONCLUDING REMARKS

This chapter focused on a theoretical demonstration of the potential of combining technological innovations and qualitative methods to evaluate occupant behaviour in buildings, as well as to include it in the loop of building control. This concept relies on the fact that modern buildings can be considered Cyber-Physical Systems. Such a characteristic is enabled by innovative technologies that combine physical features with cyber representations, as well as exchange information between them. However, by focusing specifically, on objective measures, the outcomes achieved might be little accepted by the occupants. Therefore, it was concluded that several methods commonly used in the social sciences might provide stakeholders of the building sector with ways to understand occupants' preferences and behaviours. By carefully and continually assessing occupants' views regarding their buildings, such rich information can improve the performance of technology-driven systems. Thus, an in-depth understanding of occupant-related aspects can evolve modern buildings by turning them into Cyber-Physical-Social Systems (CPSS). By combining technical and social dimensions in this new generation of buildings, occupants' satisfaction and energy use can be improved. Importantly, understanding occupants' preferences in buildings should not be a bothering task as stakeholders may apply available resources and tailor experiments to local contexts. Indeed, by enriching data collection and presentation, more professionals can access previous outcomes and adapt their practices towards achieving comfortable and energy-efficient buildings.

CONSENT FOR PUBLICATION

Not Applicable.

CONFLICT OF INTEREST

The author confirms that this chapter contents have no conflict of interest.

ACKNOWLEDGEMENTS

This study was financed in part by the Coordenação de Aperfeiçoamento de Pessoal de Nível Superior – Brasil (CAPES) – Finance Code 001.

REFERENCES

[1]   GABC, "2019 Global Status Report: Towards a zero-emission, efficient and resilient buildings and construction sector", *International Energy Agency,* 2019.https://www.unenvironment.org/resources/publication/2019-global-status-report-buildings-and-construction-sector

[2]   H. Yoshino, T. Hong, and N. Nord, "IEA EBC annex 53: Total energy use in buildings—Analysis and evaluation methods", *Energy Build.,* vol. 152, pp. 124-136, 2017.
[http://dx.doi.org/10.1016/j.enbuild.2017.07.038]

[3]   T. Hong, S. D'Oca, W.J.N. Turner, and S.C. Taylor-Lange, "An ontology to represent energy-related occupant behavior in buildings. Part I: Introduction to the DNAs framework", *Build. Environ.,* vol. 94, no. P1, pp. 196-205, 2015.
[http://dx.doi.org/10.1016/j.buildenv.2015.08.006]

[4]   N.E. Klepeis, W.C. Nelson, W.R. Ott, J.P. Robinson, A.M. Tsang, P. Switzer, J.V. Behar, S.C. Hern, and W.H. Engelmann, "The National Human Activity Pattern Survey (NHAPS): a resource for assessing exposure to environmental pollutants", *J. Expo. Anal. Environ. Epidemiol.,* vol. 11, no. 3, pp. 231-252, 2001.
[http://dx.doi.org/10.1038/sj.jea.7500165] [PMID: 11477521]

[5]   D. Yan, "IEA EBC Annex 66: Definition and simulation of occupant behavior in buildings", *Energy Build.,* vol. 156, pp. 258-270, 2017.
[http://dx.doi.org/10.1016/j.enbuild.2017.09.084]

[6]   W. O'Brien, "Introducing IEA EBC Annex 79: Key challenges and opportunities in the field of occupant-centric building design and operation", *Build. Environ.,* vol. 178, p. 106738, 2020.
[http://dx.doi.org/10.1016/j.buildenv.2020.106738]

[7]   D. Calì, P. Matthes, K. Huchtemann, R. Streblow, and D. Müller, "CO_2 based occupancy detection algorithm: Experimental analysis and validation for office and residential buildings", *Build. Environ.,* vol. 86, pp. 39-49, 2015.
[http://dx.doi.org/10.1016/j.buildenv.2014.12.011]

[8]   R. de Dear, J. Kim, and T. Parkinson, "Residential adaptive comfort in a humid subtropical climate—Sydney Australia", *Energy Build.,* vol. 158, pp. 1296-1305, 2018.
[http://dx.doi.org/10.1016/j.enbuild.2017.11.028]

[9]   D. Calì, R.K. Andersen, D. Müller, and B.W. Olesen, "Analysis of occupants' behavior related to the use of windows in German households", *Build. Environ.,* vol. 103, pp. 54-69, 2016.
[http://dx.doi.org/10.1016/j.buildenv.2016.03.024]

[10]   F. Naspi, M. Arnesano, L. Zampetti, F. Stazi, G.M. Revel, and M. D'Orazio, "Experimental study on occupants' interaction with windows and lights in Mediterranean offices during the non-heating season", *Build. Environ.,* vol. 127, pp. 221-238, 2018.
[http://dx.doi.org/10.1016/j.buildenv.2017.11.009]

[11]   R. Andersen, V. Fabi, J. Toftum, S.P. Corgnati, and B.W. Olesen, "Window opening behaviour modelled from measurements in Danish dwellings", *Build. Environ.,* vol. 69, pp. 101-113, 2013.
[http://dx.doi.org/10.1016/j.buildenv.2013.07.005]

[12]   F. Haldi, and D. Robinson, "On the behaviour and adaptation of office occupants", *Build. Environ.,* vol. 43, no. 12, pp. 2163-2177, 2008.
[http://dx.doi.org/10.1016/j.buildenv.2008.01.003]

[13]   Y.S. Lee, and A.M. Malkawi, "Simulating multiple occupant behaviors in buildings: An agent-based modeling approach", *Energy Build.,* vol. 69, pp. 407-416, 2014.
[http://dx.doi.org/10.1016/j.enbuild.2013.11.020]

[14]   B. Lin, Z. Wang, Y. Liu, Y. Zhu, and Q. Ouyang, "Investigation of winter indoor thermal environment and heating demand of urban residential buildings in China's hot summer - Cold winter climate region", *Build. Environ.,* vol. 101, pp. 9-18, 2016.

[http://dx.doi.org/10.1016/j.buildenv.2016.02.022]

[15]   Y. Zhang, and P. Barrett, "Factors influencing the occupants' window opening behaviour in a naturally ventilated office building", *Build. Environ.,* vol. 50, pp. 125-134, 2012.
[http://dx.doi.org/10.1016/j.buildenv.2011.10.018]

[16]   M. Schweiker, and M. Shukuya, "Comparative effects of building envelope improvements and occupant behavioural changes on the exergy consumption for heating and cooling", *Energy Policy,* vol. 38, no. 6, pp. 2976-2986, 2010.
[http://dx.doi.org/10.1016/j.enpol.2010.01.035]

[17]   V. Fabi, R.V. Andersen, and S.P. Corgnati, "Influence of occupant's heating set-point preferences on indoor environmental quality and heating demand in residential buildings", *HVAC & R Res.,* vol. 19, no. 5, pp. 635-645, 2013.
[http://dx.doi.org/10.1080/10789669.2013.789372]

[18]   M. Schweiker, F. Haldi, M. Shukuya, and D. Robinson, "Verification of stochastic models of window opening behaviour for residential buildings", *J. Build. Perform. Simul.,* vol. 5, no. 1, pp. 55-74, 2012.
[http://dx.doi.org/10.1080/19401493.2011.567422]

[19]   M.V. Bavaresco, and E. Ghisi, "A low-cost framework to establish internal blind control patterns and enable simulation-based user-centric design", *J. Build. Eng.,* vol. 28, p. 101077, 2020.
[http://dx.doi.org/10.1016/j.jobe.2019.101077]

[20]   W. O'Brien, K. Kapsis, and A. K. Athienitis, "Manually-operated window shade patterns in office buildings: A critical review", *Building and Environment,* vol. 60, pp. 319-338, 2013.
[http://dx.doi.org/10.1016/j.buildenv.2012.10.003]

[21]   J.K. Day, "A review of select human-building interfaces and their relationship to human behavior, energy use and occupant comfort", *Build. Environ,* vol. 178, p. 106920, 2020.
[http://dx.doi.org/10.1016/j.buildenv.2020.106920]

[22]   M. Schweiker, "Review of multi□domain approaches to indoor environmental perception and behaviour", *Build. Environ.,* no. Mar, p. 106804, 2020.
[http://dx.doi.org/10.1016/j.buildenv.2020.106804]

[23]   A. Heydarian, "What drives our behaviors in buildings? A review on occupant interactions with building systems from the lens of behavioral theories", *Build. Environ.,* vol. 179, p. 106928, 2020.
[http://dx.doi.org/10.1016/j.buildenv.2020.106928]

[24]   M.V. Bavaresco, S. D'Oca, E. Ghisi, and R. Lamberts, "Methods used in social sciences that suit energy research: A literature review on qualitative methods to assess the human dimension of energy use in buildings", *Energy Build.,* vol. 209, p. 109702, 2020.
[http://dx.doi.org/10.1016/j.enbuild.2019.109702]

[25]   T. Hong, D. Yan, and S. D'Oca, *Build. Environ.,* vol. 114, pp. 518-530, 2017.
[http://dx.doi.org/10.1016/j.buildenv.2016.12.006]

[26]   P. De Wilde, "The gap between predicted and measured energy performance of buildings: A framework for investigation", *Autom. Construct.,* vol. 41, pp. 40-49, 2014.
[http://dx.doi.org/10.1016/j.autcon.2014.02.009]

[27]   C. Turner, and M. Frankel, *Energy Performance of LEED ® for New Construction Buildings*, 2008.

[28]   G.R. Newsham, S. Mancini, and B.J. Birt, "Do LEED-certified buildings save energy? Yes, but...", *Energy Build.,* vol. 41, no. 8, pp. 897-905, 2009.
[http://dx.doi.org/10.1016/j.enbuild.2009.03.014]

[29]   Z. Belafi, T. Hong, and A. Reith, "Smart building management vs. intuitive human control—Lessons learnt from an office building in Hungary", *Build. Simul.,* vol. 10, no. 6, pp. 811-828, 2017.
[http://dx.doi.org/10.1007/s12273-017-0361-4]

[30]   T. Hong, S.C. Taylor-Lange, S. D'Oca, D. Yan, and S.P. Corgnati, "Advances in research and

applications of energy-related occupant behavior in buildings", *Energy Build.,* vol. 116, pp. 694-702, 2016.
[http://dx.doi.org/10.1016/j.enbuild.2015.11.052]

[31]   B. Dong, Sensing and Data Acquisition

[32]   B.F. Balvedi, E. Ghisi, and R. Lamberts, "A review of occupant behaviour in residential buildings", *Energy Build.,* vol. 174, pp. 495-505, 2018.
[http://dx.doi.org/10.1016/j.enbuild.2018.06.049]

[33]   P.F. Pereira, N.M.M. Ramos, and M.L. Simões, "Data-driven occupant actions prediction to achieve an intelligent building", *Build. Res. Inform.,* vol. 48, no. 5, pp. 485-500, 2020.
[http://dx.doi.org/10.1080/09613218.2019.1692648]

[34]   Y. Peng, A. Rysanek, Z. Nagy, and A. Schlüter, *Using machine learning techniques for occupancy-prediction-based cooling control in office buildings,* 2018.
[http://dx.doi.org/10.1016/j.apenergy.2017.12.002]

[35]   A. Mirakhorli, and B. Dong, "Occupancy behavior based model predictive control for building indoor climate—A critical review", *Energy Build.,* vol. 129, pp. 499-513, 2016.
[http://dx.doi.org/10.1016/j.enbuild.2016.07.036]

[36]   J. Du, W. Pan, and C. Yu, "In-situ monitoring of occupant behavior in residential buildings – a timely review", *Energy Build.,* vol. 212, p. 109811, 2020.
[http://dx.doi.org/10.1016/j.enbuild.2020.109811]

[37]   M.V. Bavaresco, S. D'Oca, E. Ghisi, and R. Lamberts, "Technological innovations to assess and include the human dimension in the building-performance loop: A review", *Energy Build.,* vol. 202, p. 109365, 2019.
[http://dx.doi.org/10.1016/j.enbuild.2019.109365]

[38]   S. D'Oca, C.F. Chen, T. Hong, and Z. Belafi, "Synthesizing building physics with social psychology: An interdisciplinary framework for context and occupant behavior in office buildings", *Energy Res. Soc. Sci.,* vol. 34, pp. 240-251, 2017.
[http://dx.doi.org/10.1016/j.erss.2017.08.002]

[39]   A. Bandura, *Prentice-Hall series in social learning theory. Social foundations of thought and action: Social cognitive theory.,* 1986.

[40]   I. Ajzen, "The Theory of Planned Behavior", *Organ. Behav. Hum. Decis. Process.,* vol. 50, pp. 179-211, 1991.
[http://dx.doi.org/10.1016/0749-5978(91)90020-T]

[41]   T. Hong, "C. fei Chen, Z. Wang, and X. Xu, "Linking human-building interactions in shared offices with personality traits", *Build. Environ.,* vol. 170, 2020.
[http://dx.doi.org/10.1016/j.buildenv.2019.106602]

[42]   Z. Deme Bélafi, and A. Reith, "Interdisciplinary survey to investigate energy-related occupant behavior in offices – the Hungarian case", *Pollack Period.,* vol. 13, no. 3, pp. 41-52, 2018.
[http://dx.doi.org/10.1556/606.2018.13.3.5]

[43]   C. Chen, "Culture, conformity, and carbon? A multi-country analysis of heating and cooling practices in office buildings", *Energy Res. Soc. Sci.,* vol. 61, p. 101344, 2020.
[http://dx.doi.org/10.1016/j.erss.2019.101344]

[44]   C-F. Chen, "The impacts of building characteristics, social psychological and cultural factors on indoor environment quality productivity belief", *Build. Environ.,* vol. 185, p. 107189, 2020.
[http://dx.doi.org/10.1016/j.buildenv.2020.107189]

[45]   M.V. Bavaresco, S. D'Oca, E. Ghisi, and A.L. Pisello, "Assessing underlying effects on the choices of adaptive behaviours in offices through an interdisciplinary framework", *Build. Environ.,* vol. 181, p. 107086, 2020.
[http://dx.doi.org/10.1016/j.buildenv.2020.107086]

[46]   L. Chen, "Di. J. Cook, B. Guo, and W. Leister, "Guest Editorial - Special Issue on Situation, Activity, and Goal Awareness in Cyber-Physical Human-Machine Systems", *IEEE Trans. Hum. Mach. Syst.,* vol. 47, no. 3, pp. 305-309, 2017.
[http://dx.doi.org/10.1109/THMS.2017.2689178]

[47]   J.Y. Park, and Z. Nagy, "Comprehensive analysis of the relationship between thermal comfort and building control research - A data-driven literature review", *Renew. Sustain. Energy Rev.,* vol. 82, pp. 2664-2679, 2018.
[http://dx.doi.org/10.1016/j.rser.2017.09.102]

[48]   P. Kumar, "Indoor air quality and energy management through real-time sensing in commercial buildings", *Energy Build.,* vol. 111, pp. 145-153, 2016.
[http://dx.doi.org/10.1016/j.enbuild.2015.11.037]

[49]   K. Chandra Sahoo, and U. Chandra Pati, "IoT based intrusion detection system using PIR sensor", *2nd IEEE International Conference On Recent Trends in Electronics Information & Communication Technology (RTEICT).* 2017.

[50]   N.M. Saad, M.F. Abas, and D. Pebrianti, "Wireless PIR & D6T thermal sensor based lighting & airconditioning control device for building", *4th IET Clean Energy and Technology Conference.* 2016.
[http://dx.doi.org/10.1049/cp.2016.1283]

[51]   S. Baldi, I. Michailidis, C. Ravanis, and E.B. Kosmatopoulos, "Model-based and model-free 'plug-and-play' building energy efficient control", *Appl. Energy,* vol. 154, pp. 829-841, 2015.
[http://dx.doi.org/10.1016/j.apenergy.2015.05.081]

[52]   P. Lindahl, A.T. Avestruz, W. Thompson, E. George, B.R. Sennett, and S.B. Leeb, "A Transmitter-Receiver System for Long-Range Capacitive Sensing Applications", *IEEE Trans. Instrum. Meas.,* 2016.
[http://dx.doi.org/10.1109/TIM.2016.2575338]

[53]   J. Berry, and K. Park, "A Passive System for Quantifying Indoor Space Utilization", *ACADIA Discipline + Disruption,* 2017.

[54]   P. Skocir, P. Krivic, M. Tomeljak, M. Kusek, and G. Jezic, "Activity Detection in Smart Home Environment", *Procedia Computer Science,* 2016.
[http://dx.doi.org/10.1016/j.procs.2016.08.249]

[55]   S.S. Shetty, H.D. Chinh, M. Gupta, and S.K. Panda, "User presence estimation in multi-occupancy rooms using plug-load meters and PIR sensors", *2017 IEEE Global Communications Conference, GLOBECOM 2017 - Proceedings.* 2018.

[56]   P. Urwyler, R. Stucki, R. Muri, U.P. Mosimann, and T. Nef, "Passive wireless sensor systems can recognize activites of daily living", *Proceedings of the Annual International Conference of the IEEE Engineering in Medicine and Biology Society, EMBS.* 2015.
[http://dx.doi.org/10.1109/EMBC.2015.7320259]

[57]   Y. Li, M. Liu, and W. Sheng, "Indoor human tracking and state estimation by fusing environmental sensors and wearable sensors", *2015 IEEE International Conference on Cyber Technology in Automation, Control and Intelligent Systems, IEEE-CYBER.* 2015.
[http://dx.doi.org/10.1109/CYBER.2015.7288161]

[58]   A. Wickramasinghe, D. C. Ranasinghe, C. Fumeaux, K. D. Hill, and R. Visvanathan, *Sequence Learning with Passive RFID Sensors for Real-Time Bed-Egress Recognition in Older People,* 2017.
[http://dx.doi.org/10.1109/JBHI.2016.2576285]

[59]   R.P. O'Brien, S. Katkoori, and M.A. Rowe, "Design and implementation of an embedded system for monitoring at-home solitary Alzheimer's patients", *Midwest Symposium on Circuits and Systems.* 2015.
[http://dx.doi.org/10.1109/MWSCAS.2015.7282201]

[60]   K. Nakahata, E. Dorronzoro, N. Imamoglu, M. Sekine, K. Kita, and W. Yu, *Active Sensing for Human*

Activity Recognition by a Home Bio-monitoring Robot in a Home Living Environment, 2017. http://www.springer.com/series/11156
[http://dx.doi.org/10.1007/978-3-319-48036-7_23]

[61]   O. Shih, and A. Rowe, "Occupancy Estimation using Ultrasonic Chirps", *ACM/IEEE 6th International Conference on Cyber-Physical Systems.* 2015.
[http://dx.doi.org/10.1145/2735960.2735969]

[62]   B. Sun, M. Shimoyama, and N. Matsuhita, "Human Detection by Active Sensing Using a Laser Range Sensor and a Pan-Tilt Camera", *IEEE 15th International Workshop on Advanced Motion Control (AMC).* 2018.
[http://dx.doi.org/10.1109/AMC.2019.8371113]

[63]   R. Ma, F. Hu, and Q. Hao, "Active Compressive Sensing viaPyroelectric Infrared Sensor for Human Situation Recognition", *IEEE Trans. Syst. Man Cybern. Syst.,* 2017.
[http://dx.doi.org/10.1109/TSMC.2016.2578465]

[64]   P. Mohebbi, E. Stroulia, and I. Nikolaidis, "Indoor localization: a cost-effectiveness vs. accuracy study", *Proceedings - IEEE Symposium on Computer-Based Medical Systems.* 2017.
[http://dx.doi.org/10.1109/CBMS.2017.126]

[65]   G.E. Biccario, M. De Vittorio, and S. D'Amico, "Fluids energy harvesting system with low cut-in velocity piezoelectric MEMS", *2017 IEEE International Conference on IC Design and Technology, ICICDT 2017,* 2017.
[http://dx.doi.org/10.1109/ICICDT.2017.7993506]

[66]   A. Proto, "Wearable PVDF transducer for biomechanical energy harvesting and gait cycle detection", *IECBES 2016 - IEEE-EMBS Conference on Biomedical Engineering and Sciences.,* 2017.

[67]   K.N. Kim, J. Chun, J.W. Kim, K.Y. Lee, J.U. Park, S.W. Kim, Z.L. Wang, and J.M. Baik, "Highly Stretchable 2D Fabrics for Wearable Triboelectric Nanogenerator under Harsh Environments", *ACS Nano,* vol. 9, no. 6, pp. 6394-6400, 2015.
[http://dx.doi.org/10.1021/acsnano.5b02010] [PMID: 26051679]

[68]   W. Seung, M.K. Gupta, K.Y. Lee, K.S. Shin, J.H. Lee, T.Y. Kim, S. Kim, J. Lin, J.H. Kim, and S.W. Kim, "Nanopatterned textile-based wearable triboelectric nanogenerator", *ACS Nano,* vol. 9, no. 4, pp. 3501-3509, 2015.
[http://dx.doi.org/10.1021/nn507221f] [PMID: 25670211]

[69]   K. Leuenberger, R. Gonzenbach, S. Wachter, A. Luft, and R. Gassert, "A method to qualitatively assess arm use in stroke survivors in the home environment", *Med. Biol. Eng. Comput.,* vol. 55, no. 1, pp. 141-150, 2017.
[http://dx.doi.org/10.1007/s11517-016-1496-7] [PMID: 27106757]

[70]   W. Li, S. Zhao, N. Wu, J. Zhong, B. Wang, S. Lin, S. Chen, F. Yuan, H. Jiang, Y. Xiao, B. Hu, and J. Zhou, "Sensitivity-Enhanced Wearable Active Voiceprint Sensor Based on Cellular Polypropylene Piezoelectret", *ACS Appl. Mater. Interfaces,* vol. 9, no. 28, pp. 23716-23722, 2017.
[http://dx.doi.org/10.1021/acsami.7b05051] [PMID: 28613808]

[71]   I. Pigliautile, S. Casaccia, N. Morresi, M. Arnesano, A.L. Pisello, and G.M. Revel, "Assessing occupants ' personal attributes in relation to human perception of environmental comfort : measurement procedure and data analysis", *Build. Environ.,* vol. 177, p. 106901, 2020.
[http://dx.doi.org/10.1016/j.buildenv.2020.106901]

[72]   A. Ghahramani, "Learning occupants' workplace interactions from wearable and stationary ambient sensing systems", *Appl. Energy,* vol. 230, no. April, pp. 42-51, 2018.
[http://dx.doi.org/10.1016/j.apenergy.2018.08.096]

[73]   P. Jayathissa, M. Quintana, T. Sood, N. Nazarian, and C. Miller, "Is your clock-face cozie? A smartwatch methodology for the in-situ collection of occupant comfort data", *J. Phys. Conf. Ser.,* vol. 1343, no. 1, 2019.
[http://dx.doi.org/10.1088/1742-6596/1343/1/012145]

[74] P. Jayathissa, M. Quintana, M. Abdelrahman, and C. Miller, "Humans-as-Sensor for Buildings—Intensive Longitudinal Indoor Comfort Models", *Buildings,* vol. 174, pp. 1-22, 2020.
[http://dx.doi.org/10.3390/buildings10100174]

[75] I. Chew, D. Karunatilaka, C.P. Tan, and V. Kalavally, "Smart lighting: The way forward? Reviewing the past to shape the future", *Energy Build.,* vol. 149, pp. 180-191, 2017.
[http://dx.doi.org/10.1016/j.enbuild.2017.04.083]

[76] G. Happle, E. Wilhelm, J.A. Fonseca, and A. Schlueter, "Determining air-conditioning usage patterns in Singapore from distributed, portable sensors", *Energy Procedia,* vol. 122, pp. 313-318, 2017.
[http://dx.doi.org/10.1016/j.egypro.2017.07.328]

[77] M. Victoria Moreno, and A.F. Skarmeta, "An indoor localization system based on 3D magnetic fingerprints for smart buildings", *IEEE nternational Conference on Computing and Communication Technologies: Research, Innovation, and Vision for Future (RIVF),* pp. 186-191, 2015.

[78] F.D. Salim, "Modelling urban-scale occupant behaviour, mobility, and energy in buildings: A survey", *Build. Environ.,* vol. 183, p. 106964, 2020.
[http://dx.doi.org/10.1016/j.buildenv.2020.106964]

[79] H. Zou, Y. Zhou, J. Yang, and C.J. Spanos, "Device-free occupancy detection and crowd counting in smart buildings with WiFi-enabled IoT", *Energy Build.,* vol. 174, pp. 309-322, 2018.
[http://dx.doi.org/10.1016/j.enbuild.2018.06.040]

[80] H. Zou, Y. Zhou, J. Yang, and C.J. Spanos, "Towards occupant activity driven smart buildings viaWiFi-enabled IoT devices and deep learning", *Energy Build.,* vol. 177, pp. 12-22, 2018.
[http://dx.doi.org/10.1016/j.enbuild.2018.08.010]

[81] N.A. Zakaria, Z.Z. Abidin, N. Harum, L.C. Hau, N.S. Ali, and F.A. Jafar, "Wireless internet of things-based air quality device for smart pollution monitoring", *Int. J. Adv. Comput. Sci. Appl.,* vol. 9, no. 11, pp. 65-69, 2018.
[http://dx.doi.org/10.14569/IJACSA.2018.091110]

[82] J. Xie, H. Li, C. Li, J. Zhang, and M. Luo, "Review on occupant-centric thermal comfort sensing, predicting, and controlling", *Energy Build.,* vol. 226, p. 110392, 2020.
[http://dx.doi.org/10.1016/j.enbuild.2020.110392]

[83] A.P. Ramallo-Gonz, "000E1;lez, V. Tomat, P. J. Fern&000E1;ndez-Ruiz, M. &000C1;ngel Zamora-Izquierdo, and A. F. Skarmeta-G&000F3;mez, "Conceptualisation of an IoT framework for multi-person interaction with conditioning systems", *Energies,* vol. 13, no. 12, 2020.
[http://dx.doi.org/10.3390/en13123094]

[84] C.H. Lu, "IoT-enabled adaptive context-aware and playful cyber-physical system for everyday energy savings", *IEEE Trans. Hum. Mach. Syst.,* vol. 48, no. 4, pp. 380-391, 2018.
[http://dx.doi.org/10.1109/THMS.2018.2844119]

[85] S. Saeidi, C. Chokwitthaya, Y. Zhu, and M. Sun, *Spatial-temporal event-driven modeling for occupant behavior studies using immersive virtual environments,* 2018.
[http://dx.doi.org/10.1016/j.autcon.2018.07.019]

[86] S. Saeidi, Y. Zhu, and C. Chokwitthaya, "Application of immersive virtual environments for longitudinal data collection: a pilot study", *Construction Research Congress 2018,* Sustainable Design and Construction and Education, pp. 208-215, 2018. https://www.wunderground.com/
[http://dx.doi.org/10.1061/9780784481301.021]

[87] S. Niu, W. Pan, and Y. Zhao, "A virtual reality integrated design approach to improving occupancy information integrity for closing the building energy performance gap", *Sustain. Cities Soc.,* vol. 27, pp. 275-289, 2016.
[http://dx.doi.org/10.1016/j.scs.2016.03.010]

[88] E. Vilar, F. Rebelo, P. Noriega, J. Teles, and C. Mayhorn, "Signage *versus* environmental affordances: Is the explicit information strong enough to guide human behavior during a wayfinding task?", *Hum.*

Factors Ergon. Manuf. Serv. Ind., vol. 25, pp. 439-452, 2015.
[http://dx.doi.org/10.1002/hfm.20557]

[89] Y. Zhu, S. Saeidi, T. Rizzuto, A. Roetzel, and R. Kooima, "Potential and challenges of immersive virtual environments for occupant energy behavior modeling and validation: A literature review", *J. Build. Eng.,* vol. 19, pp. 302-319, 2018.
[http://dx.doi.org/10.1016/j.jobe.2018.05.017]

[90] A. Heydarian, J.P. Carneiro, D. Gerber, and B. Becerik-Gerber, "Immersive virtual environments, understanding the impact of design features and occupant choice upon lighting for building performance", *Build. Environ.,* vol. 89, pp. 217-228, 2015.
[http://dx.doi.org/10.1016/j.buildenv.2015.02.038]

[91] F. Stazi, F. Naspi, and M. D'Orazio, "A literature review on driving factors and contextual events influencing occupants' behaviours in buildings", *Build. Environ.,* vol. 118, pp. 40-66, 2017.
[http://dx.doi.org/10.1016/j.buildenv.2017.03.021]

[92] A. Wolff, I. Weber, B. Gill, J. Schubert, and M. Schneider, "Tackling the interplay of occupants' heating practices and building physics: Insights from a German mixed methods study", *Energy Res. Soc. Sci.,* vol. 32, pp. 65-75, 2017.
[http://dx.doi.org/10.1016/j.erss.2017.07.003]

[93] W. O'Brien, and H.B. Gunay, "The contextual factors contributing to occupants' adaptive comfort behaviors in offices - A review and proposed modeling framework", *Build. Environ.,* vol. 77, pp. 77-87, 2014.
[http://dx.doi.org/10.1016/j.buildenv.2014.03.024]

[94] B.K. Sovacool, "Diversity: Energy studies need social science", *Nature,* vol. 511, no. 7511, pp. 529-530, 2014.
[http://dx.doi.org/10.1038/511529a] [PMID: 25079540]

[95] C. Carpino, D. Mora, and M. De Simone, "On the use of questionnaire in residential buildings. A review of collected data, methodologies and objectives", *Energy Build.,* vol. 186, pp. 297-318, 2019.
[http://dx.doi.org/10.1016/j.enbuild.2018.12.021]

[96] Z. Deme Belafi, T. Hong, and A. Reith, "A critical review on questionnaire surveys in the field of energy-related occupant behaviour", *Energy Effic.,* vol. 11, no. 8, pp. 2157-2177, 2018.
[http://dx.doi.org/10.1007/s12053-018-9711-z]

[97] P.J. Lavrakas, *Encyclopedia of Survey Research Methods.* 4th ed. SAGE Publications, 2009.

[98] ASHRAE, *Standard 55 - Thermal environmental conditions for human occupancy.* ASHRAE, 2013.

[99] E.D. de Leeuw, J. Hox, and D. Dillman, *International Handbook of Survey Methodology.* vol. Vol. 19. Routledge, 2012.
[http://dx.doi.org/10.4324/9780203843123]

[100] X. Feng, D. Yan, and C. Wang, "Classification of occupant air-conditioning behavior patterns", *14th Conference of International Building Performance Simulation Association,* 2015p. 1516 1522.

[101] M.V. Bavaresco, and E. Ghisi, "Influence of user interaction with internal blinds on the energy efficiency of office buildings", *Energy Build.,* vol. 166, pp. 538-549, 2018.
[http://dx.doi.org/10.1016/j.enbuild.2018.02.011]

[102] A.S. Silva, L.S.S. Almeida, and E. Ghisi, "Decision-making process for improving thermal and energy performance of residential buildings: A case study of constructive systems in Brazil", *Energy Build.,* vol. 128, pp. 270-286, 2016.
[http://dx.doi.org/10.1016/j.enbuild.2016.06.084]

[103] C. Vogiatzi, G. Gemenetzi, L. Massou, S. Poulopoulos, S. Papaefthimiou, and E. Zervas, "Energy use and saving in residential sector and occupant behavior: A case study in Athens", *Energy Build.,* vol. 181, pp. 1-9, 2018.
[http://dx.doi.org/10.1016/j.enbuild.2018.09.039]

[104] D. Li, and X. Xu, "C. fei Chen, and C. Menassa, "Understanding energy-saving behaviors in the American workplace: A unified theory of motivation, opportunity, and ability", *Energy Res. Soc. Sci.,* vol. 51, pp. 198-209, 2019.
[http://dx.doi.org/10.1016/j.erss.2019.01.020]

[105] E.L. Hewitt, C.J. Andrews, J.A. Senick, R.E. Wener, U. Krogmann, and M. Sorensen Allacci, "Distinguishing between green building occupants reasoned and unplanned behaviours", *Build. Res. Inform.,* vol. 44, no. 2, pp. 119-134, 2016.
[http://dx.doi.org/10.1080/09613218.2015.1015854]

[106] A.K. Mishra, and M. Ramgopal, "A thermal comfort field study of naturally ventilated classrooms in Kharagpur, India", *Build. Environ.,* vol. 92, pp. 396-406, 2015.
[http://dx.doi.org/10.1016/j.buildenv.2015.05.024]

[107] M.K. Singh, S. Mahapatra, and J. Teller, "Relation between indoor thermal environment and renovation in Liege residential buildings", *Therm. Sci.,* vol. 18, no. 3, pp. 889-902, 2014.
[http://dx.doi.org/10.2298/TSCI1403889S]

[108] C. Sun, R. Zhang, S. Sharples, Y. Han, and H. Zhang, "A longitudinal study of summertime occupant behaviour and thermal comfort in office buildings in northern China", *Build. Environ.,* vol. 143, pp. 404-420, 2018.
[http://dx.doi.org/10.1016/j.buildenv.2018.07.004]

[109] M. Indraganti, D. Boussaa, S. Assadi, and E. Mostavi, "User satisfaction and energy use behavior in offices in Qatar", *Build. Serv. Eng. Res. Tech.,* vol. 39, no. 4, pp. 391-405, 2018.
[http://dx.doi.org/10.1177/0143624417751388]

[110] G.H. Lim, N. Keumala, and N.A. Ghafar, "Energy saving potential and visual comfort of task light usage for offices in Malaysia", *Energy Build.,* vol. 147, pp. 166-175, 2017.
[http://dx.doi.org/10.1016/j.enbuild.2017.05.004]

[111] J.F. Robinson, T.J. Foxon, and P.G. Taylor, "Performance gap analysis case study of a non-domestic building", *Eng. Sustain.,* vol. 169, no. 1, pp. 31-38, 2015.
[http://dx.doi.org/10.1680/ensu.14.00055]

[112] J.K. Day, and W. O'Brien, "Oh behave! Survey stories and lessons learned from building occupants in high-performance buildings", *Energy Res. Soc. Sci.,* vol. 31, pp. 11-20, 2017.
[http://dx.doi.org/10.1016/j.erss.2017.05.037]

[113] N. G. Vásquez, M. L. Felippe, F. O. R. Pereira, and A. Kuhnen, *Luminous and visual preferences of young children in their classrooms: Curtain use, artificial lighting and window views,* 2019.
[http://dx.doi.org/10.1016/j.buildenv.2019.01.049]

[114] N.G. Vásquez, R.F. Rupp, L.A. Díaz, A.G. Cardona, and D.M. Arenas, "Testing a method to assess the thermal sensation and preference of children in kindergartens", *30th International PLEA Conference: Sustainable Habitat for Developing Societies: Choosing the Way Forward - Proceedings,* 2014 p. 492 499.

[115] O. Guerra-Santin, S. Boess, T. Konstantinou, N. Romero Herrera, T. Klein, and S. Silvester, "Designing for residents: Building monitoring and co-creation in social housing renovation in the Netherlands", *Energy Res. Soc. Sci.,* vol. 32, pp. 164-179, 2017.
[http://dx.doi.org/10.1016/j.erss.2017.03.009]

[116] E. Brandt, T. Binder, and E.B-N. Sanders, Tools and techniques: Ways to engage telling, making and enacting.*Routledge International Handbook of Participatory Design.* Routledge - Taylor & Francis Group: London, 2012, pp. 145-181.
[http://dx.doi.org/10.4324/9780203108543.ch7]

[117] J. Kim, R. de Dear, T. Parkinson, and C. Candido, "Understanding patterns of adaptive comfort behaviour in the Sydney mixed-mode residential context", *Energy Build.,* vol. 141, pp. 274-283, 2017.
[http://dx.doi.org/10.1016/j.enbuild.2017.02.061]

[118]   K. Kalvelage, and M.C. Dorneich, "Using human factors to establish occupant task lists for office building simulations", *Proceedings of the Human Factors and Ergonomics Society.,* 2016 p. 450 454.
[http://dx.doi.org/10.1177/1541931213601102]

[119]   V. Haines, K. Kyriakopoulou, and C. Lawton, "End user engagement with domestic hot water heating systems: Design implications for future thermal storage technologies", *Energy Res. Soc. Sci.,* vol. 49, pp. 74-81, 2019.
[http://dx.doi.org/10.1016/j.erss.2018.10.009]

[120]   G. Rojas, W. Wagner, J. Suschek-Berger, R. Pfluger, and W. Feist, "Applying the passive house concept to a social housing project in Austria – evaluation of the indoor environment based on long-term measurements and user surveys", *Adv. Build. Energy Res.,* vol. 10, no. 1, pp. 125-148, 2016.
[http://dx.doi.org/10.1080/17512549.2015.1040072]

[121]   B. Ozarisoy, and H. Altan, "Adoption of energy design strategies for retrofitting mass housing estates in Northern Cyprus", *Sustain.,* vol. 9, no. 8, p. 1477, 2017.
[http://dx.doi.org/10.3390/su9081477]

[122]   S. Oliveira, E. Marco, B. Gething, and C. Robertson, "Exploring Energy Modelling in Architecture Logics of Investment and Risk", *Energy Procedia,* vol. 111, pp. 61-70, 2017.
[http://dx.doi.org/10.1016/j.egypro.2017.03.008]

[123]   S. Attia, S. Bilir, T. Safy, C. Struck, R. Loonen, and F. Goia, "Current trends and future challenges in the performance assessment of adaptive façade systems", *Energy Build.,* vol. 179, pp. 165-182, 2018.
[http://dx.doi.org/10.1016/j.enbuild.2018.09.017]

[124]   J. Lee, and M. Shepley, "Analysis of human factors in a building environmental assessment system in Korea: Resident perception and the G-SEED for MF scores", *Build. Environ.,* vol. 142, pp. 388-397, 2018.
[http://dx.doi.org/10.1016/j.buildenv.2018.06.044]

[125]   W. Pan, and M. Pan, "A dialectical system framework of zero carbon emission building policy for high-rise high-density cities: Perspectives from Hong Kong", *J. Clean. Prod.,* vol. 205, pp. 1-13, 2018.
[http://dx.doi.org/10.1016/j.jclepro.2018.09.025]

[126]   J. Carlander, K. Trygg, and B. Moshfegh, "Integration of Measurements and Time Diaries as Complementary Measures to Improve Resolution of BES", *Energies,* vol. 12, no. 11, p. 2072, 2019.
[http://dx.doi.org/10.3390/en12112072]

[127]   K. Vrotsou, M. Bergqvist, M. Cooper, and K. Ellegård, *PODD: A portable diary data collection system.* Proc. Work. Adv. Vis. Interfaces AVI, 2014, pp. 381-382.
[http://dx.doi.org/10.1145/2598153.2600046]

[128]   V.M. Barthelmes, R. Li, R.K. Andersen, W. Bahnfleth, S.P. Corgnati, and C. Rode, "Profiling occupant behaviour in Danish dwellings using time use survey data", *Energy Build.,* vol. 177, pp. 329-340, 2018.
[http://dx.doi.org/10.1016/j.enbuild.2018.07.044]

[129]   L. Diao, Y. Sun, Z. Chen, and J. Chen, "Modeling energy consumption in residential buildings: A bottom-up analysis based on occupant behavior pattern clustering and stochastic simulation", *Energy Build.,* vol. 147, pp. 47-66, 2017.
[http://dx.doi.org/10.1016/j.enbuild.2017.04.072]

[130]   A. Wang, R. Li, and S. You, "Development of a data driven approach to explore the energy flexibility potential of building clusters", *Appl. Energy,* vol. 232, pp. 89-100, 2018.
[http://dx.doi.org/10.1016/j.apenergy.2018.09.187]

[131]   S. Kim, S. Jung, and S.M. Baek, "A model for predicting energy usage pattern types with energy consumption information according to the behaviors of single-person households in South Korea", *Sustain.,* vol. 11, no. 1, 2019.
[http://dx.doi.org/10.3390/su11010245]

[132] G. McGill, M. Qin, and L. Oyedele, "A case study investigation of indoor air quality in UK Passivhaus dwellings", *Energy Procedia,* vol. 62, pp. 190-199, 2014.
[http://dx.doi.org/10.1016/j.egypro.2014.12.380]

[133] T. Lovett, "Designing sensor sets for capturing energy events in buildings", *Build. Environ.,* vol. 110, pp. 11-22, 2016.
[http://dx.doi.org/10.1016/j.buildenv.2016.09.004]

[134] R. Escandón, J.J. Sendra, and R. Suárez, "Energy and climate simulation in the Upper Lawn Pavilion, an experimental laboratory in the architecture of the Smithsons", *Build. Simul.,* vol. 8, no. 1, pp. 99-109, 2015.
[http://dx.doi.org/10.1007/s12273-014-0197-0]

[135] W.F.E. Preiser, "Feedback, feedforward and control: post-occupancy evaluation to the rescue", *Build. Res. Inform.,* vol. 29, no. 6, pp. 456-459, 2001.
[http://dx.doi.org/10.1080/09613210110072692]

[136] J.K. Day, S. Ruiz, W. O'Brien, and M. Schweiker, "Seeing is believing: an innovative approach to post-occupancy evaluation", *Energy Effic.,* vol. 13, no. 3, pp. 473-486, 2020.
[http://dx.doi.org/10.1007/s12053-019-09817-8]

[137] Y. Geng, W. Ji, Z. Wang, B. Lin, and Y. Zhu, "A review of operating performance in green buildings: Energy use, indoor environmental quality and occupant satisfaction", *Energy Build.,* vol. 183, pp. 500-514, 2019.
[http://dx.doi.org/10.1016/j.enbuild.2018.11.017]

[138] A. Vásquez-Hernández, and M.F. Restrepo Álvarez, "Evaluation of buildings in real conditions of use: Current situation", *J. Build. Eng.,* vol. 12, pp. 26-36, 2017.
[http://dx.doi.org/10.1016/j.jobe.2017.04.019]

[139] F.A. Mustafa, "Performance assessment of buildings viapost-occupancy evaluation: A case study of the building of the architecture and software engineering departments in Salahaddin University-Erbil, Iraq", *Front. Archit. Res.,* vol. 6, no. 3, pp. 412-429, 2017.
[http://dx.doi.org/10.1016/j.foar.2017.06.004]

[140] J.Y. Park, "A critical review of field implementations of occupant-centric building controls", *Build. Environ.,* vol. 165, no. July, p. 106351, 2019.
[http://dx.doi.org/10.1016/j.buildenv.2019.106351]

[141] A. AlHogail, and M. AlShahrani, "Building Consumer Trust to Improve Internet of Things (IoT) Technology Adoption", *International Conference on Neuroergonomics and Cognitive Engineering2 ,* 2018 p. 325 334.

[142] S.A. Sadeghi, P. Karava, I. Konstantzos, and A. Tzempelikos, "Occupant interactions with shading and lighting systems using different control interfaces: A pilot field study", *Build. Environ.,* vol. 97, pp. 177-195, 2016.
[http://dx.doi.org/10.1016/j.buildenv.2015.12.008]

[143] S. Monfils, and J.M. Hauglustaine, "Introduction of behavioral parameterization in the EPC calculation method and assessment of five typical urban houses in Wallonia, Belgium", *Sustain.,* vol. 8, no. 11, 2016.
[http://dx.doi.org/10.3390/su8111205]

[144] V.L. Castaldo, I. Pigliautile, F. Rosso, F. Cotana, F. De Giorgio, and A.L. Pisello, "How subjective and non-physical parameters affect occupants' environmental comfort perception", *Energy Build.,* vol. 178, pp. 107-129, 2018.
[http://dx.doi.org/10.1016/j.enbuild.2018.08.020]

[145] J.A. Veitch, K.E. Charles, K.M.J. Farley, and G.R. Newsham, "A model of satisfaction with open-plan office conditions: COPE field findings", *J. Environ. Psychol.,* vol. 27, no. 3, pp. 177-189, 2007.
[http://dx.doi.org/10.1016/j.jenvp.2007.04.002]

[146] A.C. Boerstra, M.G.L.C. Loomans, and J.L.M. Hensen, "Personal Control Over Indoor Climate and Productivity", *Indoor Air conference,* pp. 1-8, 2014.

[147] E.L.A. da Guarda, "The influence of climate change on renewable energy systems designed to achieve zero energy buildings in the present : A case study in the Brazilian Savannah", *Sustain. Cities Soc.,* vol. 52, p. 101843, 2020.
[http://dx.doi.org/10.1016/j.scs.2019.101843]

[148] J. Kim, "Design Automation for Smart Building Systems", *Proc. IEEE,* vol. 106, no. 9, pp. 1680-1699, 2018.
[http://dx.doi.org/10.1109/JPROC.2018.2856932]

CHAPTER 2

Monitoring Occupant Window Opening Behaviour in Buildings: A Critical Review

Shen Wei[1,*], Yan Ding[2] and Wei Yu[3, 4]

[1] *The Bartlett School of Sustainable Construction, University College London, London, United Kingdom*

[2] *School of Environmental Science and Engineering, Tianjin Key Laboratory of Indoor Air Environmental Quality Control, Tianjin University, Tianjin, 300072, China*

[3] *Joint International Research Laboratory of Green Buildings and Built Environments (Ministry of Education), Chongqing University, Chongqing 400045, China*

[4] *National Centre for International Research of Low-carbon and Green Buildings (Ministry of Science and Technology), Chongqing University, Chongqing 400045, China*

Abstract: People's behaviour can significantly impact both the energy consumption and the indoor thermal environment of the buildings, and of particular interest is their window opening behaviour. A better understanding of why, when and how occupants open windows is, therefore, essential in the quest to achieve low-carbon buildings. Many studies have sought to answer these questions based on behavioural data measured in actual buildings. This paper introduces existing methods that have been used to monitor occupant window opening behaviour in buildings based on a comprehensive review of literature, as well as for relevant influential factors, and critically discusses the advantages and disadvantages of each method. The review has identified five methods monitoring window usage (*i.e.* self-recording, electronic recording, observing by surveyors and self-estimating), and each method has its advantages and disadvantages in terms of feasible sample size, monitoring interval and duration, recognition of window states/opening angle, and the relative dynamic nature of behaviour. The aim has been to provide researchers with systematic criteria for selecting a suitable monitoring method for their specific research objectives. Additionally, the paper demonstrates the need for a standard method for monitoring relevant influential factors, as these varied considerably between existing studies with respect to the accuracy, interval and location. Such variation clearly has the potential to influence the ability to perform cross-study comparisons.

Keywords: Behavioural modelling, Buildings, Driver, Energy, Electronic measurement, Indoor air quality, Indoor environment, Monitoring, Outdoor environment, Window opening behaviour.

* **Corresponding author Shen Wei:** The Bartlett School of Sustainable Construction, University College London, London, United Kingdom; Tel: +0044 203 108 3060; E-mail: shen.wei@ucl.ac.uk

Enedir Ghisi, Ricardo Forgiarini Rupp and Pedro Fernandes Pereira (Eds.)

INTRODUCTION

The high importance of occupants' role in the performance of buildings has been demonstrated by both real measured data [1 - 7] and building performance simulation [8 - 14]. As Gram-Hanssen [15], Ben and Steemers [16] and Pisello and Asdrubali [17] have argued, how occupants use the building has a direct and significant impact on the building's indoor environment and energy consumption, no less than the building construction and building systems. Furthermore, occupants' behaviour within the buildings can also affect their comfort perceptions: when provided with a higher level of adaptive opportunities [18] (*e.g.* opening/closing windows, adjusting blind/shading positions and turning up/down thermostatic settings) to rebalance comfort, occupants displayed greater comfort acceptance [19 - 21].

In order to achieve energy efficient buildings the aspect of occupant behaviour is crucial and should not be neglected [22 - 27]. It has been demonstrated by Masoso and Grobler [28] and Guerra-Santin [29] that improper building use may cause a significant waste of energy, and user-centred building control strategy can greatly help to reduce the building's energy demand [30, 31]. Furthermore, according to Ben and Steemers [16] and Wei *et al.* [32] insufficient consideration of occupant behaviour when retrofitting/refurbishing buildings may also lead to improperly selected energy efficient measures. Aiming at a golden role proposed by Masoso and Grobler [28], *i.e.* '*If you don't need it, don't use it*', for building energy saving, a number of studies have been carried out to make occupants use buildings more energy efficiently [33 - 40], achieving low-carbon life styles [41 - 44]. Mullaly [45] has pointed out the importance of this task should be of serious concern not only by building users, but also by local and national government.

Occupant behaviour is a complex process: it is influenced by a number of factors [46 - 49] and can manifest itself in what Peng *et al.* [50], call various modes, *i.e.* time-related, environment-related and random. When introduced into building simulation models, occupant behaviour appears to be one of the biggest contributors to the gap between the predicted and actual building performance, commonly considered as performance gap [51 - 56]. This being the case, it has been argued that improved representation of occupant behaviour in building simulation is of paramount importance for the optimisation of building design for real applications [57 - 59].

Windows, as an important component of building façade, have a major impact on buildings' indoor visual environment and energy consumption [60 - 62]. In non-air-conditioned buildings, opening windows provides a useful way for occupants to adjust their indoor thermal environment [63, 64] and air quality [65, 66], *via* promoted air exchange between indoors and outdoors [67 - 70]. However when the building is heated or cooled mechanically, opening windows will cause extra energy loss [62, 71 - 74]. Wei [75] has revealed that in the past 30 years, a number of studies have been performed using engineering methods to gain a better understanding of why, when and how occupants open windows. Generally, these studies started from a field monitoring of occupant window use in buildings, concurrently with a recording of relevant environmental and non-environmental parameters, as shown in Fig. (**1**). Then the studies can be grouped into two methodological fluxes. Studies following Flux 1 were designed mainly to evaluate the impact of occupant behaviour on the building performance, such as indoor environmental quality and energy consumption, based on the real measured data from the building [68, 76, 77]. Studies based on Flux 2 were aiming for deeper research on occupant window opening behaviour, namely developing reliable behavioural models, to reduce the gap between the building simulation results and actual building performance [78 - 83]. These studies generally included a 'driver determination' stage, in which all monitored environmental and non-environmental parameters were assessed for their influence on the state of the windows. The parameters with significant influences were categorised as 'drivers'/'factors' of window opening behaviour [46, 47] and were used to build statistical models for window state predictions [84]. After a validation process, these models would be used in building performance simulation to provide more reliable simulation results to support building design and operation.

In both fluxes shown in Fig. (**1**), apparently, the monitoring stage is fundamental as it provides the raw data that is used for later analysis. Therefore, a critical understanding and controlling of the accuracies of available data collection methods are extremely important both existing data usage and future data collection. This paper, therefore, presents a critical review of existing methods that have been used to monitor occupant window opening behaviour in buildings, and discussed their advantages and disadvantages. Additionally, existing methods that have been used to monitor relevant influential factors have been critically reviewed and compared as well.

MONITORING OCCUPANT WINDOW OPENING BEHAVIOUR

In existing studies regarding occupant window opening behaviour in buildings, there are five methods that have been used to monitor the window usage (either on/off or opening angles) by building occupants:

1. Self-recording by building occupants;
2. Recording by electronic measuring devices;

3. Observing by surveyors;
4. Self-estimating by building occupants; and,
5. Camera-based estimation.

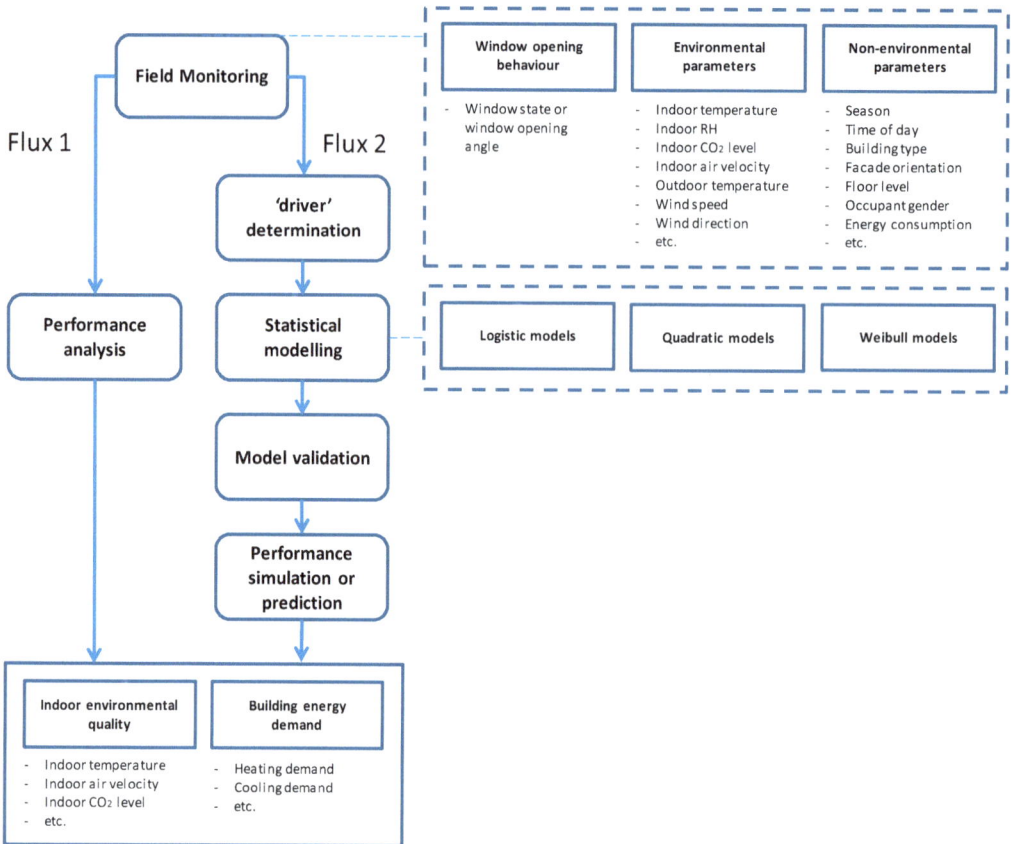

Fig. (1). Two methodological fluxes for studies of occupant window opening behaviour.

The studies reviewed in this section were all focusing on occupant adaptive behaviour in buildings with window opening behaviour as one major component of the analysis. In total, 50 studies involving this topic have been found. The papers cited in the following discussions have been taken mainly from (1) high SCI impact journals (75%), such as Energy and Buildings or Building and Environment; (2) key conferences (20%), such as the ACEEE Summer Study conference and the Windsor conference; and (3) academic technical reports or PhD thesis (5%), using keywords like 'occupant behaviour', 'adaptive behaviour', 'window behaviour' and 'window control'.

Self-Recording Window Usage By Building Occupants

The self-recording method for window usage involves asking building occupants to record the window state themselves by filling out questionnaires or logs. This method has been used in several existing studies to monitor occupants' window opening behaviour in buildings.

Haldi and Robinson [85] carried out a case study in several non-air-conditioned office buildings in Switzerland, during which the building occupants were asked to record their clothing level, activity level, thermal sensation and preference, as well as adaptive opportunities, such as opening a window, exercised during the preceding hour. Occupants recorded the information by completing a short electronic questionnaire installed on their personal computers, three or four times per day. A similar method has been adopted by Nicol and Humphreys to monitor occupants' window usage in three projects that explored thermal comfort and human behaviour in non-air-conditioned office buildings [86 - 88]. All three projects included a longitudinal and a transverse survey. The longitudinal survey collected data, including the use of windows, from building occupants three to four times a day and allowed assessment of changes in the building's environment and the occupants' response to these changes. The transverse survey was performed once a month for each participant, aiming to increase the number of samples (the number of subjects investigated) of the study. In Denmark, Andersen *et al.* [89] have asked occupants to provide information about their window usage using an online questionnaire, by answering the question *"Right now, is the window in the room you are sitting in open?"*. A similar question has been used by Liu *et al.* [90] in a field study carried out in China.

In two field surveys conducted in August and September 2003 in Hyogo and Osaka in Japan, Nakaya *et al.* [91] asked building occupants to self-record their window usage using questionnaires, at a frequency of 10 minutes. A similar questionnaire has been used by Iwashita and Akasaka [72] when evaluating the influence of occupant behaviour on natural ventilation rates and indoor air environment in summer. Beko *et al.* [66] also adopted this method to investigate the impact of indoor environment on asthma and allergy among children. In this study, the 'opening angle' of windows (closed, ajar or fully open) rather than 'window state' (open or closed) was recorded continuously by the children's parents [92]. In a study carried out in California, USA, Offermann [65] asked building occupants to self-record their use of windows, using a log placed on the glass or panel. In Germany, Hellwig *et al.* [93] carried out field studies in two schools for an estimation of the influence of occupants' window opening behaviour on the classrooms' indoor thermal environment, and the self-recording method was adopted in the study carried out in the first school.

Recording Window Usage by Electronic Measuring Devices

Electronic measuring devices provide a convenient and continuous way to record occupants' window use and they have been widely used in existing studies. Normally, an electronic measuring device includes a sensor, which can detect the window state or opening angle, and this state/angle will be recorded automatically by a data logger at a specified logging interval. Two types of sensors have been adopted by Andersen *et al.* [94] in their eight-month field study carried out in 15 Danish dwellings, so both the window opening state and opening angle can be captured. For measuring window states, a pair of magnetic contact sensors was used to determine whether the window is closed or open, and for measuring the window opening angle an accelerometer was used. Figs. (**2a and 2b**) have shown an example of each method respectively. The window state was measured by gathering and storing momentary contact events using special data loggers, such as an event data logger. The accelerometer method of measuring window position applies the principle that by measuring angular displacement in three specific spatial directions, x, y and z, the tilt and duration of an opened window can be determined. In the studies carried out by Schweiker *et al.* [95] in both Switzerland and Japan, window opening angle was measured by a potentiometer whose electrical resistance varies with the opening angle. Herkel *et al.* [81] have used two pairs of window contacts to differentiate between tilted open and fully open, one on the side and one on the bottom of the frame.

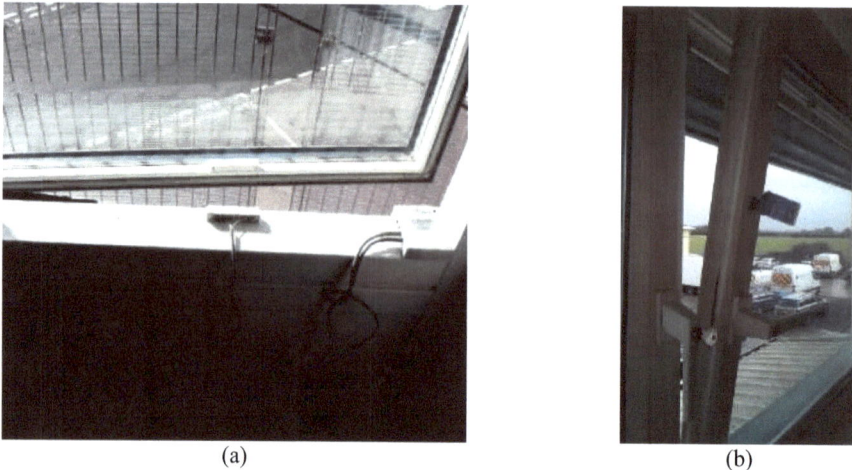

(a)   (b)

Fig. (2). Measurement of window state **(a)** and window opening angle **(b)**.

This technique has been widely adopted by other researchers as well to capture occupants' window opening behaviour in both residential and office buildings. These include studies by Yun and Steemers [79, 80], Yun *et al.* [96], Haldi and

Robinson [78], Dutton and Shao [97], Fritsch *et al.* [98], Erhorn [99], Antretter *et al.* [100], Li *et al.* [101], Weihl and Gladhart [102], Schakib-Ekbatan *et al.* [103], Jeong *et al.* [104], Wei *et al.* [105] and Bruce-Konuah [106], as well as the second of the studies already mentioned, that were carried out by Hellwig *et al.* [93].

Recording Window Usage by Surveyor Observations

Another option is for surveyors to record the state of windows themselves through observations. Using this method, the surveyors need to either develop a fully-structured questionnaire (Table **1**) before the survey so the observed window states and other important parameters can be recorded efficiently and accurately [107 - 109], or they can photograph the monitored building façade first and then decide the window states with reference to the photos after the survey [106, 110 - 112].

Table 1. A fully-structured questionnaire used by Johnson and Long [108].

Visit	Time	Number of Open Windows Per Wall	Number of Open Doors (omit Garage)	Floor Location of Openings (circle One)	Status of Car Door of Attached Garage	Likelihood of AC Operation (circle One)	Likelihood of Occupancy (circle One)	Evidence Supporting Occupancy Rating	Precip. During Last Hour	Special Conditions (Write in)
A	am pm	39 Front__ 40 Right__ 41 Left__ 42 Back__> 43 Total__	44 Front__ 45 Right__ 46 Left__ 47 Back__ 48 Total__	49 None 50 Ground 51 Upper 52 Both	53 Closed 54 Open w/vehicle 55 Open w/o vehicle	56 100% 57 >50% 58 <50% 59 0% 60 Uncertain	61 100% 62 >50% 63 <50% 64 0% 65 Uncertain	Write in:	66 Yes 67 No 68 Uncertain	
B	am pm	69 Front__ 70 Right__ 71 Left__ 72 Back__ 73 Total__	74 Front__ 75 Right__ 76 Left__ 77 Back__ 78 Total__	79 None 80 Ground 81 Upper 82 Both	83 Closed 84 Open w/vehicle 85 Open w/o vehicle	86 100% 87 >50% 88 <50% 89 0% 90 Uncertain	91 100% 92 >50% 93 <50% 94 0% 95 Uncertain	Write in:	96 Yes 97 No 98 Uncertain	

In order to develop a realistic algorithm for simulating the opening and closing of windows, Johnson and Long [108] adopted the first method to monitor the state of windows from 1100 dwellings. In their study, a fully structured questionnaire, as shown in Table **1**, was used to record the observed state of windows and other important parameters. This included questions about the time of the observation occurred, number of open windows per wall, floor location of openings *etc.* Wei *et al.* [107] have also used the first method to record the end-of-day window states in 36 cellular offices in an office building, so as to investigate the significance of a number of non-environmental factors on occupants' choice of end-of-day window states. Brundrett [109] has surveyed 123 dwellings in the UK for a year and recorded the window use conditions by observations, aiming to better understand occupants' window-opening behaviour in residential buildings.

All Warren and Parkins [111], Inkarojrit and Paliaga [112], Bruce-Knouah [106], and, Zhang and Barrett [110] have used the second method, *i.e.* photography, to record window states in office buildings. In these studies, the state of office windows was photographed up to six times per day.

Recording Window Usage by Self-Estimation from Building Occupants

Some existing studies have asked building occupants to estimate their own ventilation habits, such as the number of hours per day with open windows in the living room, so that possible window use patterns and determinants of window opening behaviour can be identified, by either filling out questionnaires [29, 113] or attending interviews [114, 115]. In the USA, Price and Sherman [76] have asked occupants of 1448 houses in California to self-estimate their use of windows, in order to better understand occupants' ventilation behaviour and their satisfaction of indoor air quality in California homes. Similarly, Encinas Pino and de Herde [114] used this method to study occupants' ventilation habits in 91 apartments in one building located in the city of Santiago, Chile. In the Netherlands, Guerra-Santin and Itard [113], and Guerra-Satin [29] have chosen this method to estimate possible influential factors of occupants' energy behaviour in buildings, including their ventilation behaviour by opening windows. In Greece, Papakostas and Sotiropoulos [115] have asked occupants of 158 families to self-estimate their habits of airing the houses, in order to obtained behavioural patterns that can be used for building performance simulation. In the UK, Gill *et al.* [2] carried out a post-occupancy evaluation in 11 low-energy dwellings in order to identify the contribution of occupant behaviour to the actual performance of buildings. In the study, occupant window opening behaviour was collected during the interview to estimate its impact on the house heating demand. In China, Huang *et al.* [73] interviewed 500 random subjects in the northern China, asking them to evaluate their window opening habits at home in winter. In Portugal, de Freitas and Geudes [116] have used this method to identify occupants' window-use behaviour and related details in 'Baixa Pombalina's' heritage buildings. In Denmark, Andersen *et al.* [117] asked occupants of five apartments to estimate how often they opened windows during a two-month period and used the data for behavioural validation.

Camera-Based Estimation

In a case study carried out in Southampton, UK, Bourikas *et al.* [118] tested the performance of using camera images to decide the state of windows, as a low-cost method. In the study, the states of 40 windows on one façade of an office building were monitored using one digital camera and the pictures taken were computationally analysed to deduce both window state (Open/Close) and opening angle. Although high accuracies were obtained for both summer (97% between 17–23 May and 7–20 June 2015) and winter (90% between 1-30 November 2015) but the method is susceptible to certain limitations, such as available level of daylight, glare issue and raindrops on the camera. As this method is still new in this research area, with only one demo study available, it will not be used in the following method comparison.

Method Comparison

As exemplified by the studies collected above, four methods have been used by researchers to monitor occupants' window opening behaviour. In this section, a comparison of these methods is made and discussed, based upon:

1. Popularity;
2. Sample size;
3. Measurement interval;
4. Measurement duration;
5. Whether measuring window state (on/off) or window opening angle;
6. Whether occupants' window opening/closing action has been analysed.

Table **2** provides relevant statistics used for the comparison. The ' – ' sign in the table means that the parameter has not been described in the literature, and 'N/A' means that the parameter is not available or feasible for that study.

Popularity

Popularity was examined by comparing the number of existing studies using each method, as shown in Fig. (**3**). The comparison reflects that, when monitoring occupant window opening behaviour in buildings, the self-recording and electronic measuring devices methods are preferred by researchers over the other two methods, namely, the surveyor-observation method and the self-estimating method.

Table 2. Overview of literature on window opening behaviour.

Measurement Method	Study	Number of Monitored Subjects	Interval	Duration	On/Off or Opening Angle	Analysis of Occupant Action
Self-recording	Haldi and Robinson 2008	60 participants in several office buildings	2 hours	3 months	on/off	Yes
	Nakaya *et al.*, 2008	31 detached houses + 31 apartments	10 minutes	2 days	on/off	Yes
	Iwashita and Akasaka, 1997	8 dwellings	10 minutes	4 days	on/off	Yes
	Andersen *et al.*, 2009	933 dwellings for summer and 636 dwellings for winter	N/A	Single time slot	on/off	No
	Liu *et al.*, 2012	148 occupants	N/A	Single time slot	on/off	No
	Raja *et al.*, 2001 (longitudinal)	219 participants	three to four	3 months	on/off	No

(Table 2) cont.....

		from 15 buildings	times a day			
	Raja *et al.*, 2001 (transverse)	909 participants from more than 25 buildings	1 month	18 months	on/off	No
	Nicol *et al.*, 1999	846 participants from 25 buildings	1 month	16 months	on/off	No
	McCartney and Nicol, 2002 (transverse)	840 participants	1 month	12 months	on/off	No
	McCartney and Nicol, 2002 (longitudinal)	120 participants	up to four times a day	3 months	on/off	No
	Beko *et al.*, 2011	500 dwellings	real-time	2.5 days	opening angle	No
	Offermann, 2009	108 dwellings	real-time	1 week	on/off	No
Self-recording_continued	Hellwig *et al.* 2008 (School No.1)	4 classrooms in 4 buildings	real-time	2 weeks	on/off	No

(Table 2) cont.....

	Haldi and Robinson, 2009	14 offices in 1 building	real-time	86 months	on/off	Yes
	Schakib-Ekbatan *et al.*, 2015	35 rooms in 1 building	real-time	72 months	on/off	No
	Schweiker *et al.*, 2011 (apartments)	3 apartments in 2 buildings	real-time	6 months	opening angle	Yes
	Schweiker *et al.*, 2011 (dormitory)	24 students in winter and 19 students in summer	real-time	2.5 months	opening angle	Yes
Electronic Measuring devices	Dutton and Shao	2 classrooms	real-time	14 months	on/off	Yes
	Andersen *et al.*, 2013	15 dwellings	real-time	8 months	opening angle	Yes
	Yun and Steemers, 2008	6 offices in 2 buildings	real-time	3 months	on/off	Yes
	Yun and Steemers, 2010	3 offices in 2 buildings	real-time	6 months	on/off	Yes
	Yun *et al.*, 2012	4 offices in 1 building	real-time	11 months	on/off	Yes
	Fritsch *et al.*, 1990	4 offices in 1 building	30 minutes	one winter	opening angle	Yes

(Table 2) cont.....

	Herkel *et al.*, 2008	21 offices in 1 building	1 minute	13 months	opening angle	Yes
	Erhorn, 1986	24 flats	30 minutes	12 months	on/off	No
	Antretter *et al.*, 2011	17 dwellings	real-time	24 months	on/off	Yes
	Li *et al.*, 2015	5 offices	real-time	2 months	on/off	No
	Weihl and Gladhart, 1990 (Phase 1)	7 dwellings	real-time	24 months	on/off	No
Electronic Measuring devices_continued	Weihl and Gladhart, 1990 (Phase 2)	10 dwellings	real-time	12 months	on/off	No
	Hellwig *et al.*, 2008 (School No.2)	5 classrooms in 4 buildings	1 minute	8 months	on/off	No
	Wei *et al.*, 2015	5 offices	1 minute	9 months	on/off	Yes
	Jeong *et al.* 2016	20 housing units	real-time	7 months	on/off	No
	Bruce-Konuah, 2014 (Study 1)	7 offices in 2 buildings	real-time	10 months	on/off	Yes

(Table 2) cont.....

Surveyor observation	Brundrett, 1997	123 dwellings	Twice a day	12 months	on/off	No
	Wei *et al.*, 2013	36 offices in 1 building	1 day	8 months	on/off	No
	Zhang and Barrett, 2012	333 rooms in 1 building	Twice a day	16 months	on/off	No
	Warren and Parkins, 1984	196 offices in 5 buildings	Twice a day	3 months	opening angle	No
	Johnson and Long, 2005	1100 dwellings	–	16 months	on/off	Yes
	Inkarojrit and Paliaga, 2004	270 windows	Four times a day	9 days	on/off	No
	Bruce-Konuah, 2014 (Study 2)	2004 windows	2.5 hours	Six weeks	on/off	Yes
Self-estimating	Guerra-Santin and Itard, 2010	313 dwellings	N/A	Single time slot	on/off	No
	Guerra-Santin, 2013	4727 dwellings	N/A	Single time slot	on/off	No
	Encinas Pino and de Herde, 2011	91 apartments	N/A	Single time slot	on/off	No

(Table 2) cont.....

Self-estimating_continued	Papakostas and Sotiropoulos, 1997	158 families	N/A	Single time slot	on/off	No
	Gill *et al.*, 2010	11 dwellings	N/A	Single time slot	on/off	No
	Price and Sherman, 2006	1448 homes	N/A	Single time slot	on/off	No
	Huang *et al.* 2014	500 subjects	N/A	Single time slot	opening angle	No
	Andersen *et al.* 2016	5 apartments	N/A	Single time slot	on/off	No
	de Freitas and Guedes, 2015	249 subjects	N/A	Single time slot	on/off	No
Camera-based estimation	Bourikas *et al.* 2018	40 windows	N/A	Summer: 21 days Winter: 30 days	on/off opening angle	No

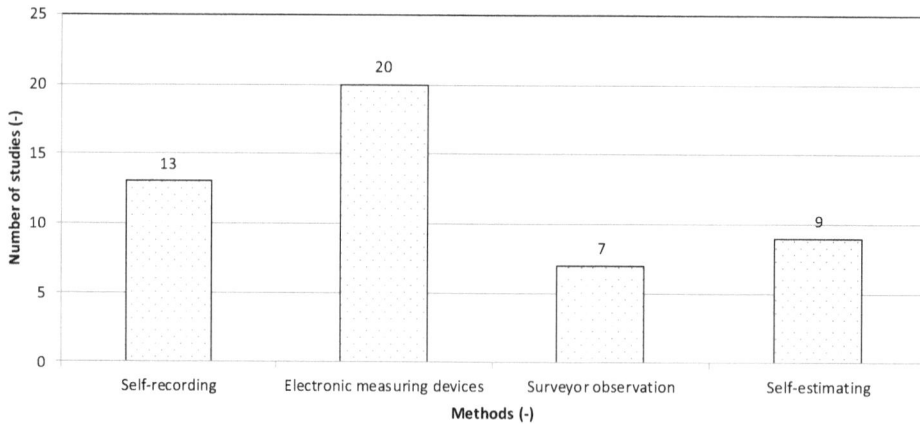

Fig. (3). Popularity comparison between methods capturing occupants' window opening behaviour.

Sample Size

The sample size of a study reflects the generalizability of its findings to a larger population [88]. Table **3** lists relevant statistics from existing studies using each method, including the average value, minimum value, maximum value and standard deviation (SD). From the table, it can be seen that studies using electronic measuring devices have a much smaller sample size than those using the other three methods, according to the average values. This is possibly because of the cost intensive nature of this method [102]. However, when comparing the SD of each method, it seems that studies using the electronic measuring devices have much more consistent sample sizes between studies than those using the other three methods. Additionally, although asking the monitored occupants to self-record or self-evaluate their window opening behaviour can help to increase the sample size of the study, researchers [65, 119] have expressed their worries on the accuracy of these methods.

Table 3. Statistics with respect to the number of monitored subjects.

Statistics/Method	Self-Recording	Electronic Measuring Devices	Surveyor Observation	Self-Estimating
Minimum	4	2	36	5
Maximum	933	35	2004	4727
Average	366	12	580	834
SD	380	9	720	1526

Measurement Interval

The measurement interval defines whether the dynamic behaviour of opening/closing windows can be captured [120]. Based on the statistics provided in Table **2**, it can be seen that using electronic devices provides the ability to capture occupants' window opening/closing actions with high frequency (the maximum interval adopted in the studies was 30 minutes). The surveyor observation method, however, had to be carried out less frequently, with a minimum available interval of 2.5 hours. The various measurement intervals of the self-recording method (with a minimum of 'instant' to a maximum interval of 1 month) appear to be much more random than the other two methods. Furthermore, a closer look at the studies using this method reveals that such studies with short-interval monitoring were generally more rapid than that of the long-interval monitoring: the former required more occupant-involvement and potential 'survey-fatigue' [121] and thus resulted in short monitoring duration.

When using the self-estimating method, occupants were asked to estimate their window opening behaviour only once by either filling out questionnaires or attending interviews so the measurement interval is not available for this method.

Measurement Duration

The measurement duration reflects how well the occupants' behavioural response to the changes in the building's environment can be captured, as described by Raja *et al.* [88]. For this parameter, the existing studies have been categorised into three groups regarding their measurement duration: 'shorter than or equal to one month', 'longer than one month and less than or equal to 12 months' and 'longer than 12 months'. Fig. (**4**) lists the percentages of studies in each duration group for different measurement methods.

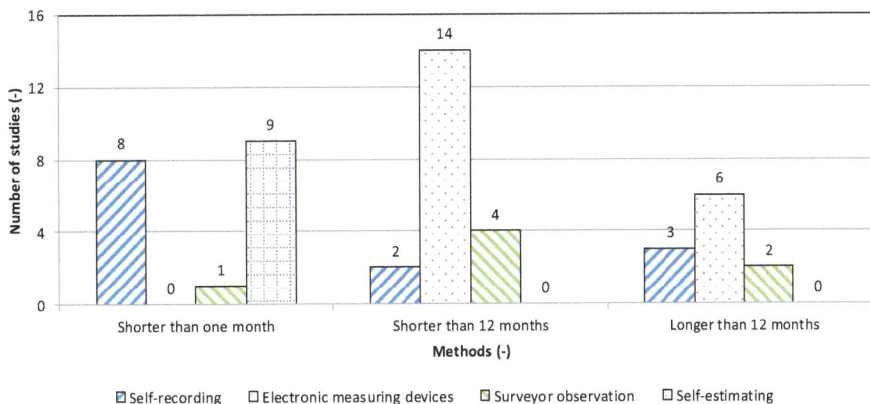

Fig. (4). Number of studies in each duration group for different measurement methods.

Fig. (**4**) reflects that the electronic measuring devices method and the surveyor observation method provide a possibility of long-term (longer than 1 year) monitoring of occupants' window opening behaviour in buildings, as they need less involvement from the monitored occupants. The three studies with long monitoring duration using the self-recording method were all coming from the transverse studies carried out by Nicol and Humphreys, and they all have a one-month measurement interval to minimise the influence from survey-fatigue to the participants. Due to the advantages in both measurement interval and measurement duration, the electronic measuring device method was preferred by researchers who wanted to model and simulate occupant window behaviour in buildings (see Flux 2 described in Fig. (**1**)). To better capture occupant window opening behaviour, most existing studies have carried out monitoring of longer than one month. Some studies using the self-recording method were shorter than one month, as in those studies the occupants were asked to record their use of windows at a high intensity, *e.g.* with an interval of less than 10 minutes (see Table **2**). All studies asking occupants to self-estimate their window opening behaviour were single measurement so all of them were shorter than one month.

Whether Measuring Window State or Window Opening Angle

Some of the studies focused on the window state, that is, whether the window was open or closed, while some captured the window opening angle. Fig. (**5**) lists the percentages of studies measuring either window state or window opening angle for each measuring method.

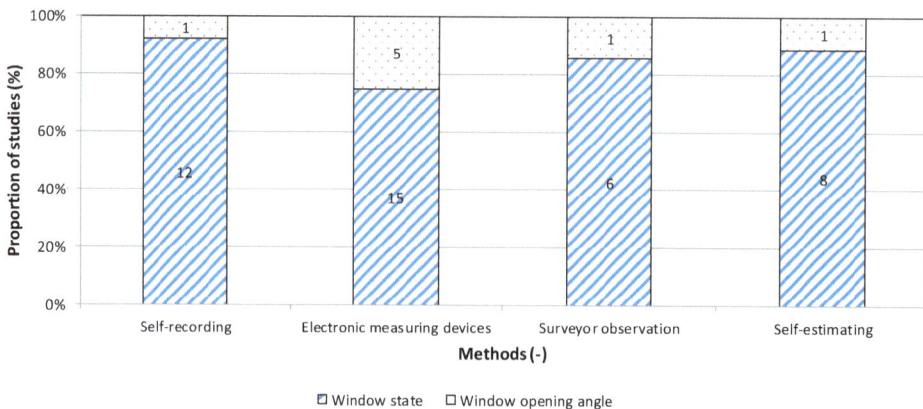

Fig. (5). Percentages of studies measuring either window state or window opening angle for each measurement method.

As seen in Fig. (**5**), the majority (39 in 47 studies) focused on the window state, rather than the window opening angle. In theory, all measurement methods were able to capture the window opening angle. However, questionnaire-based methods can only do this simply (*i.e.* by specification of 'closed, tilted open and fully open' [73, 92, 111]) while electronic measuring devices are preferred by researchers to capture window opening angle in a continuous nature.

Whether Occupants' Window Opening/Closing Action has been Analysed

Occupants' window opening behaviour is dynamic and the previous state of windows can influence occupants' decision of the current state [120]. To reflect this dynamic nature of people's behaviour, the measurement normally needs to record the time when occupants change the state of windows or alternatively, record the state of windows at a high frequency. Fig. (6) compares the studies with and without analysis of this dynamic nature for each measurement method.

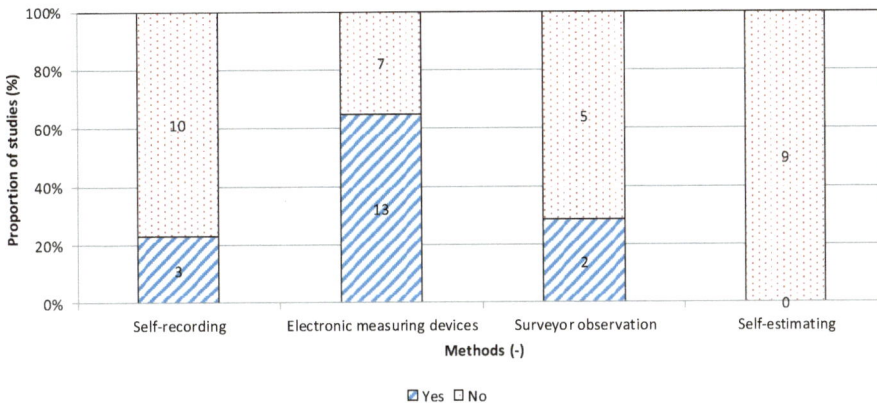

Fig. (6). Percentages of studies about whether analysing window actions for each measurement method.

Compared to the other three methods, the electronic measuring devices method provides a much higher potential for analysing the dynamic nature of occupants' window opening behaviour, due to its high measurement frequency. To capture this using the self-recording method needs a high frequency of recording window states [72, 91] or asking occupants to provide details of window opening actions during a specific period of time, *e.g.* during the preceding one hour [85]. Johnson and Long [108] have catered for the phenomenon by used the surveyor observation method based on a revisit of monitored houses within 1 hour.

CAPTURING IMPORTANT INFLUENTIAL FACTORS

Identifying influential factors is important for a better understanding of why, when and how occupants open windows, so they need to be collected along with occupants' window opening behaviour. This section reviews how these influential factors were collected in the existing studies.

Both Wei [75] and Fabi *et al.* [46] have carried out a review of influential factors on occupants' window opening behaviour in buildings, and their results are combined in Table **4**. For further analysis, these factors are classified into six groups, namely, 'outdoor environmental factors', 'indoor environmental factors', 'building- and system-related factors', 'occupant-related factors', 'time-dependent factors' and 'other factors'.

Table 4. Factors affecting window opening behaviour in buildings.

Outdoor Environmental Factors	Outdoor temperature, Wind speed, Solar radiation, Rain and Outdoor air pollution
Indoor Environmental Factors	Indoor temperature and CO_2 concentration
Building- and System-related Factors	Dwelling type, Room type, Room orientation, Ventilation type, Heating system, Window type, Floor level and Shared offices
Occupant-related Factors	Occupant age, Occupant gender, Ownership of the property and Smoking behaviour
Time-dependent Factors	Time of day and Season
Other Factors	Previous window state and room occupancy

Outdoor Environmental Factors

Outdoor environmental factors were generally monitored by a weather station with specific types of sensors for each factor. In existing studies, however, the location of the weather station during the monitoring period was not identical: some studies put the weather station on the roof of the investigated building in order to collect local meteorological data for the analysis [78, 81, 96, 105, 107]; some studies used the weather data measured from a 'nearby' weather station [85, 91], with a distance between the weather station and the investigated building varying from 280m [106] to 23 miles (about 37014m) [65]; some studies employed a remote weather station that was handheld by the experimenters during the survey time for a measurement of the onsite local weather conditions [108].

When putting the weather station on the roof of the investigated building, some researchers have considered the influence from the heat generated from the building itself on the temperature readings, so the local weather station was installed a few metres higher than the roof level [75, 97]. Schakib-Ekbatan *et al.* [103] have suggested that even when the meteorological data was measured on the roof of the investigated building, it ignored the microclimatic difference among different building façades. When using the weather data collected from the nearby public weather stations for the behavioural analysis, Andersen *et al.* [89] have suggested that such data did not reflect the local meteorological conditions well, especially in respect of wind speed. To deal with this issue, Haldi and Robinson [78] have used a linear regression method to rectify the meteorological data measured from a weather station 7.7 km away from the investigated building. From the work, they found a satisfactory agreement between measured temperature and relative humidity at the two locations, but not wind speed and direction. Therefore, they used a coarse representation of wind speed and direction in their analysis, by considering four levels of wind intensity defined by the observed quartiles of wind speed and direction in the weather station. With respect to measurement interval, this parameter was also not identical between existing studies as well. For example, the meteorological data used by Andersen *et al.* [89], Antretter *et al.* [100], Bruce-Knouah [106] and Offermann [65] were taken every hour; the ones used by Yun and Steemers [80] every 30 minutes; the ones used by Schweiker *et al.* [95], Andersen *et al.* [94] and Haldi and Robison [78] every 10 minutes; the ones adopted by Pan *et al.* [122] every 5 minutes; and the one used by Herkel *et al.* [81] at 10-minute intervals. In two studies carried out by the same researchers, the meteorological data were measured at two different locations using different intervals (one every 16 minutes and another one every hour) [88].

In most of the existing studies, the measurement accuracy of outdoor climatic data was not provided so this parameter is not compared in this paper.

Indoor Environmental Factors

As with outdoor environmental factors, indoor environmental factors were also measured using specific sensors and then recorded by data loggers. For example, thermocouples [72, 93], thermistors [91, 106, 107] and PT100s [78, 81, 95, 97] have been used to measure indoor temperature, and Tong [123] has discussed the advantages and disadvantages of using each sensor to measure indoor temperature. Non-dispersive Infrared (NDIR) sensors have been used to measure indoor CO_2 concentrations [65, 92 - 94], and its principle has been introduced by Hodgkinson *et al.* [124].

Indoor temperature has been measured in most studies regarding window opening behaviour in buildings, but the monitoring location differed significantly. Haldi and Robinson [85], Zhang and Barrett [110], Yun *et al.* [96], Nicol *et al.* [87], McCartney and Nicol [86], Pan *et al.* [122] and Raja *et al.* [88] placed temperature sensors near each participant's workstation. Nakaya *et al.* [91], and, Iwashita and Akasaka [72] carried out their temperature measurement at the centre of the zones to be measured. Wei *et al.* [107] measured indoor air temperature in cellular offices under the occupants' desks at the abdomen level. Bruce-Knouah [106] measured indoor environmental parameters on the desk of building occupants at a height of 0.76 m above the floor. Andersen *et al.* [94] and Herkel *et al.* [81] mounted their sensors on internal walls, though Herkel *et al.* have argued that using this method to measure temperature will not prevent the influence of the wall, so the measured temperature should be considered as operative temperature, rather than air temperature. Yun and Steemers [79, 80] have measured temperatures at two different locations in the monitored offices, one on the workstation and another on the book shelves, and used the average value for the analysis. In the two studies carried out by Hellwig *et al.* [93], indoor temperature was measured at different locations. In the first study, the temperature was measured at four heights, namely 0.1, 0.6, 1.1 and 1.7m, following the recommended values in ISO 7726 [125] for thermal comfort studies. In the second study, it was only measured at two heights, namely, 1.7 and 2.0m, on the interior walls. Liu *et al.* [90] also measured the indoor environmental parameters according to the requirement of ISO 7726. For large rooms, such as classrooms, Dutton and Shao [97] have measured this parameter at three locations in the room and used their average for the analysis. Although the location of measuring indoor air temperature varied among existing studies, there was a golden rule that the chosen location should minimise the chance of direct sunshine and the influence from heating/cooling resources, such as windows and heaters, in the room [79, 91, 93, 94, 106, 107].

Indoor CO_2 concentration has been monitored in some studies and the measurement locations were either at the locale of the monitored subjects [86, 92, 96] or at the centre of the room [93] or on the room interior wall [93]. As with their measurement of air temperature, Dutton and Shao [97] obtained their CO_2 readings at three different locations in classrooms.

Table **5** summarises the accuracy and interval of measuring indoor temperature and CO_2 concentration in the studies. Based on the statistics provided in Table **5**, it can be seen that the accuracy and interval of measurements of both indoor temperature and CO_2 concentration, vary between the studies, although the variation was not great. For temperature measurement, the highest accuracy was $\pm0.1°C$ and the lowest was $\pm0.5°C$, with an average of $\pm0.3°C$ and SD of $0.1°C$. The longest interval between measurements was 45 minutes and the shortest interval was 1 minute, with an average of 7 minutes and SD of 10 minutes. For measurement of CO_2 concentration, the highest accuracy was $\pm1\%$ of measurement range and the lowest accuracy was $\pm5\%$, with an average of $\pm3\%$ and SD of $\pm1.6\%$.

Table 5. Measurement accuracy and interval for indoor environmental factors in existing studies.

Study	Indoor Temperature		CO$_2$ Concentration	
	Accuracy	Interval	Accuracy	Interval
Haldi and Robinson, 2008	–	45 minutes	–	–
Nakaya et al, 2008	±0.3°C	5 minutes	–	–
Raja *et al.*, 2001	±0.3°C	15 minutes	–	–
Beko *et al.*, 2011	±0.4°C	5 minutes	±2% of range or ±2% of reading	5 minutes
Offermann, 2009	±0.5°C	1 minute	±3% of range or ±50ppm	1 minute
Schweiker *et al.* 2011(apartments)	±0.2°C	1 minute	–	–
Schweiker *et al.* 2011(dormitory)	±0.3°C	1 minute	–	–
Dutton and Shao, 2010	±0.2°C	2 minutes	±1% of range	2 minutes
Andersen *et al.*, 2013	±0.4°C	10 minutes	±2% of range or ±2% of reading	10 minutes
Wei *et al.*, 2013	±0.5°C	10 minutes	–	–
Pan *et al.*, 2015	±0.4°C	5 minutes	–	–
Yun and Steemers, 2008 and 2010	–	10 minutes	–	–
Yun *et al.*, 2011	±0.4°C	10 minutes	±5% of range or ±50ppm	10 minutes
Herkel *et al.*, 2008	–	1 minute	–	–
Hellwig *et al.* 2008 (School No.1)	±0.3°C	1 minute	±3% of range	1 minute
Hellwig *et al.* 2008 (School No.2)	±0.1°C	1 minute	±5% of range	1 minute
Wei *et al.* 2015	±0.4°C	5 minutes	–	–
Bruce-Konuah	±0.4°C	5 minutes	±5% of range or ±50ppm	5 minutes

Building- and System-Related Factors and Occupant-Related Factors

In the various studies reviewed, Building- and System- Related Factors and Occupant-Related Factors were normally collected before or after the survey and remained constant over the whole survey period. The most common method used to collect this information was questionnaires filled out by either the building occupants or surveyors, as illustrated by Andersen *et al.* [89], Yun and Steemers [80], Warren and Parkins [111], Guerra-Santin and Itard [113], Pino and Herde [114], Papakostas and Sotiropoulos [115], Price and Sherman [76], and Huang *et al.* [73]. Another method, used by Zhang and Barrett [110] and Beko *et al.* [92], was to use house inspections by experimenters. Guerra-Santin [29] collected building- and system-related factors using inspections and obtained occupant-related factors using questionnaires. Andersen *et al.* [89] have used questionnaires to collect occupant-related factors while collected building- and system- related factors from an official database.

Time-Dependent Factors and Other Factors

Time-dependent factors generally include the time of day (*i.e.* first arrival, intermediate and last departure) [64, 78, 81] and season (*i.e.* summer, transitional or winter) [81, 101]. The time of day was generally captured by occupancy sensors [78, 81], or from occupant behaviours indoors, such as opening and closing of office doors [79]. If occupants' window operation was recorded using data loggers, the season can be easily determined by the built-in timer in the data logger [81, 96]. If the survey was conducted by observations or interviews, this information can be recorded by either the building occupants or the experimenters in the questionnaires [67, 107, 110].

Other Factors include the previous window state and occupancy conditions. In existing studies, the previous window state has been measured using two methods. One was to ask building occupants to record their window operation (closing an opened window, or opening a closed window) rather than window states [85] and another was to use data loggers to make a continuous monitoring of window states [78, 80, 81, 122]. The occupancy of the room/dwelling has been determined using several methods in existing studies. Movement sensors, either passive infrared sensors [78] or ultrasonic sensors [81], have been used as a popular method to capture room occupancy. Some studies have asked building occupants to self-estimate their occupancy conditions after the survey [80, 102, 113] or self-record their occupancy conditions during the survey time [72, 91 - 93, 111]. Experimenter observation was another method that has been used to monitor occupancy in buildings, for obtaining either directly observed occupancy [86 - 88, 107, 110] or deduced occupancy based on reasonable evidence [108]. Andersen *et al.* [94] have used the CO_2 concentration and the instant window state to indicate the occupancy of rooms. Additionally, if both the bedroom and the living room were unoccupied, they assumed that the whole dwelling was unoccupied. Yun and Steemers [79] have used the state of office doors to determine the daily occupancy of offices: the time of first door opening of each day was used as the start of daily occupancy and the time of last door closing was used as the end of daily occupancy, and Bruce-Konuah [106] has adopted this method in one of his studies.

CONCLUSION

Existing studies have provided sufficient evidence to demonstrate the great impact of occupants' adaptive behaviour (*e.g.* opening/closing windows and turning up/down thermostatic setting) upon building performance. In non-air-conditioned buildings, opening/closing windows is an important adaptive opportunity for occupants to adjust the indoor environment, through increasing/decreasing the air change between indoors and outdoors. Therefore, it is important to gain a better understanding of why, when and how occupants open windows. In the past several decades, a number of studies have been performed, focusing on occupant window opening behaviour in buildings. For these studies, a good method of capturing occupant window opening behaviour and relevant influential factors has been fundamental for the later data analysis. In order to provide guidance on designing future window behaviour studies, this paper has introduced a thorough review of existing methods that have been used to capture occupants' window use in buildings, as well as relevant influential factors, and critically discussed their advantages and disadvantages.

Essentially, there are four methods of data capture, namely: self-recording by building occupants, recording by electronic measuring devices, observing by surveyors and self-estimating by building occupants, for occupant window opening behaviour in buildings. The use of electronic measuring devices has proved most useful for continuous monitoring window states/opening angles, though this method can only be used to monitor a small set of occupants due to its cost. Self-recording and self-estimating methods allow an increase in the number of samples monitored, but there is a lack of confidence in their accuracy. The surveyor observation method can also help to increase the sample size, but this method generally cannot be used for high frequency measurement, making it hard to capture occupants' dynamic behaviour in opening/closing windows. The self-recording method is much more flexible with respect to measurement interval and measurement duration, but there seems to be a conflict between these two parameters when using this method: that is, a higher frequency of behavioural monitoring results in a shorter monitoring period, and vice versa. Most studies to-date have focused on monitoring the state of windows: further studies exploring the opening angle of windows would be highly valuable. As all measurement methods have their advantages and disadvantages, the choice of method should be carried out carefully by the researchers before the study, based on the objective of the evaluation.

Regarding the measurement of relevant influential factors, it seems that a standard methodology is required, because the measurement location, measurement accuracy and measurement interval have been found to vary significantly between existing studies.

The review work also reveals that existing studies mostly collected data from one case study building or from buildings in one country. The inconsistency in monitoring methodologies may diminish the value of performing cross-comparison studies between various countries or climates (these may have various climates and cultures that may have influences on occupants' behaviour), which are important for justifying the portability and comparability of the developed models – if the raw data were varying due to the nature of different monitoring methods applied, how convince the comparison results between models can be?

In summary, a standard methodology of capturing occupants' window opening behaviour in buildings and relevant influential factors is urgently required, and this issue is being discussed currently in the ongoing international project, IEA ANNEX 66 [126]. This standard methodology should consider the measurement intervals, accuracy, resolution and location of important parameters usable for better understanding occupants' window behaviour, as well as how the developed behavioural models can best fit dynamic building simulations (*e.g.* using existing parameters in current building simulation tools to model behaviour and how to best match the behavioural models with the simulation time step). This will be helpful for better understanding occupants' window opening behaviour in actual buildings and can also contribute to bridging the gap between the simulated performance and the actual performance of buildings. Additionally, an accurate capture of occupants' daily window use can also help identify improper window uses in real buildings, which may cause great energy waste [127]. Based on this, suitable measures can be applied to promote the building's energy performance regarding to occupants' window behaviour, such as automatic window control systems [128 - 130] and occupant behaviour education [34, 38, 131].

CONSENT FOR PUBLICATION

Not Applicable.

CONFLICT OF INTEREST

The author confirms that this chapter contents have no conflict of interest.

ACKNOWLEDGEMENT

Declared none.

REFERENCES

[1]   K. Steemers, and G.Y. Yun, "Household energy consumption: a study of the role of occupants", *Build. Res. Inform.*, vol. 37, no. 5-6, pp. 625-637, 2009.
[http://dx.doi.org/10.1080/09613210903186661]

[2]   Z.M. Gill, "Low-energy dwellings: the contribution of behaviours to actual performance", *Build. Res. Inform.*, vol. 38, no. 5, pp. 491-508, 2010.
[http://dx.doi.org/10.1080/09613218.2010.505371]

[3]   K. Gram-Hanssen, "Residential heat comfort practices: understanding users", *Build. Res. Inform.*, vol. 38, no. 2, pp. 175-186, 2010.
[http://dx.doi.org/10.1080/09613210903541527]

[4]   O. Guerra Santin, L. Itard, and H. Visscher, "The effect of occupancy and building characteristics on energy use for space and water heating in Dutch residential stock", *Energy Build.*, vol. 41, no. 11, pp.

1223-1232, 2009.
[http://dx.doi.org/10.1016/j.enbuild.2009.07.002]

[5]   R. Haas, H. Auer, and P. Biermayr, "The impact of consumer behavior on residential energy demand for space heating", *Energy Build.,* vol. 27, no. 2, pp. 195-205, 1998.
[http://dx.doi.org/10.1016/S0378-7788(97)00034-0]

[6]   Z. Yu, "A systematic procedure to study the influence of occupant behavior on building energy consumption", *Energy Build.,* vol. 43, no. 6, pp. 1409-1417, 2011.
[http://dx.doi.org/10.1016/j.enbuild.2011.02.002]

[7]   M. Mulville, K. Jones, and G. Huebner, "The potential for energy reduction in UK commercial offices through effective management and behaviour change", *Architectural Engineering and Design Management,* vol. 10, no. 1-2, pp. 79-90, 2014.
[http://dx.doi.org/10.1080/17452007.2013.837250]

[8]   M. Bonte, F. Thellier, and B. Lartigue, "Impact of occupant's actions on energy building performance and thermal sensation", *Energy Build.,* vol. 76, pp. 219-227, 2014.
[http://dx.doi.org/10.1016/j.enbuild.2014.02.068]

[9]   A. Mavrogianni, "The impact of occupancy patterns, occupant-controlled ventilation and shading on indoor overheating risk in domestic environments", *Build. Environ.,* vol. 78, pp. 183-198, 2014.
[http://dx.doi.org/10.1016/j.buildenv.2014.04.008]

[10]   V. Fabi, R.V. Andersen, and S.P. Corgnati, "Influence of occupant's heating set-point preferences on indoor environmental quality and heating demand in residential buildings", *HVAC & R Res.,* vol. 19, no. 5, pp. 635-645, 2013.

[11]   A.S. Silva, and E. Ghisi, "Uncertainty analysis of user behaviour and physical parameters in residential building performance simulation", *Energy Build.,* vol. 76, pp. 381-391, 2014.
[http://dx.doi.org/10.1016/j.enbuild.2014.03.001]

[12]   Y.S. Lee, and A.M. Malkawi, "Simulating multiple occupant behaviors in buildings: An agent-based modeling approach", *Energy Build.,* vol. 69, pp. 407-416, 2014.
[http://dx.doi.org/10.1016/j.enbuild.2013.11.020]

[13]   A. Roetzel, "Occupant behaviour simulation for cellular offices in early design stages—Architectural and modelling considerations", *Build. Simul.,* vol. 8, no. 2, pp. 211-224, 2015.
[http://dx.doi.org/10.1007/s12273-014-0203-6]

[14]   T. de Meester, "Impacts of occupant behaviours on residential heating consumption for detached houses in a temperate climate in the northern part of Europe", *Energy Build.,* vol. 57, pp. 313-323, 2013.
[http://dx.doi.org/10.1016/j.enbuild.2012.11.005]

[15]   K. Gram-Hanssen, "Households' energy use - which is the more important: efficient technologies or user practices", *World Renewable Energy Congress 2011* Linkoping, Sweden. 2011.
[http://dx.doi.org/10.3384/ecp11057992]

[16]   H. Ben, and K. Steemers, "Energy retrofit and occupant behaviour in protected housing: A case study of the Brunswick Centre in London", *Energy Build.,* vol. 80, pp. 120-130, 2014.
[http://dx.doi.org/10.1016/j.enbuild.2014.05.019]

[17]   A.L. Pisello, and F. Asdrubali, "Human-based energy retrofits in residential buildings: A cost-effective alternative to traditional physical strategies", *Appl. Energy,* vol. 133, pp. 224-235, 2014.
[http://dx.doi.org/10.1016/j.apenergy.2014.07.049]

[18]   G.S. Brager, and R.J. de Dear, "Thermal adaptation in the built environment: a literature review", *Energy Build.,* vol. 27, no. 1, pp. 83-96, 1998.
[http://dx.doi.org/10.1016/S0378-7788(97)00053-4]

[19]   P. Xue, C.M. Mak, and H.D. Cheung, "The effects of daylighting and human behavior on luminous comfort in residential buildings: A questionnaire survey", *Build. Environ.,* vol. 81, pp. 51-59, 2014.

[http://dx.doi.org/10.1016/j.buildenv.2014.06.011]

[20]   M. Luo, "Can personal control influence human thermal comfort? A field study in residential buildings in China in winter", *Energy Build.,* vol. 72, pp. 411-418, 2014.
[http://dx.doi.org/10.1016/j.enbuild.2013.12.057]

[21]   R.J. de Dear, and G.S. Brager, "Thermal comfort in naturally ventilated buildings: revisions to ASHRAE Standard 55", *Energy Build.,* vol. 34, no. 6, pp. 549-561, 2002.
[http://dx.doi.org/10.1016/S0378-7788(02)00005-1]

[22]   A. Kashif, "Simulating the dynamics of occupant behaviour for power management in residential buildings", *Energy Build.,* vol. 56, pp. 85-93, 2013.
[http://dx.doi.org/10.1016/j.enbuild.2012.09.042]

[23]   T.A. Nguyen, and M. Aiello, "Energy intelligent buildings based on user activity: A survey", *Energy Build.,* vol. 56, pp. 244-257, 2013.
[http://dx.doi.org/10.1016/j.enbuild.2012.09.005]

[24]   M.A.R. Lopes, C.H. Antunes, and N. Martins, "Energy behaviours as promoters of energy efficiency: A 21st century review", *Renew. Sustain. Energy Rev.,* vol. 16, no. 6, pp. 4095-4104, 2012.
[http://dx.doi.org/10.1016/j.rser.2012.03.034]

[25]   J.K. Day, and D.E. Gunderson, "Understanding high performance buildings: The link between occupant knowledge of passive design systems, corresponding behaviors, occupant comfort and environmental satisfaction", *Build. Environ.,* vol. 84, pp. 114-124, 2015.
[http://dx.doi.org/10.1016/j.buildenv.2014.11.003]

[26]   G.M. Huebner, J. Cooper, and K. Jones, "Domestic energy consumption—What role do comfort, habit, and knowledge about the heating system play?", *Energy Build.,* vol. 66, pp. 626-636, 2013.
[http://dx.doi.org/10.1016/j.enbuild.2013.07.043]

[27]   H. Darby, "Influence of occupants' behaviour on energy and carbon emission reduction in a higher education building in the UK", *Intelligent Buildings International,* vol. 8, no. 3, pp. 157-175, 2016.
[http://dx.doi.org/10.1080/17508975.2016.1139535]

[28]   O.T. Masoso, and L.J. Grobler, "The dark side of occupants' behaviour on building energy use", *Energy Build.,* vol. 42, no. 2, pp. 173-177, 2010.
[http://dx.doi.org/10.1016/j.enbuild.2009.08.009]

[29]   O. Guerra Santin, "Occupant behaviour in energy efficient dwellings: evidence of a rebound effect", *J. Housing Built Environ.,* vol. 28, no. 2, pp. 311-327, 2013.
[http://dx.doi.org/10.1007/s10901-012-9297-2]

[30]   R. Yang, and L. Wang, "Development of multi-agent system for building energy and comfort management based on occupant behaviors", *Energy Build.,* vol. 56, pp. 1-7, 2013.
[http://dx.doi.org/10.1016/j.enbuild.2012.10.025]

[31]   L. Klein, "Coordinating occupant behavior for building energy and comfort management using multi-agent systems", *Autom. Construct.,* vol. 22, pp. 525-536, 2012.
[http://dx.doi.org/10.1016/j.autcon.2011.11.012]

[32]   S. Wei, "Impact of occupant behaviour on the energy-saving potential of retrofit measures for a public building in the UK", *Intelligent Buildings International,* vol. 9, no. 2, pp. 97-106, 2017.
[http://dx.doi.org/10.1080/17508975.2016.1139538]

[33]   H. Staats, E. van Leeuwen, and A. Wit, "A longitudinal study of informational interventions to save energy in an office building", *J. Appl. Behav. Anal.,* vol. 33, no. 1, pp. 101-104, 2000.
[http://dx.doi.org/10.1901/jaba.2000.33-101] [PMID: 10738959]

[34]   J. Goodhew, S. Pahl, T. Auburn, and S. Goodhew, "Making Heat Visible: Promoting Energy Conservation Behaviors Through Thermal Imaging", *Environ. Behav.,* vol. 47, no. 10, pp. 1059-1088, 2015.
[http://dx.doi.org/10.1177/0013916514546218] [PMID: 26635418]

[35]   R. Parnell, and O.P. Larsen, "Informing the Development of Domestic Energy Efficiency Initiatives: An Everyday Householder-Centered Framework", *Environ. Behav.,* vol. 37, no. 6, pp. 787-807, 2005.
[http://dx.doi.org/10.1177/0013916504274008]

[36]   W. Abrahamse, "The effect of tailored information, goal setting, and tailored feedback on household energy use, energy-related behaviors, and behavioral antecedents", *J. Environ. Psychol.,* vol. 27, no. 4, pp. 265-276, 2007.
[http://dx.doi.org/10.1016/j.jenvp.2007.08.002]

[37]   W. Abrahamse, "A review of intervention studies aimed at household energy conservation", *J. Environ. Psychol.,* vol. 25, no. 3, pp. 273-291, 2005.
[http://dx.doi.org/10.1016/j.jenvp.2005.08.002]

[38]   Z. Yu, "A methodology for identifying and improving occupant behavior in residential buildings", *Energy,* vol. 36, no. 11, pp. 6596-6608, 2011.
[http://dx.doi.org/10.1016/j.energy.2011.09.002]

[39]   R. Gulbinas, and J.E. Taylor, "Effects of real-time eco-feedback and organizational network dynamics on energy efficient behavior in commercial buildings", *Energy Build.,* vol. 84, pp. 493-500, 2014.
[http://dx.doi.org/10.1016/j.enbuild.2014.08.017]

[40]   A.R. Carrico, and M. Riemer, "Motivating energy conservation in the workplace: An evaluation of the use of group-level feedback and peer education", *J. Environ. Psychol.,* vol. 31, no. 1, pp. 1-13, 2011.
[http://dx.doi.org/10.1016/j.jenvp.2010.11.004]

[41]   P.C. Stern, "New Environmental Theories: Toward a Coherent Theory of Environmentally Significant Behavior", *J. Soc. Issues,* vol. 56, no. 3, pp. 407-424, 2000.
[http://dx.doi.org/10.1111/0022-4537.00175]

[42]   Y. Kaluarachchi, and K. Jones, "Promoting low-carbon home adaptations and behavioural change in the older community", *Architectural Engineering and Design Management,* vol. 10, no. 1-2, pp. 131-145, 2014.
[http://dx.doi.org/10.1080/17452007.2013.837242]

[43]   M.M. Agha-Hossein, "Providing persuasive feedback through interactive posters to motivate energy-saving behaviours", *Intelligent Buildings International,* vol. 7, no. 1, pp. 16-35, 2015.
[http://dx.doi.org/10.1080/17508975.2014.960357]

[44]   X. Zhang, "Smart meter and in-home display for energy savings in residential buildings: a pilot investigation in Shanghai, China", *Intelligent Buildings International,* vol. 11, no. 1, pp. 4-26, 2019.
[http://dx.doi.org/10.1080/17508975.2016.1213694]

[45]   C. Mullaly, "Home energy use behaviour: a necessary component of successful local government home energy conservation (LGHEC) programs", *Energy Policy,* vol. 26, no. 14, pp. 1041-1052, 1998.
[http://dx.doi.org/10.1016/S0301-4215(98)00046-9]

[46]   V. Fabi, "Occupants' window opening behaviour: A literature review of factors influencing occupant behaviour and models", *Build. Environ.,* vol. 58, pp. 188-198, 2012.
[http://dx.doi.org/10.1016/j.buildenv.2012.07.009]

[47]   S. Wei, R. Jones, and P. de Wilde, "Driving factors for occupant-controlled space heating in residential buildings", *Energy Build.,* vol. 70, pp. 36-44, 2014.
[http://dx.doi.org/10.1016/j.enbuild.2013.11.001]

[48]   W. O'Brien, and H.B. Gunay, "The contextual factors contributing to occupants' adaptive comfort behaviors in offices – A review and proposed modeling framework", *Build. Environ.,* vol. 77, pp. 77-87, 2014.
[http://dx.doi.org/10.1016/j.buildenv.2014.03.024]

[49]   W. O'Brien, K. Kapsis, and A.K. Athienitis, "Manually-operated window shade patterns in office buildings: A critical review", *Build. Environ.,* vol. 60, pp. 319-338, 2013.
[http://dx.doi.org/10.1016/j.buildenv.2012.10.003]

[50] C. Peng, "Quantitative description and simulation of human behavior in residential buildings", *Build. Simul.,* vol. 5, no. 2, pp. 85-94, 2012.
[http://dx.doi.org/10.1007/s12273-011-0049-0]

[51] E. Azar, and C.C. Menassa, "A comprehensive analysis of the impact of occupancy parameters in energy simulation of office buildings", *Energy Build.,* vol. 55, pp. 841-853, 2012.
[http://dx.doi.org/10.1016/j.enbuild.2012.10.002]

[52] A.C. Menezes, "Predicted vs. actual energy performance of non-domestic buildings: Using post-occupancy evaluation data to reduce the performance gap", *Appl. Energy,* vol. 97, pp. 355-364, 2012.
[http://dx.doi.org/10.1016/j.apenergy.2011.11.075]

[53] P. de Wilde, "The gap between predicted and measured energy performance of buildings: A framework for investigation", *Autom. Construct.,* vol. 41, pp. 40-49, 2014.
[http://dx.doi.org/10.1016/j.autcon.2014.02.009]

[54] D. Daly, P. Cooper, and Z. Ma, "Understanding the risks and uncertainties introduced by common assumptions in energy simulations for Australian commercial buildings", *Energy Build.,* vol. 75, pp. 382-393, 2014.
[http://dx.doi.org/10.1016/j.enbuild.2014.02.028]

[55] S. D'Oca, "Effect of thermostat and window opening occupant behavior models on energy use in homes", *Build. Simul.,* vol. 7, no. 6, pp. 683-694, 2014.
[http://dx.doi.org/10.1007/s12273-014-0191-6]

[56] L. Daniel, V. Soebarto, and T. Williamson, "House energy rating schemes and low energy dwellings: The impact of occupant behaviours in Australia", *Energy Build.,* vol. 88, pp. 34-44, 2015.
[http://dx.doi.org/10.1016/j.enbuild.2014.11.060]

[57] P. Hoes, "User behavior in whole building simulation", *Energy Build.,* vol. 41, no. 3, pp. 295-302, 2009.
[http://dx.doi.org/10.1016/j.enbuild.2008.09.008]

[58] F. Haldi, and D. Robinson, "The impact of occupants' behaviour on building energy demand", *J. Build. Perform. Simul.,* vol. 4, no. 4, pp. 323-338, 2011.
[http://dx.doi.org/10.1080/19401493.2011.558213]

[59] J. Virote, and R. Neves-Silva, "Stochastic models for building energy prediction based on occupant behavior assessment", *Energy Build.,* vol. 53, pp. 183-193, 2012.
[http://dx.doi.org/10.1016/j.enbuild.2012.06.001]

[60] C.E. Ochoa, "Considerations on design optimization criteria for windows providing low energy consumption and high visual comfort", *Appl. Energy,* vol. 95, pp. 238-245, 2012.
[http://dx.doi.org/10.1016/j.apenergy.2012.02.042]

[61] W.J. Hee, "The role of window glazing on daylighting and energy saving in buildings", *Renew. Sustain. Energy Rev.,* vol. 42, pp. 323-343, 2015.
[http://dx.doi.org/10.1016/j.rser.2014.09.020]

[62] L. Wang, and S. Greenberg, "Window operation and impacts on building energy consumption", *Energy Build.,* vol. 92, pp. 313-321, 2015.
[http://dx.doi.org/10.1016/j.enbuild.2015.01.060]

[63] S.M. Porritt, "Ranking of interventions to reduce dwelling overheating during heat waves", *Energy Build.,* vol. 55, pp. 16-27, 2012.
[http://dx.doi.org/10.1016/j.enbuild.2012.01.043]

[64] G.Y. Yun, P. Tuohy, and K. Steemers, "Thermal performance of a naturally ventilated building using a combined algorithm of probabilistic occupant behaviour and deterministic heat and mass balance models", *Energy Build.,* vol. 41, no. 5, pp. 489-499, 2009.
[http://dx.doi.org/10.1016/j.enbuild.2008.11.013]

[65] F.J. Offermann, *Ventilation and indoor air quality in new homes.* California Energy Commission, 2009.

[66] G. Bekö, "Ventilation rates in the bedrooms of 500 Danish children", *Build. Environ.*, vol. 45, no. 10, pp. 2289-2295, 2010.
[http://dx.doi.org/10.1016/j.buildenv.2010.04.014]

[67] L.A. Wallace, S.J. Emmerich, and C. Howard-Reed, "Continuous measurements of air change rates in an occupied house for 1 year: the effect of temperature, wind, fans, and windows", *J. Expo. Anal. Environ. Epidemiol.*, vol. 12, no. 4, pp. 296-306, 2002.
[http://dx.doi.org/10.1038/sj.jea.7500229] [PMID: 12087436]

[68] D. Marr, "The influence of opening windows and doors on the natural ventilation rate of a residential building", *HVAC & R Res.*, vol. 18, no. 1-2, pp. 195-203, 2012.

[69] C. Howard-Reed, L.A. Wallace, and W.R. Ott, "The effect of opening windows on air change rates in two homes", *J. Air Waste Manag. Assoc.*, vol. 52, no. 2, pp. 147-159, 2002.
[http://dx.doi.org/10.1080/10473289.2002.10470775] [PMID: 15143789]

[70] B. Kvisgaard, P.F. Collet, and J. Kure, *Research on fresh-air change rate: 1 Occupants' influence on air-change 1985.* Commission of European Communities, 1985.

[71] J.S. Weihl, "Monitored residential ventilation behaviour: a seasonal analysis",

[72] G. Iwashita, and H. Akasaka, "The effects of human behavior on natural ventilation rate and indoor air environment in summer — a field study in southern Japan", *Energy Build.*, vol. 25, no. 3, pp. 195-205, 1997.
[http://dx.doi.org/10.1016/S0378-7788(96)00994-2]

[73] K. Huang, "Opening window issue of residential buildings in winter in north China: A case study in Shenyang", *Energy Build.*, vol. 84, pp. 567-574, 2014.
[http://dx.doi.org/10.1016/j.enbuild.2014.09.005]

[74] S. Wei, "Using building performance simulation to save residential space heating energy: A pilot testing", *Windsor Conference.* 2014.

[75] S. Wei, Preference-based modelling and prediction of occupants' window behaviour in non-ai--conditioned office buildings.*School of Civil and Building Engineering.* Loughborough University, 2014.

[76] P.N. Price, and M.H. Sherman, *Ventilation behavior and household characteristics in new California houses.* LBNL, 2006.
[http://dx.doi.org/10.2172/883796]

[77] Y. Qi, "Large-scale and long-term monitoring of the thermal environments and adaptive behaviors in Chinese urban residential buildings", *Build. Environ.*, vol. 168, p. 106524, 2020.
[http://dx.doi.org/10.1016/j.buildenv.2019.106524]

[78] F. Haldi, and D. Robinson, "Interactions with window openings by office occupants", *Build. Environ.*, vol. 44, no. 12, pp. 2378-2395, 2009.
[http://dx.doi.org/10.1016/j.buildenv.2009.03.025]

[79] G.Y. Yun, and K. Steemers, "Night-time naturally ventilated offices: Statistical simulations of window-use patterns from field monitoring", *Sol. Energy*, vol. 84, no. 7, pp. 1216-1231, 2010.
[http://dx.doi.org/10.1016/j.solener.2010.03.029]

[80] G.Y. Yun, and K. Steemers, "Time-dependent occupant behaviour models of window control in summer", *Build. Environ.*, vol. 43, no. 9, pp. 1471-1482, 2008.
[http://dx.doi.org/10.1016/j.buildenv.2007.08.001]

[81] S. Herkel, U. Knapp, and J. Pfafferott, "Towards a model of user behaviour regarding the manual control of windows in office buildings", *Build. Environ.*, vol. 43, no. 4, pp. 588-600, 2008.
[http://dx.doi.org/10.1016/j.buildenv.2006.06.031]

[82] H. Mo, "Developing window behavior models for residential buildings using XGBoost algorithm", *Energy Build.,* vol. 205, p. 109564, 2019.
[http://dx.doi.org/10.1016/j.enbuild.2019.109564]

[83] S. Pan, "A model based on Gauss Distribution for predicting window behavior in building", *Build. Environ.,* vol. 149, pp. 210-219, 2019.
[http://dx.doi.org/10.1016/j.buildenv.2018.12.008]

[84] H.B. Gunay, W. O'Brien, and I. Beausoleil-Morrison, "A critical review of observation studies, modeling, and simulation of adaptive occupant behaviors in offices", *Build. Environ.,* vol. 70, pp. 31-47, 2013.
[http://dx.doi.org/10.1016/j.buildenv.2013.07.020]

[85] F. Haldi, and D. Robinson, "On the behaviour and adaptation of office occupants", *Build. Environ.,* vol. 43, no. 12, pp. 2163-2177, 2008.
[http://dx.doi.org/10.1016/j.buildenv.2008.01.003]

[86] K.J. McCartney, and J. Fergus Nicol, "Developing an adaptive control algorithm for Europe", *Energy Build.,* vol. 34, no. 6, pp. 623-635, 2002.
[http://dx.doi.org/10.1016/S0378-7788(02)00013-0]

[87] F.J. Nicol, *Climatic variations in comfortable temperatures: the Pakistan projects.,* 1999.
[http://dx.doi.org/10.1016/S0378-7788(99)00011-0]

[88] I.A. Raja, "Thermal comfort: use of controls in naturally ventilated buildings", *Energy Build.,* vol. 33, no. 3, pp. 235-244, 2001.
[http://dx.doi.org/10.1016/S0378-7788(00)00087-6]

[89] R.V. Andersen, "Survey of occupant behaviour and control of indoor environment in Danish dwellings", *Energy Build.,* vol. 41, no. 1, pp. 11-16, 2009.
[http://dx.doi.org/10.1016/j.enbuild.2008.07.004]

[90] J. Liu, "Occupants' behavioural adaptation in workplaces with non-central heating and cooling systems", *Appl. Therm. Eng.,* vol. 35, pp. 40-54, 2012.
[http://dx.doi.org/10.1016/j.applthermaleng.2011.09.037]

[91] T. Nakaya, N. Matsubara, and Y. Kurazumi, "Use of occupant behaviour to control the indoor climate in Japanese residences", *Windsor Conference,* 2008

[92] G. Bekö, J. Toftum, and G. Clausen, "Modeling ventilation rates in bedrooms based on building characteristics and occupant behavior", *Build. Environ.,* vol. 46, no. 11, pp. 2230-2237, 2011.
[http://dx.doi.org/10.1016/j.buildenv.2011.05.002]

[93] R.T. Hellwig, "The use of windows as controls for indoor environmental conditions in schools", *Windsor Conference* Windsor, UK 2008.

[94] R. Andersen, "Window opening behaviour modelled from measurements in Danish dwellings", *Build. Environ.,* vol. 69, no. 0, pp. 101-113, 2013.
[http://dx.doi.org/10.1016/j.buildenv.2013.07.005]

[95] M. Schweiker, "Verification of stochastic models of window opening behaviour for residential buildings", *J. Build. Perform. Simul.,* vol. 5, no. 1, pp. 55-74, 2011.
[http://dx.doi.org/10.1080/19401493.2011.567422]

[96] G.Y. Yun, H. Kim, and J.T. Kim, "Thermal and non-thermal stimuli for the use of windows in offices", *Indoor Built Environ.,* vol. 21, pp. 109-121, 2012.
[http://dx.doi.org/10.1177/1420326X11420012]

[97] S. Dutton, and L. Shao, "Window opening behaviour in a naturally ventilated school", *SimBuild 2010 Conference* New York City, New York, USA 2010.

[98] R. Fritsch, "A stochastic model of user behaviour regarding ventilation", *Build. Environ.,* vol. 25, no. 2, pp. 173-181, 1990.

[http://dx.doi.org/10.1016/0360-1323(90)90030-U]

[99] E. Erhorn, "Influence of the meteorological conditions on the inhabitants' behaviour in dwellings with mechanical ventilation", *7th AIC Conference,* 1986 p. 11.1 11.15.

[100] F. Antretter, C. Mayer, and U. Wellisch, "An approach for a statistical model for the user behaviour regarding window ventilation in residential buildings", *12th Conference of International Building Performance Simulation Association,* 2011 p. 1678 1685.

[101] N. Li, "Probability of occupant operation of windows during transition seasons in office buildings", *Renew. Energy,* vol. 73, pp. 84-91, 2015.
[http://dx.doi.org/10.1016/j.renene.2014.05.065]

[102] J.S. Weihl, and P.M. Gladhart, "Occupant behavior and successful energy conservation: Findings and implications of behavioral monitoring", *ACEEE Summer Study Conference on Energy Efficiency in Buildings* 1990.

[103] K. Schakib-Ekbatan, "Does the occupant behavior match the energy concept of the building? – Analysis of a German naturally ventilated office building", *Build. Environ.,* vol. 84, pp. 142-150, 2015.
[http://dx.doi.org/10.1016/j.buildenv.2014.10.018]

[104] B. Jeong, J-W. Jeong, and J.S. Park, "Occupant behavior regarding the manual control of windows in residential buildings", *Energy Build.,* vol. 127, pp. 206-216, 2016.
[http://dx.doi.org/10.1016/j.enbuild.2016.05.097]

[105] S. Wei, "Analysis of factors influencing the modelling of occupant window opening behaviour in an office building in Beijing, China", *Building Simulation Conference* 2015.

[106] A. Bruce-Konuah, Occupant window opening behaviour: the relative importance of temperature and carbon dioxide in university office buildings.*Department of Civil and Structural Engineering.* University of Sheffield Sheffield, 2014.

[107] S. Wei, R. Buswell, and D. Loveday, "Factors affecting 'end-of-day' window position in a non-ai--conditioned office building", *Energy Build.,* vol. 62, no. 0, pp. 87-96, 2013.
[http://dx.doi.org/10.1016/j.enbuild.2013.02.060]

[108] T. Johnson, and T. Long, *Determining the frequency of open windows in residences: a pilot study in Durham, North Carolina during varying temperature conditions.* Print, 2005, pp. 1053-4245.

[109] G.W. Brundrett, "Ventilation: A behavioural approach", *Int. J. Energy Res.,* vol. 1, no. 4, pp. 289-298, 1977.
[http://dx.doi.org/10.1002/er.4440010403]

[110] Y. Zhang, and P. Barrett, "Factors influencing the occupants' window opening behaviour in a naturally ventilated office building", *Build. Environ.,* vol. 50, no. 0, pp. 125-134, 2012.
[http://dx.doi.org/10.1016/j.buildenv.2011.10.018]

[111] P.R. Warren, and L.M. Parkins, "Window-opening behaviour in office buildings", *ASHRAE Trans.,* vol. 90, no. Part 1B, pp. 1056-1076, 1984.

[112] V. Inkarojrit, and G. Paliaga, "Indoor climatic influences on the operation of windows in a naturally ventilated building", *Plea2004 - the 21th conference on passive and low energy architecture* Eindhoven, The Netherlands 2004.

[113] O. Guerra-Santin, and L. Itard, "Occupants' behaviour: determinants and effects on residential heating consumption", *Build. Res. Inform.,* vol. 38, no. 3, pp. 318-338, 2010.
[http://dx.doi.org/10.1080/09613211003661074]

[114] F. Encinas Pino, and A. de Herde, "Definition of occupant behaviour patterns with respect to ventilation for apartments from the real estate market in Santiago de Chile", *Sustainable Cities and Society,* vol. 1, no. 1, pp. 38-44, 2011.
[http://dx.doi.org/10.1016/j.scs.2010.08.005]

[115] K.T. Papakostas, and B.A. Sotiropoulos, "Occupational and energy behaviour patterns in Greek residences", *Energy Build.,* vol. 26, no. 2, pp. 207-213, 1997.
[http://dx.doi.org/10.1016/S0378-7788(97)00002-9]

[116] P. Nunes de Freitas, and M.C. Guedes, "The use of windows as environmental control in "Baixa Pombalina's" heritage buildings", *Renew. Energy,* vol. 73, pp. 92-98, 2015.
[http://dx.doi.org/10.1016/j.renene.2014.08.029]

[117] R.K. Andersen, V. Fabi, and S.P. Corgnati, "Predicted and actual indoor environmental quality: Verification of occupants' behaviour models in residential buildings", *Energy Build.,* vol. 127, pp. 105-115, 2016.
[http://dx.doi.org/10.1016/j.enbuild.2016.05.074]

[118] L. Bourikas, "Camera-based window-opening estimation in a naturally ventilated office", *Build. Res. Inform.,* vol. 46, no. 2, pp. 148-163, 2018.
[http://dx.doi.org/10.1080/09613218.2016.1245951]

[119] S. Gauthier, and D. Shipworth, "Behavioural responses to cold thermal discomfort", *Build. Res. Inform.,* vol. 43, no. 3, pp. 355-370, 2015.
[http://dx.doi.org/10.1080/09613218.2015.1003277]

[120] F. Haldi, and D. Robinson, "A comparison of alternative approaches for the modelling of window opening and closing behaviour", *Proceedings of Conference: Air Conditioning and the Low Carbon Cooling Challenge* Cumberland Lodge, Windsor, UK 2008.

[121] F.J. Nicol, and K.J. McCartney, "SCATS: Final Report-Public", *Oxford Brookes University,* 2001.

[122] S. Pan, "A study on influential factors of occupant window-opening behavior in an office building in China", *Build. Environ.,* vol. 133, pp. 41-50, 2018.
[http://dx.doi.org/10.1016/j.buildenv.2018.02.008]

[123] A. Tong, "Improving the accuracy of temperature measurements", *Sens. Rev.,* vol. 21, no. 3, pp. 193-198, 2001.
[http://dx.doi.org/10.1108/02602280110398044]

[124] J. Hodgkinson, "Non-dispersive infra-red (NDIR) measurement of carbon dioxide at 4.2μm in a compact and optically efficient sensor", *Sens. Actuators B Chem.,* vol. 186, pp. 580-588, 2013.
[http://dx.doi.org/10.1016/j.snb.2013.06.006]

[125] *International Standard 7726: Ergonomics of the thermal environment - Instruments for measuring physical quantities.* International Standard Organization, Geneva: Switzerland, 1998.

[126] *ANNEX 66 Project Homepage.,* 2015.http://www.annex66.org/

[127] S. Pan, "Improper Window Use in Office Buildings: Findings from a Longitudinal Study in Beijing, China", *Energy Procedia,* vol. 88, pp. 761-767, 2016.
[http://dx.doi.org/10.1016/j.egypro.2016.06.104]

[128] J. Cai, "A general multi-agent control approach for building energy system optimization", *Energy Build.,* vol. 127, pp. 337-351, 2016.
[http://dx.doi.org/10.1016/j.enbuild.2016.05.040]

[129] T. Labeodan, "On the application of multi-agent systems in buildings for improved building operations, performance and smart grid interaction – A survey", *Renew. Sustain. Energy Rev.,* vol. 50, pp. 1405-1414, 2015.
[http://dx.doi.org/10.1016/j.rser.2015.05.081]

[130] S. Firląg, "Control algorithms for dynamic windows for residential buildings", *Energy Build.,* vol. 109, pp. 157-173, 2015.
[http://dx.doi.org/10.1016/j.enbuild.2015.09.069]

[131] S. Wei, R. Jones, and P. de Wilde, "Extending the UK's green deal with the consideration of occupant behaviour", *Conference of Building Simulation and Optimization* UCL, London 2014.

CHAPTER 3

Supporting the Decision-making Process of Building Users in the Selection of Energy-Efficient Heating Solutions by Identifying and Evaluating Co-benefits

Ricardo Barbosa[1,*] and **Manuela Almeida**[1]

[1] *ISISE, Department of Civil Engineering, University of Minho, Guimarães, Portugal*

Abstract: Space heating is responsible for a significant share of the energy consumed in European households, and the replacement of appliances with more efficient alternatives can be decisive for meeting the targets set by the European Union for 2030 and 2050. Although an estimated 60% of the heating stock consists of inefficient equipment, users are often not aware of this inefficiency and associated costs. Also, changing and improving the heating systems have been systematically associated with a wide range of effects, such as thermal comfort and improved air quality, which are often termed as co-benefits or ancillary benefits. Previous research has shown that co-benefits can be decisive when users choose a heating solution. This chapter reports on the results obtained in a study conducted in the scope of the EU H2020 HARP research project, in which an international qualitative survey was used to identify, quantify and evaluate the co-benefits associated with heating solutions, to clarify the relevance of the co-benefits in the decision-making process of building users. The results suggest that both the degree of relevance and the willingness to pay for co-benefits vary significantly amongst different national contexts.

Keywords: Building Systems, Co-Benefits, Contingent Valuation, Degree of Relevance, Economic Valuation, Heating, Qualitative survey, Willingness to Pay.

INTRODUCTION

Climate change is being recognised as one of the biggest challenges of the century, and there are considerable problems to be tackled in the next decade. One of the most well-defined issues is related to residential energy consumption, which is very significant, and it is causing determinant impacts in terms of carbon

* **Corresponding author Ricardo Barbosa:** Department of Civil Engineering, University of Minho, Campus de Azurém, 4800-058 Guimarães, Portugal; Tel: +351964291410; E-mail: ricardobarbosa@civil.uminho.pt

Enedir Ghisi, Ricardo Forgiarini Rupp and Pedro Fernandes Pereira (Eds.)

emissions. Energy demand in the building sector is responsible for 40% of the European Union energy consumption, and 85% of it is used for heating and domestic hot water [1].

Furthermore, heating and cooling in the European territory are still heavily dependent on the use of fossil fuels. In fact, only around 18% of the primary energy consumption for these uses are originating from renewable sources. The most used energy source for heating and cooling is natural gas (46%). Coal (15%), biomass (11%), fuel oil (10%) and nuclear energy (7%) are also used in abundance for the heating and cooling of buildings and industry in Europe [2]. There is, therefore, the need to increase the energy efficiency in this sector as heating demand has been a central problem in the majority of the European territory. In this context, it is objectively recognised that the replacement and retrofit of old heating equipment should be promoted to fulfill EU climate and energy goals, namely regarding full carbon neutrality until 2050 [3]. In particular, in the built environment, the implementation of sustainable and energy-efficient heating and cooling systems, mainly based on renewable sources, can play a determinant role in bringing these buildings closer to net-zero energy and emissions [4].

Following that direction, the European Union has been demonstrating the political will to support that transition. In 2016, as part of the Sustainable Energy Security Package, the European Commission proposed an EU Heating and Cooling strategy [5] which includes measures that should be addressed in future regulations and policies to improve energy efficiency, promote energy renewable energy sources and tackle climate change. The strategy identifies the following areas as priorities for legislative and policy actions: i) Facilitating building renovations; ii) Increasing the share of renewables; iii) Reusing the energy waste from industry and iv) Getting consumers and industry involved. In addition, the Energy Efficiency Directive (revised in 2019) [6] intends to motivate and drive the Member States to periodically assess the energy-efficiency of heating and cooling infrastructure with the objective of promoting continuous improvement.

Given the European stock of installed appliances (126 million space heaters), of which about half has an efficiency of 60% or less [5], there is considerable room for improvement. However, the current European building stock is the property of millions of different owners, which are known to have different perspectives on the investments that should be made in buildings. Private homeowners, for example, that in most cases do not have the technical expertise and capital availability to make investments, have very heterogeneous investment criteria and investment contexts, presenting a future behaviour that is not ruled by pure rationality being, therefore, very hard to envision [7].

The literature also shows that the adoption of sustainable and energy-efficient heating systems is a complex issue. Research studies indicate that public and political supports are very much needed to support a consistent adoption of this equipment, which has to be addressed from several scales and using different perspectives [8]. Factors that can incentivise the adoption of energy-efficient systems include favourable economic conditions, previous knowledge and literacy in engineering or energy-related subjects and awareness of the associated advantages [9]. On the other hand, barriers to the adoption of this type of technologies include not only habits and perceived behavioural control [10], but also issues related to investment and hidden costs, poor payback time perceived and the unperceived benefits of such an intervention [7, 11].

To deal with these barriers, there is evidence that incentives, such as subsidies, are known to be effective [12], in particular in the so-called problem-triggered interventions [13]. For homeowners driven by opportunities (as in the scope of a planned building renovation), additional factors, such as operational convenience (*e.g.* ease of use) [13] or even relative independence from fossil fuels, seem to be significant [14]. The most common argument for promoting investments in energy efficiency, such as the replacement of heating systems, is related to the energy and economic savings potentially achieved. This engineered-based perspective overlooks the results of several studies that point out that energy savings are often not the main motivation in the decision-making process. For example, improving the "indoor climate" is a known decision-making trigger for interventions that promote energy-efficiency, such as replacing a heating appliance or improving the insulation of the house (*e.g.,* [15]). In fact, besides energy and economic savings, there is a wide range of other known effects, at various scales, that are related to this type of investments. These effects are often termed as "co-benefits", "ancillary benefits" or "non-economic benefits" [16] and can be determinant in promoting the change needed to successfully address climate change and its effects [17].

However, co-benefits, because of their subjective nature, are more difficult to quantify than objective indicators, such as savings, and are consequently often ignored and rarely measured, quantified, or monetised. Despite this difficulty, identifying and quantifying the relevance of co-benefits regarding energy-efficiency investments, can bring additional knowledge and support the decision-making of building users and energy consumers, as seen in other studies addressing different contexts [18, 19]. Following the traditional engineering-based approach, energy-efficiency improvements at the private or household scale are usually evaluated by a trade-off between the savings resulting from the use of operational energy and the investment cost of such improvements. Or, in a more evolved approach, by considering the balance between the energy use and the global costs of the renovation intervention during the life cycle of the building (*e.g* [20]). Although this can be considered a useful approach, this kind of assessment can disregard other potential benefits from interventions to improve energy-efficiency and underestimate the real value of such improvements [16]. In this context, there are several examples in the literature addressing the potential economic value of these additional benefits, but they are predominately focused on a societal co-benefits perspective (*e.g* [21, 22]).

The work presented in this chapter was developed under the European Commission H2020 funded HARP project and intends to characterise, from a private perspective, the relevance of the different co-benefits and their economic valuation in five different countries – Portugal, Italy, France, Germany and Spain. The HARP project [23] - Heating Appliances Retrofit Planning - aims at raising consumers' awareness of the opportunities that underlay the planned replacement of their old and inefficient heating appliances. This is done by supporting the consumer in identifying the energy (in)efficiency of their current heating equipment and the savings opportunities that result from its replacement with a more energy-efficient solution. The mission is to accelerate the European replacement rate of heating systems, actively contributing to the reduction of energy demand in buildings, in line with the energy-efficiency targets set by the European Union.

The chapter is organised into five sections. The next section presents an initial overview of the co-benefits. The following section introduces the methodology of the study. Then, the main results regarding the degree of relevance and the willingness to pay for co-benefits are presented. In the following section, the results from each country are detailed and compared, which is followed by a discussion of the potential and challenges of using these results to support the decision-making process. The chapter ends with the concluding notes.

CO-BENEFITS: ADDING VALUE TO ENERGY EFFICIENCY

Energy efficiency can be defined as the "ratio of energy consumed to the output produced or service performed" [24]. The main objective of promoting energy efficiency measures and policies is to reduce energy demand and conserve energy. This is done mainly by improving the efficiency of products and processes. Thanks to the effort in promoting energy efficiency since the 1970s, it has been possible to avoid energy use in a very significant way. According to the International Energy Agency (IEA), in some member countries, in 2010, for example, the avoided energy was comparable to any single supply-side resource such as oil and coal [24]. However, due to increasing energy demand, global economic growth and the pressure to limit greenhouse gas emissions, these efforts are not enough, and a continuous improvement in energy efficiency is more necessary than ever.

The impact of energy efficiency policies is generally quantified in terms of reduced energy demand. Because what is measured is a negative value (*i.e.* energy not consumed), energy efficiency was already considered in some studies as an intangible concept and referred to as a "hidden fuel" [25]. Consequently, despite the recognised potential, investors, consumers and policymakers do not always easily perceive the value of investments aimed at improving energy efficiency, and this obstacle can impair the implementation of policies and measures. In this context, it is important to highlight the results of the most recent report from the International Energy Agency, which shows that, in 2020, the rate of investments to improve energy efficiency is going to be lower than in previous years [26]. This trend is worrying and can seriously harm the achievement of energy and climate goals already defined by the Member States. It is, therefore, necessary to argue against the devaluation of the market for measures to promote energy efficiency, which requires a better holistic understanding of the benefits that energy efficiency can provide.

The idea that an increase in energy efficiency can provide a broad spectrum of benefits in addition to increasing energy savings and reducing carbon emissions is not new. The term co-benefits first appeared in the scientific literature around 1990. The report of the American Department of Energy (DOE) "National Impacts of the Weatherization Assistance Program in Single-Family and Small Multifamily Dwellings", published in 1993, identifies the early quantification of non-energy benefits as essential for the development of the concept. In this report, it is stated that "while there is a good deal of anecdotal evidence on the substantial benefits of low-income weatherisation in the areas of affordable housing, health and safety, these anecdotes do not support the assignment of dollar values to the benefits" [27]. However, it is consensual that the turning point for recognition of the concept was the inclusion of the term in the Third Assessment Report (AR3) of the Intergovernmental Panel on Climate Change (IPCC), published in 2001 [28]. Here, the co-benefits were defined as "intended positive side effects of a policy from ancillary benefits" or "unintended positive side effects". This definition highlights the importance of considering co-benefits in the initial phase of policy formulation and adoption. In particular, most of the initial studies addressing the subject have mainly focused on the relationship between reducing carbon emissions and improving the physical and mental human health. These societal and macroeconomic effects were later related to the reduction of human mortality, and further studies have expanded even more the range of associated effects. A widely recognised classification divides co-benefits into health effects, ecological (or environmental) effects, social effects and benefits linked to service provision [16]. Health was one of the most studied and evaluated co-benefits and was decisive for the introduction of the concept in the communication strategy of the advantages of energy efficiency. The co-benefits associated with implementing energy efficiency measures in terms of health improvement are considerable. The most recognisable include reduced symptoms of respiratory disease and lower rates of winter mortality (EWM), in particular in colder climates. Although noticeable at the individual level, improvements in health relieve pressure from public health budgets. As an example of its significance, when impacts on health and well-being are included in assessments of energy efficiency retrofit programmes, the benefit-cost ratio can be as high as 4:1 [24]. Other significant co-benefits on a macroeconomic scale include energy security, energy prices, improvement of industrial productivity and economic development. It is generally accepted that improvements in energy efficiency can have positive macroeconomic impacts through the increase in Gross Domestic Product (GDP) and employment, but there are also indirect effects that should be considered such as lower energy expenditure. Importantly, there is overwhelming scientific evidence that these additional benefits exceed the mitigation costs, including those needed to achieve the objectives set in the Paris Agreement [29], and are thus instrumental in driving more ambitious policies that can lead the society to a more sustainable lifestyle.

Although very useful for the formulation of climate policies, this view is considered limited [30] as it does not cover all situations. The co-benefits of some technologies that promote energy and material efficiency, emitting fewer pollutants and reducing the amount of greenhouse gas emissions are not restricted to the macroeconomic effects. Changes can also be seen at other scales, such as at the household level. The co-benefits associated with improvements in energy efficiency can then be analysed from two different perspectives, depending on the scale of application of these improvements and on the scale of the effects achieved by these improvements. One common perspective addresses societal or macro-economic co-benefits. In this case, the discussion is centred on quantifying the direct and indirect benefits for the economy and the environment, such as the effects on climate change, public health and productivity. This perspective is centred on policy formulation and decision-making (*e.g* [31]). On the other hand, a private perspective on co-benefits concerns primarily the building users and/or energy consumers and indicates, as possible co-benefits, effects that are directly related to personal or household levels, such as improved air quality, thermal comfort and increased value of the property in the real estate market (Fig. **1**). It is important to refer that these co-benefits can be verified in these two perspectives simultaneously. For example, improved air quality at the private level will benefit public health at the societal level as well.

It is important to highlight that the building sector always showed a strong potential to deliver significant benefits. This was further elaborated in the IPCC fourth assessment report (AR4) [32] regarding energy efficiency and utilisation of renewable energy in the built environment establishing a link between the co-benefits and sustainable development either in developed and least developed countries. Similarly, the new recast of the Energy Performance of Buildings Directive (EPBD), released in 2018 [33], urges the EU Member States to consider co-benefits when designing new policy instruments, such as in the energy certificates, as they can significantly impact at the household level.

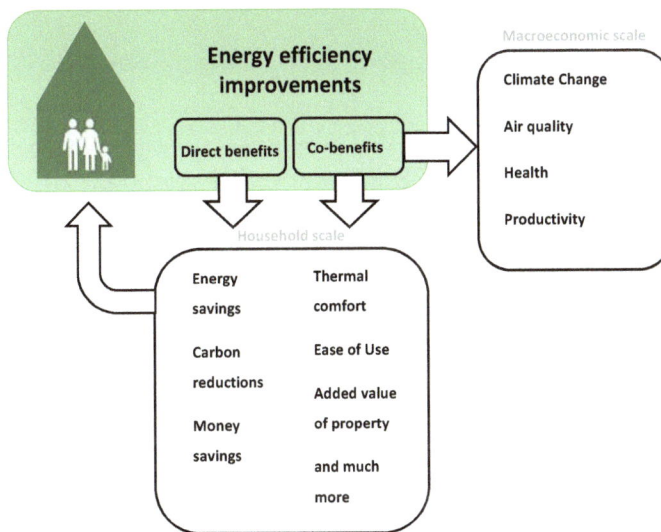

Fig. (1). Energy Efficiency and co-benefits.

In this context, the private perspective is particularly important to encourage the increase of energy efficiency in buildings, such as energy renovations and the implementation of more energy-efficient appliances. As private co-benefits relate to a value that depends mainly on the individual, the methods to determine and quantify this type of co-benefits are mainly based on self-reporting surveys to monetise the potential impacts of improving energy efficiency. Some of the methods include the simple contingent valuation, relative scaling methods and ranking based surveys.

Contingent valuation is a very straightforward approach that in its basic form consists of asking individuals what value they are willing to pay for a specific benefit. Relative scaling methods, on the other hand, consist of asking this value in relation to a base, which can be a monetary value or something of a subjective nature. For ranking based surveys, respondents are asked to attribute a ranking, comparing at least two co-benefits. It is important to note that the results of quantification using these methods point to the co-benefits assessed in a range of 9% to 43% of the energy costs saved.

ADDRESSING CO-BENEFITS THROUGH A QUALITATIVE SURVEY

The research presented in this chapter has two main objectives: to assess the relevance of the co-benefits associated with the replacement of the heating system and to quantify the potential added value provided by the co-benefits through an economic valuation.

The work performed consisted of the following methodological steps:

1. Literature review regarding the identification of co-benefits;
2. Preparation of a survey regarding the indication of the relevance of the co-benefits and their economic valuation;
3. Assessment of the relevance of the co-benefits and economic valuation analysis.

To identify the most important co-benefits to be addressed in the study, a critical review [34] of relevant literature was performed: a structured keyword-based search was used on two databases containing peer-reviewed scientific papers, books and reports – Google Scholar and Web of Science. Keywords used were "co-benefits", "non-energy benefits", "ancillary benefits", combined with terms such as "energy-efficient heating" "renewable heating", "heating systems", "energy renovation", "building renovation", "building rehabilitation", "building retrofit" and "building upgrading". The literature review allowed to preliminarily collect 23 co-benefits, found to be important for consumers when considering the replacement of their heating appliances. In the scope of this study, co-benefits are defined as additional benefits to the consumer arising from the specific characteristics (technical and physical) of the heating (production) system.

Following the literature review, the list of initially collected co-benefits was submitted to the judgment of experts to narrow the number of co-benefits that could be the most influential in the national contexts under study. The experts were chosen from the consortium of the project, which consists of eighteen partners with strong expertise in the energy field, and in particular in the heating and cooling sector. Partners in the HARP consortium included not only national energy agencies but also industry associations and research institutions. Table **1** presents the typology of the co-benefits considered in this study, as well as a brief explanation of the intended effect to be analysed.

Table 1. Typology of the co-benefits associated to the replacement of the heating equipment.

Co-benefits	Description
Thermal comfort	Higher thermal comfort due to more adequate room temperatures and relative humidity
Air quality	Improved indoor air quality, meaning reducing harmful gases, particulates, microbial contaminants (which can cause mould), or other stressor that induce adverse health conditions
Aesthetics	Aesthetic improvement of the building after the implementation of the heating solution
Ease of use /control by user	Ease of use and control of the heating solution by the users (*e.g.* automatic thermostat controls, easier filter changes, faster hot water delivery, *etc.*)
Added value into the market	Improvement of the real estate market value of the property after the implementation of the heating solution
Impact on useful area	Increase or reduction of useful area of the dwelling after the implementation of the heating solution
Independence from energy prices	Reduction of exposure to energy price fluctuations to maintain the desired level of thermal comfort
Reduction of environmental impact	Improved environmental performance regarding energy and associated carbon emissions (*e.g.* avoidance of the use of fossil fuels as energy source)

In the second methodological step of the study, an international survey was developed to be disseminated in the five countries where the project is being implemented - Portugal, Italy, France, Germany and Spain. In this survey, a question was prepared for the assessment of the co-benefits where respondents were asked to fill an ascending 7-point numerical scale corresponding to the relevance of the co-benefit in the decision-making process of choosing a heating solution. In the same survey, questions regarding the economic valuation of the same co-benefits were also asked. These questions were designed as a matrix according to a contingent valuation method (CV) to investigate the Willingness to Pay (WTP) of the consumers (Fig. **2**). An WTP analysis is a very common approach used to measure the value of non-market goods. It makes use of what is designated as a stated preference because it asks directly to respondents to place a monetary value based on a hypothetical scenario. WTP has been used for the monetisation of non-energy benefits in other contexts (*e.g.,* [35]). In this case, the scenario is posed in the following general question: *"Were you willing to invest an additional value for an energy-efficient heating solution, if it allows obtaining [co-benefit]?"*.

"Were you willing to invest an additional value for an energy-efficient heating solution, if it allows obtaining [co-benefit]?"	No	Up to 100€	Between 100€ and 500€	More than 500€
A comfortable indoor temperature during the heating season				
Better air quality				
Operating the equipment more easily				
Being independent from energy prices				
A more aesthetically pleasant equipment				
More useful living area				
A higher value of the property in the real estate market				
A reduced environmental impact				

Fig. (2). Matrix for data collection regarding Willingness to Pay for co-benefits in the international survey.

The survey was disseminated online and promoted using social media channels and institutional networks between November 2019 and February 2020. The objective was to obtain a representative sample of the population in various countries.

A random sampling methodology by country was considered, based on the assumption of infinite population size, since the exact number of consumers owning inefficient heating equipment is unknown. The sample size was, therefore calculated based on eq. (1). Regarding the level of precision (d), a value of 5% was used [36].

$$n = \frac{Z^2 p*q}{d^2} = \frac{1.96^2*0.5*0.5}{0.05^2} = 385 \tag{1}$$

where Z is the standard normal distribution for the $(1-\alpha/2)$ level, d is the precision, p is the prevalence, and $q=(1-p)$.

In this study, (p) stands for the proportion of the population that evidences the characteristic under evaluation. In this case, it refers to the proportion of the population that is willing to change its heating system. The value was calculated from the prevalence rates calculated in other studies [37] and is indicated in Table 2. The resulting representative survey sample is composed of 6044 complete responses. The sample consists of 73% male respondents and 26% female respondents. Most of the respondents have a bachelor (42%) or a master's degree (29%). Although there is a variation in the response rate for each country, the results were analysed, taking into account weighted average values for age, gender and country subsamples. Therefore, all completed responses were considered.

Table 2. Survey 1 - representative sample size per country.

Country	P (Prevalence)	N (The Necessary Number of Complete Responses)	Total Responses	Number of Complete Responses
France	0.19	237	453	411
Germany	0.12	163	300	179
Italy	0.22	264	649	387
Portugal	0.18	227	519	331
Spain	0.19	237	9531	4736
All		1128	11452	6044

The analysis carried out took into account that from the private perspective, differences in national contexts strongly affect the decision of adopting sustainable and energy-efficient heating technologies. There is evidence that not only culture and policy-driven investments are important. Differences can also be strongly related to the stage of diffusion of innovation in each country [7]. These factors are not only important for understanding the decision-making process at the household level but also influence the relevance of the different co-benefits in that process [8]. On the other hand, there is a need to deepen the knowledge in each country in relation to known individual factors. The decision on whether to adopt energy-efficient technologies depends heavily on several factors, such as ownership, gender, access to capital or income, as well as education and age [37 - 40]. For example, there is evidence that younger households are more likely to attribute importance to environmental benefits and that older households place greater value on the economic savings generated by improvements in energy efficiency. In this study, the degree of relevance was distinguished by country.

Additionally, the differences between the most consistent known factors presented in studies on the adoption of improvements in energy efficiency – gender, income and age - were detailed by each country participating in the analysis.

DEGREE OF RELEVANCE AND WILLINGNESS TO PAY FOR CO-BENEFITS

The degree of relevance of the co-benefits was defined according to the answers of the respondents in the 7-points scale question, where 1 means no relevance at all and 7 means the maximum degree possible. The degree of relevance (Fig. **3**) is not related to any specific heating solution and intends to collect information on the perceptions and preferences of the consumers. The results indicate that there are significant differences in the relevance of different co-benefits, according to the geographical context.

DEGREE OF RELEVANCE OF CO-BENEFITS

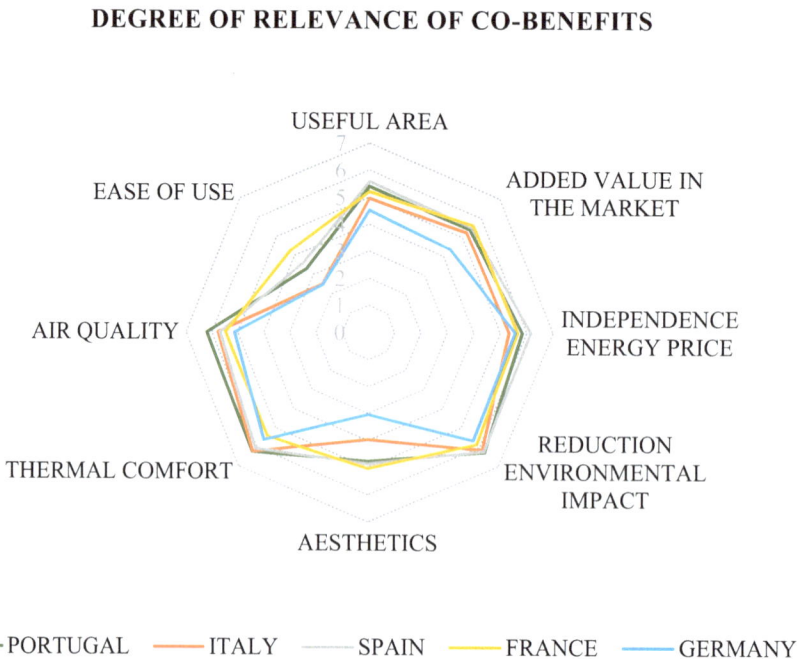

Fig. (3). Degree of relevance of co-benefits per country.

The diagram in Fig. (**3**) highlights the differences in terms of countries from what is considered relevant in terms of co-benefits. For Portugal, for example, according to the results, the most relevant co-benefits are the ones related to *thermal comfort, air quality* and *reduction of environmental impact*. Although these co-benefits are highly valued in many countries (such as in Italy or Spain), responses from France indicate that the most relevant co-benefit is the *added value in the market*. For Spain, alongside with *thermal comfort*, there is strong evidence that *independence from energy prices* is highly relevant. According to results from the survey, *independence from energy prices* is also the most relevant co-benefits in Germany. The results also indicate that in Italy and Germany, the *ease of use* is the least relevant co-benefit. *Aesthetics* is the co-benefit that consistently presents the lowest values in terms of relevance in all countries analysed, with average values ranging from 3 (Germany) to 5 (France).

WILLINGNESS TO PAY FOR CO-BENEFITS

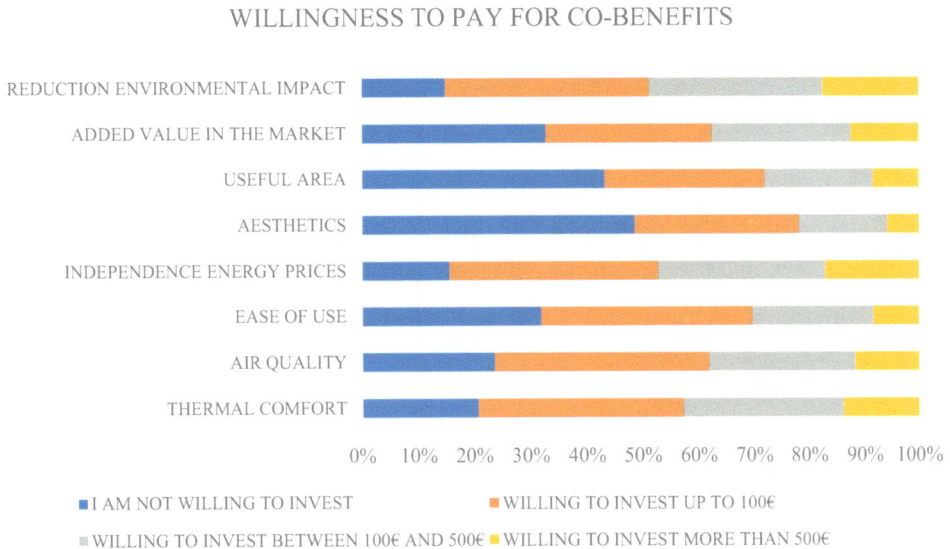

Fig. (4). Willingnesss to pay for co-benefits (all countries).

Regarding the Willingness to Pay (WTP), the results suggest that at least 15% of the respondents are not willing to invest in any co-benefit Fig. (**4**), which is significant. Additionally, an important share of respondents (34% on average) is willing to pay only an additional of 100 euros per each co-benefit. In particular, co-benefits such as *air quality, ease of use, independence from energy prices, thermal comfort* and *reduction of environmental impact* are the most valued in this tier. In the tier corresponding to the willingness to pay between 100 and 500 euros, the most valued co-benefit is *independence from energy prices* (30%), closely followed by *thermal comfort* and *air quality*, according to the data collected. Importantly, as expected, fewer respondents are willing to pay more than 500 euros for the co-benefits. The *reduction of environmental impact* and *independence from energy prices* are indicated as the most valued co-benefits in this tier. In opposition, *aesthetics* and *useful area* are the least valued. Consistently, most respondents (49%) indicated that they are not willing to invest any additional value in these two co-benefits (*aesthetics* and *useful area*) when associated with a heating solution.

COMPARING THE PERSPECTIVES OF NATIONAL CONTEXTS ON CO-BENEFITS

This section presents the most important aspects found in the analysis performed by country, with the objective of highlighting their particularities. For that purpose, the results were differentiated between the factors of age, gender and income range, which were already considered important in the owners' decision-making process in other studies (*e.g.,* [38,39]).

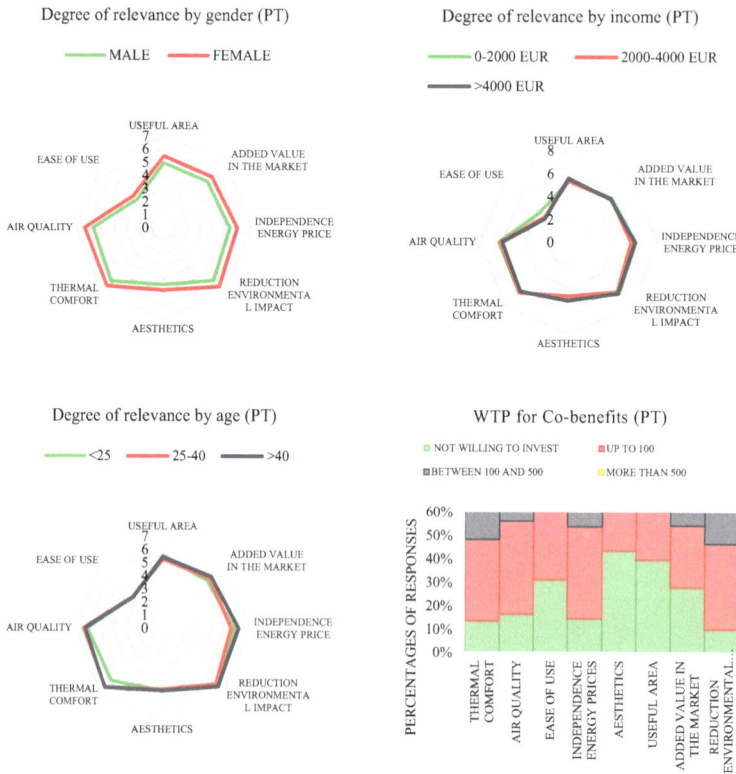

Fig. (5). Co-benefits – Degree of relevance and WTP for Portugal.

Portugal

Detailed results from Portugal (Fig. **5**) show that, in general, female respondents value co-benefits more than male respondents, although the responses were aligned in terms of importance. In Portugal, independently of the perspective under analysis, the most relevant co-benefits are *thermal comfort*, *air quality* and *reduction of environmental impact*. It is also noticeable that differences in income present only slight distinctions in how co-benefits are considered relevant. For example, there is a subtle decrease concerning *ease of use* for respondents with a higher level of income. Respondents who reported the highest level of monthly income considered in the survey (>4000 euros) also attribute a higher relevance to *aesthetics*. Respondents on the lower tier in age (<25) indicated *thermal comfort* as being less relevant than other ages tiers. In opposition, responses coming from the middle tier (between 25 and 40 years old) indicate *independence from energy prices* and *added value in the market* as being the least relevant. However, these are subtle differences between these two co-benefits (an average difference of 0.57 in the 7-point scale range).

In terms of the economic valuation, 40% and 44% of the respondents in Portugal respectively indicated that they were not willing to invest an additional value in *aesthetics* and changes in the *useful area*. At the level of willingness to pay up to 100 euros, the highest percentage of responses are on *air quality* (40%) and *independence from energy prices* (39%). Respondents also indicated that they are willing to invest an additional value between 100 and 500 euros in *reduction of the environmental impact* (40%) and *thermal comfort* (38%), as well as in *air quality* (33%). When it comes to the highest level of WTP (>500 euros), the highest percentages correspond to *reduction of environmental impacts* (13%), *thermal comfort* (12%) and *added value in the market* (12%).

Spain

Results concerning responses from Spain (Fig. **6**) indicate that the most relevant co-benefits at the country level are *reduction of environmental impacts* (average value of 6.15), *independence from energy prices* (average value of 6.1) and *thermal comfort* (average value of 6.0). In terms of gender, male respondents in Spain attribute less relevance to co-benefits, in particular to the *useful area* (average difference of -0.43), *air quality* (average difference of -0.40) and *aesthetics* (average difference of -0.26). When income levels were considered, a coincidence in almost every co-benefit for all income levels is noticeable. However, results concerning *ease of use* (which is the least relevant co-benefit in this context) present differences, namely regarding the first level of income (<2000 euros) attributing more relevance to this co-benefit than to the second (average difference of -0.14) and the third (average difference of -0.43) levels of monthly income.

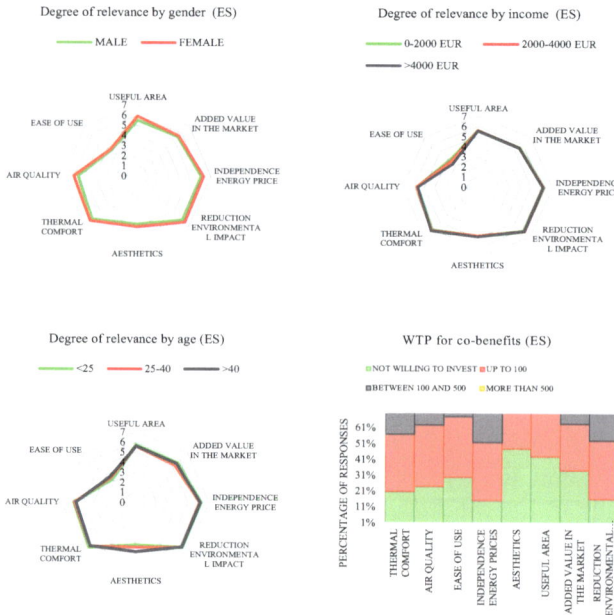

Fig. (6). Co-benefits – Degree of relevance and WTP for Spain.

When age is analysed for this national context, results indicate that *ease of use* and *aesthetics* are considered as having a higher relevance for respondents over 40 years old (average value of 4,89) than for the respondents under 25 years old (average value of 4,22). Younger respondents (<25 years old) from Spain indicated *independence from energy prices, reduction of environmental impacts, air quality* and *thermal comfort* as the most relevant co-benefits.

Concerning willingness to pay, Spanish respondents indicated that they are not willing to invest an additional value in co-benefits providing improved *aesthetics* (48%) and more *useful area* (43%). A willingness to pay up to 100 euros is evidenced for *air quality* (39%) *ease of use* (38%), *independence from energy prices* (37%) and *reduction of environmental impacts* (37%). *Independence from energy prices* (31%) and *reduction of environmental impacts* (30%) are also indicated as the most significant amongst respondents willing to pay between 100 and 500 euros, as well as in the highest level of willingness to pay (> 500 euros).

Italy

In Italy (Fig. **7**), similarly to Spain, *ease of use* (average value of 2.5) and *aesthetics* (average value of 3.9) are considered the least relevant co-benefits. On the other hand, in this country, *thermal comfort* (average value of 6.2) and *reduction of environmental impacts* (average value of 6.1) are indicated as the most relevant. As in other national contexts, in Italy, female respondents also give slightly higher relevance to co-benefits, with exception to *added value in the market*.

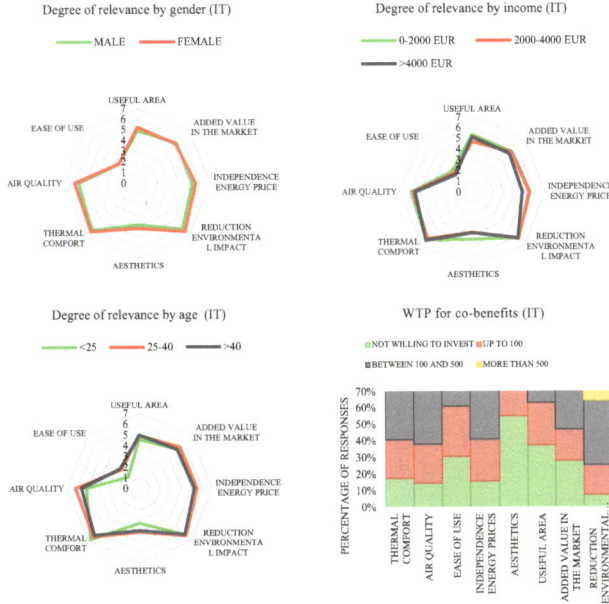

Fig. (7). Co-benefits – Degree of relevance and WTP for Italy.

Results regarding the level of income present interesting differences for Italy. Respondents within the first level of monthly income (<2000 euros) are the ones indicating *thermal comfort* (average value of 6.3) and *reduction of environmental impact* (average value of 6.1) as the most relevant co-benefits. In opposition, at this level of income, respondents also indicated that *ease of use* (average value of 2.6) and *aesthetics* (average value of 4.4) are the least relevant. At the level of income between 2000 and 4000 euros, as well as for the highest level of income (>4000 euros), results consistently point in the same direction, with small differences on the indicated degree of relevance, in particular regarding *aesthetics*. Concerning this co-benefit, respondents in the two highest levels of income in Italy reported it as being the least relevant (in fact even less relevant than for the respondents in the first level of income (<2000 euros per month)).

Respondents under 25 years old indicated *thermal comfort* as the most relevant co-benefit (average value of 6.8) and *ease of use* as the least relevant (average value of 1.5). On the other hand, respondents within 25 and 40 years old reported *air quality* as the most relevant (average value of 6.1). *Thermal comfort* is also considered the most relevant co-benefit for respondents over 40 years old (average value of 6.2).

Regarding the willingness to pay, there is a significantly high percentage of responses (55%) indicating no willingness to invest an additional value in *aesthetics,* when acquiring a new heating solution. Respondents are willing to pay an additional value up to 100 euros primarily for *ease of use* (30%) and *added value in the market* (26%). Responses indicated that *thermal comfort* is the co-benefit most people are willing to pay an additional value between 100 and 500 euros (30%) alongside with *independence from energy prices.* Most of the respondents willing to invest more than 500 euros indicated the *reduction of environmental impacts* (36%) and *thermal comfort* (29%) as the preferred co-benefits.

France

In France (Fig. **8**), the *reduction of environmental impact* is the most relevant co-benefit (average value of 5.8) and *ease of use* the least relevant (average value of 4.2). Results indicate that, in terms of gender, male respondents value *ease of use* more than female respondents (a difference of 0.5 on average).

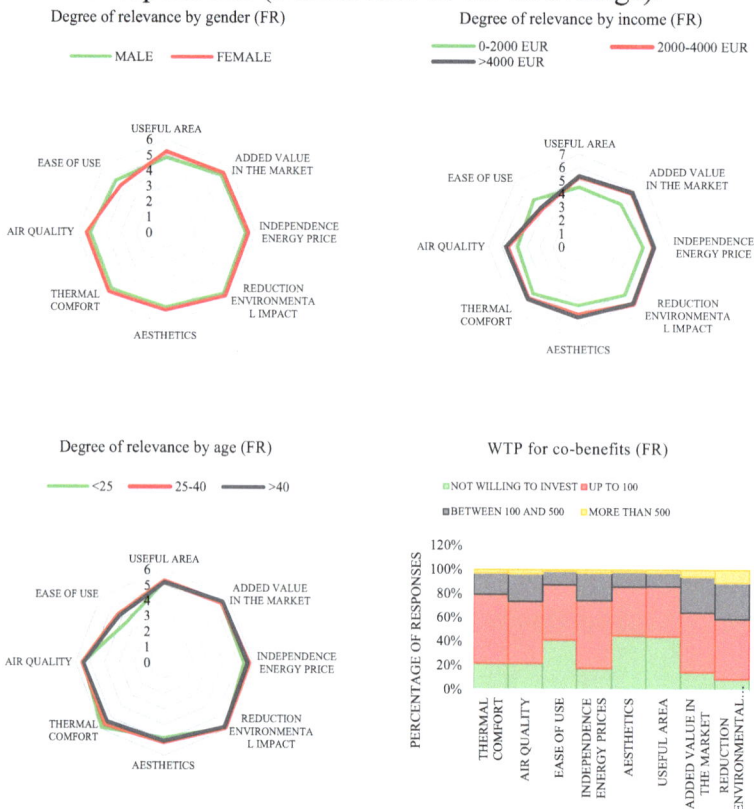

Fig. (8). Co-benefits – Degree of relevance and WTP for France.

Interesting results regarding income indicate a coincidence in responses in relation to the levels of monthly income between 2000 and 4000 euros and more than 4000 euros. However, for respondents with a level of income equal or less than 2000 euros/month, the relevance of co-benefits is consistently lower (average of 0.9 difference in responses), with the exception of *ease of use*, which was identified as more relevant for the lower income level respondents (average of 0.8 difference) than for the respondents with higher levels of income.

Generally, results from French respondents are consistent for all the age tiers considered in the analysis. However, younger respondents indicated the highest value for *thermal comfort* (average value of 5.9) and the lowest value for *ease of use* (average value of 3.58).

In terms of willingness to pay, responses point out to a high percentage of people not willing to invest in *aesthetics* (45%) and change in *useful area* (44%). Most of the respondents indicated that they are willing to pay an additional value of up to 100 euros for several co-benefits. In this context, *thermal comfort* (58%), *independence from energy prices* (57%) and *air quality* (52%) gathered the majority of the responses. Regarding the responses reporting a willingness to pay between 100 and 500 euros, most of the respondents indicated the *reduction of environmental impact* (30%). The same answer can be extracted from the results in the upper level of willingness to pay (>500 euros) although there are very few responses for this tier in this national context.

Germany

In Germany (Fig. **9**), both the *reduction of environmental impacts* and *thermal comfort* present an average value of 5.6 in the degree of relevance, which is the highest value given by the respondents. In this country, the least relevant co-benefit is *ease of use* (average value of 2.4).

The most significant differences in responses regarding the degree of relevance by gender in Germany are in *air quality* and *thermal comfort*. In both co-benefits, male respondents indicated to be of less relevance (an average difference of 0.7 and 0.8, respectively) when compared to the responses of female respondents. However, male respondents considered the *independence from energy prices* to be more relevant than female respondents (an average difference of 0.3 in the degree).

In terms of income levels in Germany, respondents within the lowest level of income (<2000 euros) generally indicated a lower level of relevance for all co-benefits. Respondents within the middle level of monthly income (between 2000 and 4000 euros) indicated *air quality* (average value of 5.5) and *useful area* (average value of 4.8), as well as *aesthetics* (average value of 3.4) the most relevant co-benefits. For respondents in Germany earning more than 4000 euros/monthly, the most relevant co-benefits are the *reduction of environmental impact* (average value of 5.8), *independence from energy prices* (average value of 5.7) and *added value in the market* (average value of 4.6).

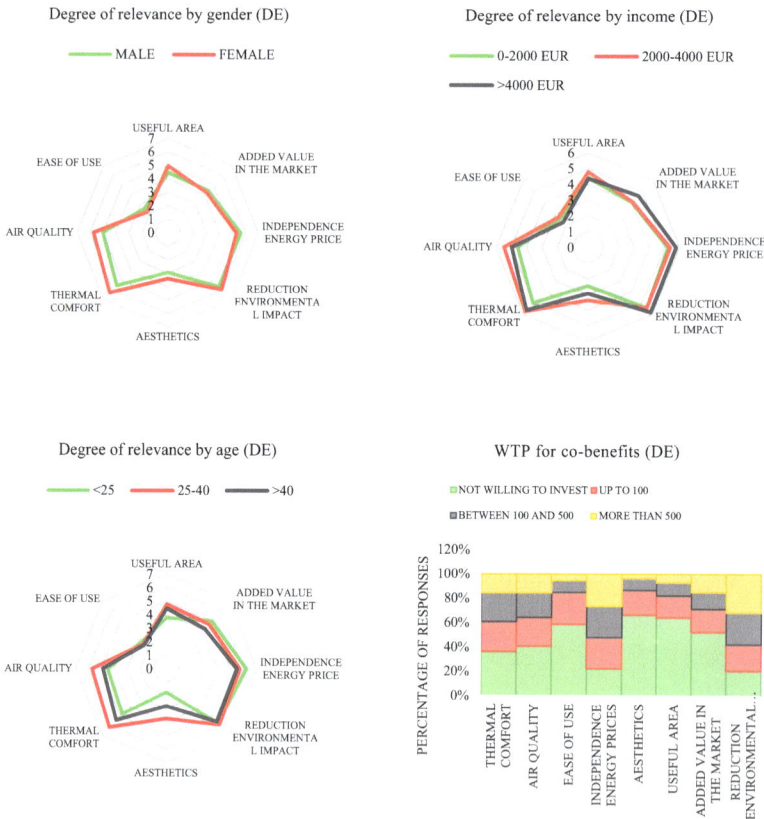

Fig. (9). Co-benefits – Degree of relevance and WTP for Germany.

Responses also varied significantly in terms of the respondents' age. For the youngest (<25 years old) German respondents, the most relevant co-benefits are the *independence from energy prices* (average value of 6.2) and the *reduction of environmental impact* (average value of 5.7). In opposition, the youngest respondents attributed the least relevance to *aesthetics* (average of 1.7). Respondents in the middle tier regarding age (between 25 and 40 years old), indicated that the most relevant co-benefits are the *reduction of environmental impact* (average value of 5.8), *thermal comfort* (average value of 6.1) and *air quality* (average value of 5.7). Despite an average difference of 0.4 in the degree of relevance, these are also the co-benefits indicated as most relevant for respondents over 40.

Regarding the willingness to pay, responses indicated *aesthetics* (66%) and *useful area* (64%) as the co-benefits with the highest percentage of people not willing to invest. Responses regarding other levels of WTP are lower when compared to other countries in terms of percentage.

However, 26% of the respondents indicated that they are willing to invest an additional value of up to 100 euros in the *reduction of environmental impact, independence from energy prices* and *ease of use*.

Respondents also indicated they were willing to pay between 100 and 500 euros for *reducing the environmental impact* (26%), *independence from energy prices* (25%) and *thermal comfort* (23%). *Reduction of the environmental impact* was the co-benefit indicated by the highest percentage of respondents (32%) willing to pay more than 500 euros.

CO-BENEFITS AS A SUPPORT FOR DECISION-MAKING

Using information such as that collected in this study to motivate consumers to replace their heating systems or improve energy-efficiency in their homes can be complex. On the one hand, there is evidence that the adoption of these measures can be strongly influenced by factors such as economic availability and knowledge on the subject, but habit and climate and cultural contexts can have a determinant influence on the decision. There is an obvious need to go deeper to better understand the differences demonstrated by the results in the previous sections by undertaking further research on the subject, which may shed some additional light on the subject. There is, nevertheless, an intrinsic value of the knowledge gained about co-benefits to support decision-making.

Co-benefits can support the adoption of energy-efficient technologies in decision-making from different perspectives. From the policy making perspective, there is potential to reduce the known barriers by supporting the communication and information for consumers and building users, as well as identifying key messages and "selling arguments". Despite a recognisable limited understanding by policymakers of the implications of using a "co-benefits approach" to address multiple climate goals [41], this type of approach is increasingly being used to communicate additional benefits at a societal level. For example, the World Bank has provided a tool that can support decision-makers regarding urban sustainability and there is a significant focus on local benefits such as improved air quality and health outcomes [42]. The European Environment Agency had also focused on this kind of information to communicate the advantages of boosting energy-efficiency [43]. However, this approach is not widely used at the private level, despite existing evidence that these additional advantages can significantly influence the decision-making process at the household level. Evidence from research, point out that these advantages are key, particularly in situations driven by opportunities, such as building renovations. In a building renovation process, the homeowners seem to need a solid decision basis to compare technologies [13]. Despite the wide range of information on the co-benefits that can support the adoption of investments in energy-efficient at the household level, there are limited communication initiatives focusing on this perspective [44]. In this context, the HARP project [23] integrated this information in a user-friendly online application that can advise consumers on the advantages of replacing a heating system at the household. There is, also a very promising venue regarding the next generation of Energy Performance Certificates and the possibility of promoting additional information on the implementation of renovation measures as hinted in the 2018 recast of the Energy Performance of Buildings Directive [33]. In this context, results from studies such as the one reported in this chapter can help to tailor this information to the specificities of each country and to what is considered relevant for building users and energy consumers. A previous work [21] has demonstrated that the co-benefits arguments are most effective when aligned with the values and priorities of the target stakeholders, who can be policymakers or building users, because there is a strong possibility that they do not consider energy efficiency as a benefit in itself. There is, therefore, the urgency of recognising different sets of actors, different cultural contexts, with specific values, which can lead to different and alternative understandings of the benefits.

Although not as directly as an information campaign or communication instrument, the economic valuation of these co-benefits can also be of key importance in motivating building users to promote energy-efficiency at the household. One of the questions that can be highlighted in the analysis performed in this study concerns the relationship between the relevance of the co-benefits and the economic value that consumers are willing to pay for them. In Fig. (**10**), the results of the international survey were used to demonstrate the relationship between the degree of relevance (X-axis) and the Willingness to Pay (Y-axis). The size of the circles is indicative of the number of responses that corresponded to the relationship of these two distinctive factors.

Generally, the results from the survey indicate that, although there is a clear relationship between the degree of relevance and the willingness to pay for a co-benefit when purchasing a new heating solution, this relationship varied significantly depending on the co-benefits. As an example, Table **3** summarises the relationships that aggregate the largest number of responses.

DEGREE OF RELEVANCE AND WTP

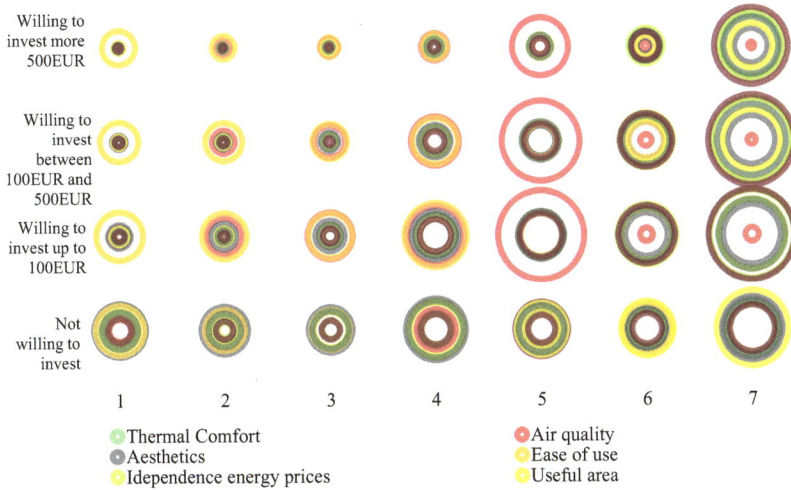

Fig. (10). Relation between Degree of Relevance and Willingness to Pay.

Table 3. Summary of the highest aggregated responses in relation to the Degree of Relevance and the WTP.

Co-benefits	Thermal Comfort	Air Quality	Aesthetics	Ease of Use	Useful Area	Added Value in Market	Independence Energy Prices	Reduction Enviromental Impact
Relation Relevance/ WTP	7/ <100€	7/<100€	7/0€	4/ <100€	7/<100€	7/100-500€	7/ <100€	7/<100€

The higher aggregation of collected responses indicated that the co-benefits were, in general, considered of high relevance (7 in most cases), but respondents are not willing to pay more than 100 euros for the co-benefit when purchasing a new heating system. Exceptions were observed in *aesthetics* that, although considered relevant, most responses indicated that users were not willing to invest in it. Another exception was observed in *added value in the market,* where the respondents indicated that they were willing to invest a value between 100 and 500 euros.

Looking at the relation presented in Fig. (**10**) in detail, it is clear that the results of the relationship between Degree of Relevance and Willingness to Pay were mostly concentrated on the co-benefits rated as the most relevant (values between 5, 6 and 7 in the X-axis) and middle level in terms of value invested (investments lower than 500 euros in the Y-axis). There is a significant relationship between the degree of relevance and WTP concerning the *reduction of environmental impact* co-benefit. Most responses regarding this co-benefit indicated a high relevance (degree of 7) and a significant willingness to pay. In particular, there is a large set of responses on the tier responding that are willing to invest up to 100 euros (1206 responses) and between 100 and 500 euros (1198 responses), although there were also some responses indicating the greater willingness to pay more than 500 euros (868 responses). Notably, there was quite a significant number of responses indicating that while some co-benefits were relevant to some respondents, there is no willingness to pay an additional value for them. Clear examples of this relationship could be found in the *added value in the market*, where 406 respondents indicated a degree of relevance 7 but showed no willingness to invest an additional value in this co-benefit. This disparity has been observed in research and is commonly referred to in the literature as asymmetric behaviour [45]. The contextual effects, such as economic and psychological factors, as well as the familiarity of the respondents with the concepts presented in the survey usually explain it. For decision-making purposes, this disparity highlights the need for comprehensive analyses of the factors influencing the gap between the degree of relevance and the willingness to pay. It also highlights the intensity of the policy and the long-term investments made in each country that may explain some of the differences, due to the way they were promoted [7]. On the other hand, it is also important to underline the importance of models to support decision-making in economic assessments, specific to each country, which must be robust and validated by several studies and methodologies. In that context, the second question of interest regarding the economic valuation performed in this study is that the contingent valuation used to obtain a willingness to pay can give a first estimate of the values to be considered for the co-benefits considering the replacement of heating appliances. It can be useful regarding the formulation of cost-benefit analyses and the expansion of other models integrating co-benefits, such as the one developed to support the choice of cost-optimal energy renovation measures in buildings [19]. Such an approach can give buildings users and energy consumers the perspective of the real value obtained from investments in energy-efficiency beyond energy and economic savings. However, the use of these results in the cost-benefit analysis should consider limitations and shortcomings that have to be recognised. One clear example is the difficulty of classifying co-benefits, since overlaps occur (*e.g.* comfort and air quality). In addition, although the achievement of co-benefits is widespread, its intrinsic value is influenced by contextual settings and location specificities [16], which can hinder the successful implementation of national policy based on these quantifications. Therefore, there is a recognised need for further research validating quantifications at various scales (*e.g.* regional). The method used for collecting the Willingness to Pay in this study – contingent valuation – has its own advantages and limitations that must also be addressed and recognised. The technique is used consistently because it is the most straightforward estimation method and admits that respondents are not fully aware of the real energy costs in their household, in opposition to scaling techniques that include such necessity. However, it depends on the respondents' level of abstraction, which has to be related to a hypothetical situation. This issue can lead to overvalued estimates and the appearance of outliers, which in this study are minimised with the use of a weighted average approach, as explained. Despite these limitations and possible inaccuracies in estimating the value and relevance of co-benefits, monetised co-benefits and their use in cost-benefit analyses can be useful in informing policy decision-making structures and helping to define prioritisation of encouraged policies and technologies.

CONCLUSION

The work reported in this chapter aimed at identifying the relevance of the different co-benefits associated with the replacement of heating appliances in the decision-making process of consumers. It was also a goal to make an economic valuation of these co-benefits based on a contingent valuation method. The data collected allowed a better understanding of the difference in the relevance attributed to certain co-benefits in different national contexts, as well as to understand the specificities of each country.

The co-benefits indicated as the most relevant were *thermal comfort, air quality* and *reduction of environmental impact*. However, there are particularities in each context. Results from France indicated that the most relevant co-benefit is the *added value in the market* (for the building). For Spain, results indicated that, alongside *with thermal comfort,* the *independence from energy prices* is a highly relevant co-benefit when considering the replacement of a heating appliance. The *independence from energy prices* is also the most relevant co-benefit in Germany, as indicated by responses in this study.

The results of this analysis point out that, although there is a clear relationship between the degree of relevance and the willingness to pay an additional value when purchasing a heating solution, this relationship varies significantly depending on the co-benefits. Notably, there is a significant number of responses indicating that, although some co-benefits were reported as relevant, there is no willingness to pay an additional value for them. Clear examples of this relationship were found concerning *added value in the market,* for example. In opposition, there is a significant relationship between the degree of relevance and the willingness to pay concerning the *reduction of environmental impact* co-benefit. Most responses to this co-benefit indicated a high relevance, as well as a significant willingness to invest an additional value in it. At least 15% of the respondents answered that were not willing to invest any value in any of the co-benefits. On the other hand, a significant share of the respondents indicated that they were willing to pay an additional 100 euros to achieve certain benefits such as *air quality, ease of use, independence from energy prices, thermal comfort and reduction of environmental impact.* Importantly, in the tier corresponding to the willingness to pay between 100 and 500 euros, the most valued co-benefit is *independence from energy prices*, closely followed by *thermal comfort* and *air quality.* Additionally, *the reduction of environmental impact* and *independence from energy prices* are indicated as the co-benefits most valued by respondents, some indicating that they are willing to pay more than 500 euros.

In conclusion, the results highlighted the importance of the relevant economic, social and cultural differences between the five countries analysed. The information collected can be used to support decision-making. Two different views were discussed in this chapter. On the one hand, the information generated by studies such as the one presented here can help in communicating the benefits of energy efficiency programs both at societal and private level, tailoring information campaigns to the specificities and relevance of each national context. On the other hand, this study also shows the importance of the economic valuation of co-benefits, as a key element for more robust cost-benefit analyses, by providing information for building users and energy consumers about the real value obtained with investments in energy efficiency beyond energy and cost savings.

CONSENT FOR PUBLICATION

Not Applicable.

CONFLICT OF INTEREST

The author confirms that this chapter contents have no conflict of interest.

ACKNOWLEDGEMENTS

The authors would like to thank the HARP project partners for supporting the dissemination of the survey. This project has received funding from the European Union's Horizon 2020 research and innovation programme under grant agreement No 847049. This work was partly financed by FCT / MCTES through national funds (PIDDAC) under the R&D Unit Institute for Sustainability and Innovation in Structural Engineering (ISISE), under reference UIDB / 04029/2020.

REFERENCES

[1]   Eurostat, *Energy consumption in households - Statistics Explained,* 2017. https://ec.europa.eu/eurostat/statistics-explained/index.php?title=Energy_consumption_in_households#Energy_consumption_in_households_by_type_of_end-use

[2]   European Commission, *EUROPA - The EU strategy on heating and cooling | SETIS - European Commission,* 2020. https://setis.ec.europa.eu/publications/setis-magazine/low-carbon-heat-ng-cooling/eu-strategy-heating-and-cooling

[3]   N. Bertelsen, and B. Vad Mathiesen, "EU-28 Residential Heat Supply and Consumption: Historical Development and Status", *Energies,* vol. 13, no. 8, p. 1894, 2020.
[http://dx.doi.org/10.3390/en13081894]

[4]   J. Terés-Zubiaga, "Cost-effective building renovation at district level combining energy efficiency & renewables – Methodology assessment proposed in IEA-Annex 75 and a demonstration case study", *Energy Build.,* vol. 224, p. 110280, 2020.
[http://dx.doi.org/10.1016/j.enbuild.2020.110280]

[5]   European Commission, *An EU Strategy on Heating and Cooling,* 2020. https://eur-lex.europa.eu/lega--content/EN/TXT/PDF/?uri=CELEX:52016DC0051&from=EN

[6]   European Commission, *COMMISSION DELEGATED REGULATION (EU) 2019/ 826 - of 4 March 2019 - amending Annexes VIII and IX to Directive 2012/ 27/ EU of the European Parliament and of the Council on the contents of comprehensive assessments of the potential for efficient heating and,* 2019. https://eur-lex.europa.eu/legal-content/EN/TXT/PDF/?uri=CELEX:32019R0826&from=EN

[7]   E. Heiskanen, and K. Matschoss, *Understanding the uneven diffusion of building-scale renewable energy systems: A review of household, local and country level factors in diverse European countries,* 2017.
[http://dx.doi.org/10.1016/j.rser.2016.11.027]

[8]   C.C. Michelsen, and R. Madlener, "Homeowners' preferences for adopting innovative residential heating systems: A discrete choice analysis for Germany", *Energy Econ.,* vol. 34, no. 5, pp. 1271-1283, 2012.
[http://dx.doi.org/10.1016/j.eneco.2012.06.009]

[9]   S. Karytsas, and H. Theodoropoulou, "Public awareness and willingness to adopt ground source heat pumps for domestic heating and cooling", *Renew. Sustain. Energy Rev.,* vol. 34, pp. 49-57, 2014.
[http://dx.doi.org/10.1016/j.rser.2014.02.008]

[10]   B. M. Sopha, and C. A. Kloeckner, *Psychological factors in the diffusion of sustainable technology: a study of Norwegian households' adoption of wood pellet heating.,* 2010. Accessed: Oct. 24, 2020.

[11]   J. Dadzie, G. Runeson, G. Ding, and F. Bondinuba, "Barriers to Adoption of Sustainable Technologies for Energy-Efficient Building Upgrade—Semi-Structured Interviews", *Buildings,* vol. 8, no. 4, p. 57, 2018.
[http://dx.doi.org/10.3390/buildings8040057]

[12]   A. Alberini, and A. Bigano, "How effective are energy-efficiency incentive programs? Evidence from Italian homeowners", *Energy Econ.,* vol. 52, pp. S76-S85, 2015.
[http://dx.doi.org/10.1016/j.eneco.2015.08.021]

[13]   M. Hecher, S. Hatzl, C. Knoeri, and A. Posch, "The trigger matters: The decision-making process for heating systems in the residential building sector", *Energy Policy,* vol. 102, pp. 288-306, 2017.
[http://dx.doi.org/10.1016/j.enpol.2016.12.004]

[14]   C.C. Michelsen, and R. Madlener, "Switching from fossil fuel to renewables in residential heating systems: An empirical study of homeowners' decisions in Germany", *Energy Policy,* vol. 89, pp. 95-105, 2016.
[http://dx.doi.org/10.1016/j.enpol.2015.11.018]

[15]   K.E. Thomsen, "Energy consumption and indoor climate in a residential building before and after comprehensive energy retrofitting", *Energy Build.,* vol. 123, pp. 8-16, 2016.
[http://dx.doi.org/10.1016/j.enbuild.2016.04.049]

[16]   D. Ürge-Vorsatz, A. Novikova, and M. Sharmina, *Counting good: quantifying the co-benefits of improved efficiency in buildings,* 2009.   http://psb.vermont.gov/sites/psb/files/projects/EEU/screening/Urge-Vorsatz__Counting_Good.pdf

[17]   P.G. Bain, "Co-benefits of addressing climate change can motivate action around the world", *Nat. Clim. Chang.,* vol. 6, no. 2, pp. 154-157, 2016.
[http://dx.doi.org/10.1038/nclimate2814]

[18]   T. Buso, F. Dell'Anna, C. Becchio, M.C. Bottero, and S.P. Corgnati, "Of comfort and cost: Examining indoor comfort conditions and guests' valuations in Italian hotel rooms", *Energy Res. Soc. Sci.,* vol. 32, pp. 94-111, 2017.
[http://dx.doi.org/10.1016/j.erss.2017.01.006]

[19]   M. Ferreira, M. Almeida, and A. Rodrigues, "Impact of co-benefits on the assessment of energy related building renovation with a nearly-zero energy target", *Energy Build.,* vol. 152, pp. 587-601, 2017.
[http://dx.doi.org/10.1016/j.enbuild.2017.07.066]

[20]   M. Almeida, and M. Ferreira, "IEA EBC Annex56 vision for cost effective energy and carbon emissions optimisation in building renovation", *Energy Procedia,* pp. 2409-2414, 2015.
[http://dx.doi.org/10.1016/j.egypro.2015.11.206]

[21]   T. Fawcett, and G. Killip, "Re-thinking energy efficiency in European policy: Practitioners' use of 'multiple benefits' arguments", *J. Clean. Prod.,* vol. 210, pp. 1171-1179, 2019.
[http://dx.doi.org/10.1016/j.jclepro.2018.11.026]

[22]   A. Kamal, S. G. Al-Ghamdi, and M. Koc, "Revaluing the costs and benefits of energy efficiency: A systematic review", *Energy Research and Social Science,* vol. 54, Elsevier Ltd, pp. 68-84, 2019.
[http://dx.doi.org/10.1016/j.erss.2019.03.012]

[23]   HARP, *HARP - Heating Appliances Retrofit Planning,* 2019. https://heating-retrofit.eu/

[24]   International Energy Agency, *Capturing the Multiple Benefits of Energy Efficiency: Executive Summary.* Capturing the Multiple Benefits of Energy Efficiency, 2014, pp. 18-25.

[25]   M. Yang, and X. Yu, "Energy efficiency becomes first fuel", *Green Energy and Technology,* vol. 142, pp. 11-18, 2015.
[http://dx.doi.org/10.1007/978-1-4471-6666-5_2]

[26]   International Energy Agency, *Energy Efficiency 2020,* 2020.

[27]   M. A. Brown, L. G. Berry, R. A. Balzer, and E. Faby, "National impacts of the Weatherization Assistance Program in single-family and small multifamily dwellings", *Oak Ridge, TN (United States),* 1993.
[http://dx.doi.org/10.2172/10179419]

[28]   IPCC, *Climate Change 2001: The Scientific Basis. Contribution of Working Group I to the Third Assessment Report of the Intergovernmental Panel on Climate Change.* Cambridge University Press: Cambridge, United Kingdom and New York, NY, USA, 2001.

[29]   J. Sampedro, S.J. Smith, I. Arto, M. González-Eguino, A. Markandya, K.M. Mulvaney, C. Pizarro-Irizar, and R. Van Dingenen, "Health co-benefits and mitigation costs as per the Paris Agreement under different technological pathways for energy supply", *Environ. Int.,* vol. 136, p. 105513, 2020.
[http://dx.doi.org/10.1016/j.envint.2020.105513] [PMID: 32006762]

[30]   E. Jochem, and R. Madlener, "The Forgotten Benefits of Climate Change Mitigation: Innovation, Technological Leapfrogging, Employment, and Sustainab", *WORKING PARTY ON GLOBAL AND STRUCTURAL POLICIES OECD Workshop on the Benefits of Climate Policy: Improving Information*

for Policy Makers 2003.

[31]   G. Heffner, and N. Campbell, *Evaluating the co-benefits of low-income energy-efficiency programmes,* 2011. Accessed: Oct. 16, 2020.

[32]   S. Solomon, D. Qin, M. Manning, Z. Chen, M. Marquis, and K.B. Averyt, IPCC, "Summary for Policymakers. In: Climate Change 2007: The Physical Science Basis", *Contribution of Working Group I to the Fourth Assessment Report of the Intergovernmental Panel on Climate Change,* Cambridge University Press: Cambridge, United Kingdom and New York, NY, USA, 2007.

[33]   EU, "Directive (EU) 2018/844 of the European Parliament and of the Council of 30 May 2018 amending Directive 2010/31/EU on the energy performance of buildings and Directive 2012/27/EU on energy efficiency", *Official Journal of the European Union,* vol. 2018, no. May, pp. 75-91, 2018.

[34]   M. J. Grant, and A. Booth, "A typology of reviews: An analysis of 14 review types and associated methodologies", *Health Information and Libraries Journal,* vol. 26, John Wiley & Sons, Ltd, no. 2, pp. 91-108, 2009. Jun. 01, 2009.
[http://dx.doi.org/10.1111/j.1471-1842.2009.00848.x]

[35]   L.A. Skumatz, and J. Gardner, *Methods and Results for Measuring Non-Energy Benefits in the Commercial and Industrial Sectors,* 2020. https://www.aceee.org/files/proceedings/2005/data/papers/ SS05_Panel06_Paper15.pdf

[36]   L. Naing, T. Winn, and B. N. Rusli, *Practical Issues in Calculating the Sample Size for Prevalence Studies,* 2020.

[37]   Eurogas, "Eurogas: Energy Survey," 2019. Accessed: Nov. 09, 2020. [Online].,

[38]   E. Zvingilaite, and H. Klinge Jacobsen, "Heat savings and heat generation technologies: Modelling of residential investment behaviour with local health costs", *Energy Policy,* vol. 77, pp. 31-45, 2015.
[http://dx.doi.org/10.1016/j.enpol.2014.11.032]

[39]   B. Mills, and J. Schleich, "Residential energy-efficient technology adoption, energy conservation, knowledge, and attitudes: An analysis of European countries", *Energy Policy,* vol. 49, pp. 616-628, 2012.
[http://dx.doi.org/10.1016/j.enpol.2012.07.008]

[40]   J. Schleich, "Energy efficient technology adoption in low-income households in the European Union – What is the evidence?", *Energy Policy,* vol. 125, pp. 196-206, 2019.
[http://dx.doi.org/10.1016/j.enpol.2018.10.061]

[41]   S.M. Karim, S. Thompson, and P. Williams, "Co-benefits of Low Carbon Policies in the Built Environment: An Investigation into the Adoption of Co-benefits by Australian Local Government", *Procedia Eng.,* vol. 180, pp. 890-900, 2017.
[http://dx.doi.org/10.1016/j.proeng.2017.04.250]

[42]   World   Bank,   *The   CURB   Tool: Climate   Action   for   Urban   Sustainability,*   2016. https://www.worldbank.org/en/topic/urbandevelopment/brief/the-curb-to- l-climate-action-for-urban-sustainability

[43]   European Environment Agency, *Interview — Energy efficiency benefits us all —,* 2017. https://www.eea.europa.eu/signals/signals-2017/articles/interview-2014-energy-efficiency-benefits

[44]   *US Department of Energy, 2016.Better Buildings Residential Network Peer Exchange Call Series: Do You Hear Me Now? Communicating the Value of Non-Energy Benefits (101),* 2016. Accessed: Nov. 06, 2020.

[45]   S. Banfi, M. Farsi, M. Filippini, and M. Jakob, "Willingness to pay for energy-saving measures in residential buildings", *Energy Econ.,* vol. 30, no. 2, pp. 503-516, 2008.
[http://dx.doi.org/10.1016/j.eneco.2006.06.001]

CHAPTER 4

The Impact of Occupants in Thermal Comfort and Energy Efficiency in Buildings

António Ruano[1,2,*], Karol Bot[1] and Maria da Graça Ruano[1,3]

[1] *Universidade do Algarve, 8005-294 Faro, Portugal*

[2] *IDMEC, Instituto Superior Técnico, Universidade de Lisboa, 1049-001 Lisboa, Portugal*

[3] *CISUC, University of Coimbra, 3030-290 Coimbra, Portugal*

Abstract: The chapter reviews the impact of occupancy in buildings, in particular in thermal comfort and energy efficiency. Concerning the first issue, this chapter will first propose a means to estimate occupancy, the impact of occupation in thermal comfort measured by the Predicted Mean Vote index, and its use for real-time control of HVAC equipment. All data used are measured data of a real university building under normal occupation. The effect of occupancy in energy efficiency will focus on the residential segment, using data of a recent installation of a data acquisition and control system in a household located in the south of Portugal. This work shows that the impact of occupancy in electricity consumption becomes more evident as the electric energy is being desegregated and that the availability of this information by the occupants can be used to improve energy efficiency. Moreover, the use of occupation in the design of electric consumption forecasting methods will also be discussed.

Keywords: Artificial neural networks, Computational learning, Data acquisition, Energy efficiency, Electricity consumption, Forecasting models, HVAC, Multi-objective genetic algorithms, Occupation, Thermal comfort.

INTRODUCTION

Industrialisation and globalisation powered a rapid economic development that was accompanied by progressively increasing energy consumption [1]. Three sectors of the economy stand out for the significant amount of energy they consume. They are the sectors of industry, transport, and buildings - with buildings representing the largest portion. For example, in the United States, the built environment consumes 35% of total primary energy, and about 45% of electricity, mainly through the use of Heating, Ventilating and Air Conditioning

---

* **Corresponding author Antonio Ruano:** Universidade do Algarve, 8005-294 Faro, Portugal and IDMEC, Instituto Superior Técnico, Universidade de Lisboa, 1049-001 Lisboa, Portugal; Tel: +351 289800912; Fax: +351 289 800 066; E-mail: aruano@ualg.pt

Enedir Ghisi, Ricardo Forgiarini Rupp and Pedro Fernandes Pereira (Eds.)

(HVAC) systems [2]. It is, therefore, of fundamental importance to decrease energy usage in residential and service building and, simultaneously, maintain the thermal comfort of its occupants. This can be obtained by integrating efficient passive solutions in the architecture and construction of buildings, the employment of renewable energy, and the use of efficient Home Energy Management Systems (HEMS), with a special emphasis on the control of existing HVAC systems.

Despite all advances obtained in these areas in the last decade, the maximum capacity for the use of intelligence in building systems, however, remains fallow, due to the complexity and variety of the systems, in addition to the frequent question of suboptimal control strategies [3]. Various reviews reported intelligent optimisation and control strategies in buildings [4], advanced control schemes [5], energy intelligent buildings [6] and passive and energy-efficient designs [7, 8]. Occupancy and occupant behaviour have a considerable impact on the operation and performance of the building. In [9], the motivations of occupant behaviour in a residential context are categorised. Detection of occupant actions through the usage of indoor environment monitoring systems is presented in [10].

This chapter deals with building thermal comfort, HVAC control and home energy efficiency, focusing on the impact of occupation are these three areas. The two former topics are discussed in the next section, while energy efficiency will be dealt with the subsequent section. Conclusions and future research directions are discussed in the final section of this chapter.

THERMAL COMFORT AND HVAC CONTROL

This section introduces the most used metrics for thermal comfort evaluation and proceeds to the estimation of occupancy in buildings. Subsequently, the impact that occupancy has in thermal comfort is highlighted, and a method to use the thermal comfort index for HVAC control is proposed, incorporating occupation estimation.

Thermal Comfort Metrics

The thermal sensation of occupants in buildings varies not only with the climate conditions in the rooms but also with the activity the occupants are performing, their clothing, and also between individuals. Because of that, the American Society of Heating, Refrigerating and Air Conditioning Engineers (ASHRAE) proposed a thermal sensation scale (Table **1**) to quantify the thermal sensation of people [11].

Table 1. The ASHRAE Thermal Sensation Scale.

Cold	Cool	Slightly Cold	Neutral	Slightly Warm	Warm	Hot
-3	-2	-1	0	1	2	3

The Predicted Mean Vote (PMV) index was presented by Fanger [12], aiming to allow the prediction of the average vote of a large group of persons, on the thermal sensation scale. The PMV relies upon six components: mean radiant temperature (), clothing insulation (I_{cl}), metabolic rate (M), air temperature (), air velocity () and air humidity (H_{ai}). Please note that there are other thermal comfort indexes (see, for instance [13]) but it is the one most used. It is computed utilising a heat-balance equation given by [14], as eq. (1):

$$PMV = \left(0.303e^{-0.036M} + 0.028\right)L \tag{1}$$

where L represents the thermal load in the human body, characterised as the variation between the internal heat production and the heat loss occurring when the human is in a thermal balance, as described in eq. (2):

$$L = (M - W) - 0.0014M\left(34 - T_{ai}\right) - 3.05*10^{-3}\left(5733 - 6.99\left(M - W\right) - P_{ai}\right) -$$
$$-0.42\left(M - W - 58.15\right) - 1.72*10^{-5}M\left(5867 - P_{ai}\right) - 0.0014M\left(34 - T_{ai}\right) - \tag{2}$$
$$-3.96*10^{-8}f_{cl}\left[\left(T_{cl} + 273\right)^4 - \left(\overline{T}_r + 273\right)^4\right] - f_{cl}h_c\left(T_{cl} - T_{ai}\right)$$

where W and M are the external work and metabolic rate, respectively, both in W/m², P_{ai} is the fractional water vapour pressure in Pascal. Both temperatures T_{ai} and are given in degrees Celsius. The h_{cl}, the convective heat transfer coefficient, and, clothing surface temperature (both in °C), are given by eq. (3) and eq. (4):

$$h_{cl} = \begin{cases} h_c^* & if\ h_c^* > 12.1\sqrt{V_{ai}} \\ 12.1\sqrt{V_{ai}} & if\ h_c^* < 12.1\sqrt{V_{ai}} \end{cases} \tag{3}$$
$$h_c^* = 2.38\left(T_{cl} - T_{ai}\right)^{1/4}$$

$$T_{cl} = 35.7 - 0.028\left(M - W\right) -$$
$$-0.1555I_{cl}\left[39.6*10^{-9}\left[\left(f_{cl} + 273\right)^4 - \left(\overline{T}_r + 273\right)^4\right] + f_{cl}h_c\left(T_{cl} - T_{ai}\right)\right] \tag{4}$$

Until a certain degree of accuracy is obtained, eq. (2) and eq. (3) are calculated recursively. Finally, in (2) and (3), f_{cl}, (ratio of body surface area covered by clothes to the naked surface area), is given by eq. (5):

$$f_{cl} = \begin{cases} 1.00 + 1.29 I_{cl} & if \ I_{cl} \leq 0.078 \\ 1.05 + 0.645 I_{cl} & if \ I_{cl} > 0.078 \end{cases} \tag{5}$$

To determine the PMV in practice, reference values of I_{cl} and M are acquired from tables existent in several books related to HVAC systems, as is presented in [11]. V_{ai} can be measured or obtained indirectly through the HVAC specifications. P_{ai} is connected to, employing Antoine's equation:

$$P_{ai} = 10 H_{ai} e^{16.6536 - \frac{4030.188}{T_{ai} + 235}} \tag{6}$$

The last variable needed to compute is the mean radiant temperature (), essential to the PMV calculation. It may be described as the uniform temperature of an imaginary enclosure wherein radiant heat transfer from the human body equals the radiant heat transfer in the actual nonuniform enclosure [11]. Its value may be calculated employing distinct methods [11]:

- From the plane radiant temperature in six opposite directions, weighted according to the projected area elements for a person;
- Through the usage of a black globe thermometer placed in the geometrical centre of the room. The following equation may be used to determine the globe temperature, denoted by (T_g):

$$\bar{T}_r = \left[\left(T_g + 273 \right)^4 + \frac{1.1 \times 10^8 V_{ai}^{0.6}}{\varepsilon D^{0.4}} \left(T_g - T_{ai} \right) \right]^{1/4} - 273 \tag{7}$$

where the globe emissivity coefficient and the globe diameter in meters are denoted by ε and D, respectively.

Finally, thermal comfort can be expressed in three classes: A, if -0.3<PMV<0.3; B, if -0.5<PMV<0.5; and C, if -0.7<PMV<0.7. The most used class is B. Obviously, considering HVAC control, class A implies the largest energy consumption, whilst class C the least.

PMV Computation

As it can be concluded from what has been said, to compute PMV for specific values of M, I_{cl} and, it is needed to acquire, H_{ai} and T_g, or the temperature (T_w). Additionally, assuming that these values are available, the computation of the actual PMV must also be considered.

Starting from this last point, PMV can be obtained by the solution of (2-7). However, iterative computations are used in this formulation, being responsible for a variable execution time. If PMV is used in a model-based predictive HVAC control scheme, as it will be explained later on, hundreds or thousands of PMV evaluations may be needed to be computed within a single sampling period.

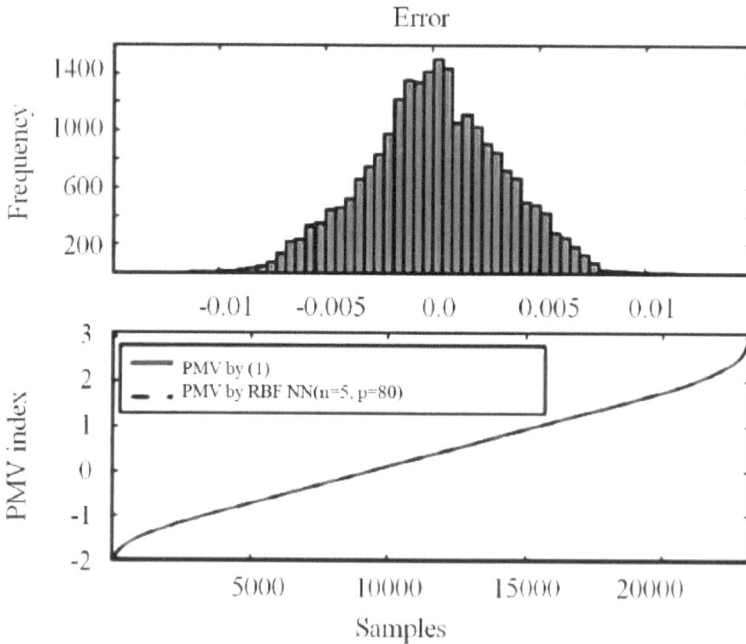

Fig. (1). Top plot: Histogram of error; bottom plot: PMV index fitting on validation data set employing a five neuron RBF model [20].

For this reason, it is of fundamental importance to employ a method for rapid, constant execution time, to the computation of the PMV index. Some computational learning approaches were proposed for PMV approximation [15 - 18]. Computational learning methods are based on a sequence of observations of variables/parameters measured at successive points in time [19], not requiring the internal states information of the system under analysis to being able to model it.

Our research group used Radial Basis Function Neural Networks (RBFs) to achieve this goal [20]. RBFs typically have a single hidden layer, where the activation function of the neurons is radial-type functions, typically Gaussian functions. The output of an RBF is just a linear combination of hidden layer outputs. The average error of 0.0014 and maximum absolute error of 0.0075 was achieved with a RBF with only five neurons. Fig. (**1**) presents the results obtained in the study [20].

The relative performance of the obtained results may be seen in Fig. (**2**). As it is shown, those results were achieved with a saving of 55 times in computation time, compared with the standard iterative computation.

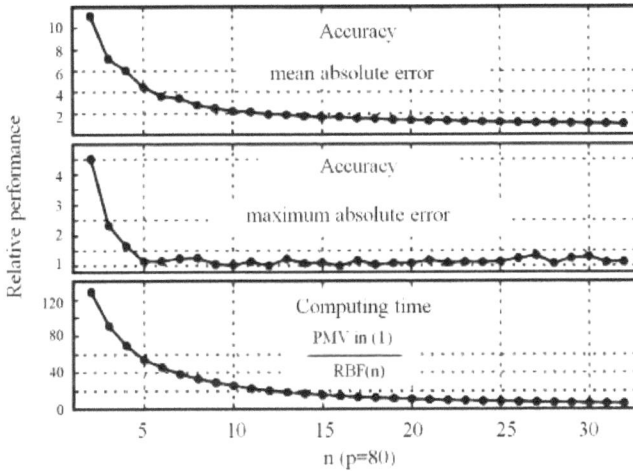

Fig. (2). Relative performance regarding computing time and accuracy [20].

Going now to the data acquisition procedure, as the room air temperature and relative humidity are quantities easy to measure, we shall focus on the computation of the mean radiant temperature. Both methods presented in the previous section were used by the team. In an early work [21], a black globe thermometer was employed, as shown in Fig. (**3**). The room used is a small lecture theatre located in Building 8 of the University of Algarve, Gambelas Campus, Faro, Portugal. As can be seen, the ball is not located in the middle height of the room, as it was not convenient for the occupants. Additionally, it is undoubtedly not an esthetical solution.

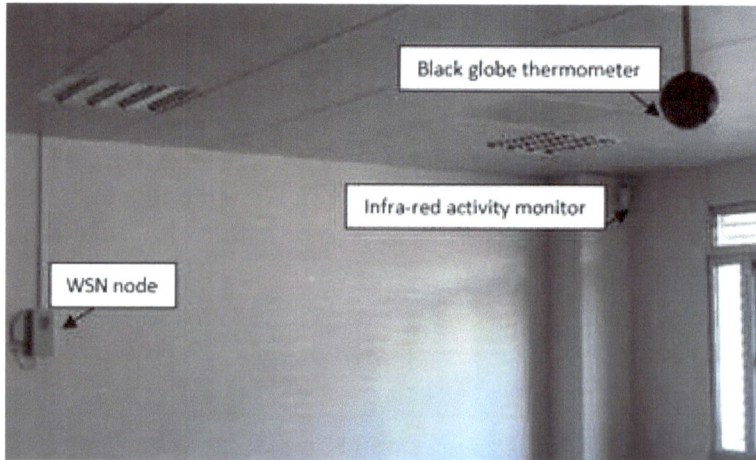

Fig. (3). A black-globe thermometer located in a small lecture theatre.

For these reasons, in a subsequent project, the second method was employed. Several experiments were carried out measuring the temperature of the six constructive surfaces in distinct classrooms and surfaces. The conclusions achieved in [22] show that the temperature of the ceiling estimates well the value of, both in summer and winter, and this is the method presently employed. This means that to compute the PMV, besides measuring and H_{ai}, it is necessary to acquire the ceiling temperature T_w.

The first two variables were previously measured, together with the state of windows/doors, by a wireless sensor node (WSN, please see Fig. (3), implemented with a Tmote Sky platform, an IEEE 802.15.4 compliant device that uses the TinyOs operating system [23]. As can be seen, although communication is wireless, cables are employed to power the system. An important finding was that this platform, as well as off-the-shell WSN, was not suitable for buildings energy management due to numerous factors, as listed below:

- economic: they are not cheap, which is a problem in their application, particularly for large buildings with many spaces;
- ergonomic: the nodes are excessively large for incorporation with the desired components.
- redundancy: their generic-purpose nature lead the architecture to consist of numerous needless components for unique functions;
- consumption of energy: due to its consumption, frequent battery replacement would be required, or connection to the electric network would be needed;
- engineering: an extensive quantity of work is needed to achieve the integration of specific sensors; especially in wiring for power-constrained available

placement locations;
• maintenance: changing batteries in large developments are not feasible.

To address these points and with focus also in providing a less intrusive platform for the occupants of the room, it was designed, implemented and tested Self-Powered Wireless Sensor (SPWS) nodes [24]. Very briefly describing, an 850 mAh lithium polymer battery is used by the SPWS applied. The EMS circuitry aims the implementation of the usage purpose of regulating voltage levels and the selection of one of three modes of operation: charging the battery while powering from the USB; powering from Universal Serial Bus (USB) port, or battery both charging and powering.

The SPWS circuit (Fig. **4**) is designed around a microcontroller Microchip XLP (eXtreme Low Power) that allows deep-sleep and sleep states, enabling the energy-efficient design of the operational duty-cycle. The microcontroller is complemented by an IEEE 802.15.4 appropriate transceiver to enable wireless transmission of data. The microcontroller performs distinct tasks, according to the type of node, as listed below:

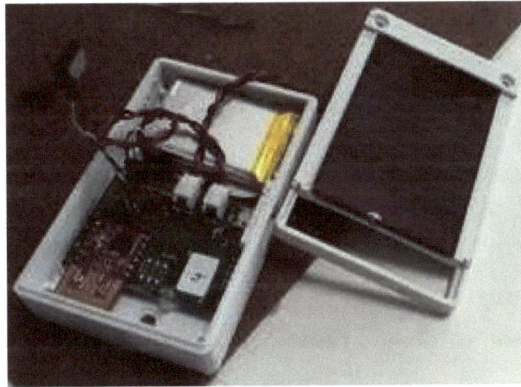

Fig. (4). Left: mounted SPWS on a wall. Right: detail of an open SPWS - PCB, battery and PV module [24].

• transmitter: reads the sensor(s) data, communicates the readings to the RF transceiver for transmission, and deep-sleeps until the next sampling time;
• receiver: repeatedly receives datagrams from the RF transceiver, extracting the sensor data sent by a transmitter or repeater node, and sending it to an Ethernet-connected device (a collector node) over the USB port;
• repeater: receives datagrams from the RF transceiver in a continuous manner, changing the necessary addressing information, and communicating the new data packet for transmission by the RF transceiver.

The following sensors were used in the SPWS (please see Fig. **4**):

- Silicon Labs Si7021; and Sensirion SHT2 sensors - employed for relative humidity and air temperature measurements;
- PT1000 superficial temperature sensor - used to measure the temperature of the walls
- Hamlin 59150-030 magnetic reed switch sensors - used to identify the state (open or closed) of fenestration surfaces;
- Panasonic EKMC1601111 Passive Infra-Red (PIR) motion sensor - employed to identify movement in rooms;
- N5AC-50108 photo-resistor from Low Power Radio Solutions – employed to measure the luminous flux,

The use of very low power components and a careful selection of the sensors made the perpetual operation of these devices possible. For more details, please see [24].

Occupancy Estimation and Detection, and Its Impact on Thermal Comfort

Occupation can be estimated or detected in several ways. For instance, Meyn and co-workers [25] propose a Sensor-Utility-Network (SUN) estimator, employing data from several sensors, historical data, room schedules and building characteristics. Cao and co-workers [26] employ cameras for occupancy estimation. For occupancy detection [27], employees WiFi probe-based ensemble classification. Also, for detection [28], presented a method based on air temperature, relative humidity and CO_2 levels of the room.

As pointed out, the SPWS introduced in the last section measure movement using a PIR sensor. Detection of occupation in a room can be obtained directly, and the level of movement (M) is correlated with how many occupants are in a room. For thermal comfort assessed with PMV, occupancy affects the inside temperature.

For predictive control of HVAC systems using PMV, models are needed to predict the evolution of, H_{ai}, and T_w. The influence of occupation in PMV is through the room temperature. In previous experiences, this research group developed models for forecasting the evolution of these variables and, using those forecasts, of PMV. Fig. (**5**) shows the prediction error (1-step-ahead) for the air temperature in a room obtained with a model which was designed with data with no occupancy. The room considered was a computer room at the University of Algarve. It is clear that the error increases between samples 20 and 35, where movement is detected using a similar PIR sensor found in the corner of Fig. (**3**).

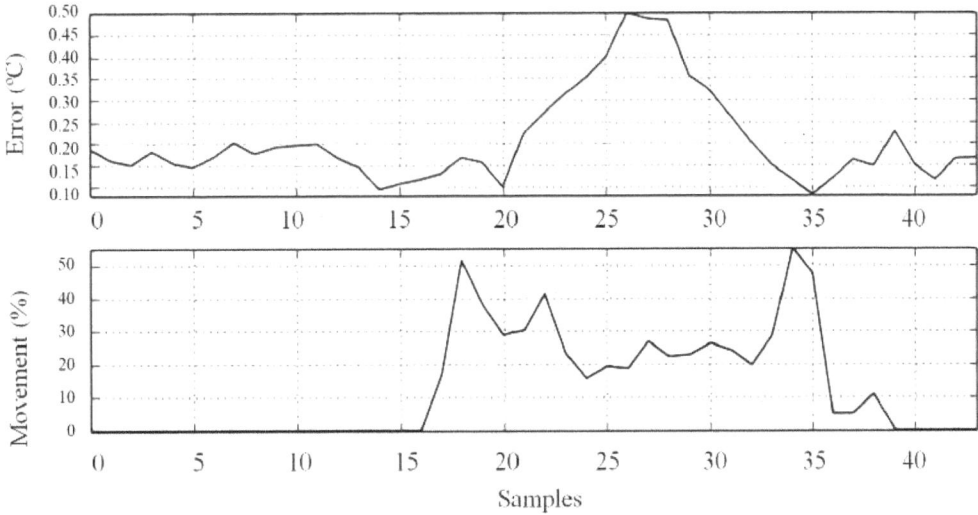

Fig. (5). Top: Absolute error signal between the measured and estimated. Bottom: Movement signal [29].

Subsequently, the model was redesigned, adding data collected in periods of occupation to the previous design data, and incorporating the movement signal. Results obtained in another experiment and presented in Fig. (6) showed that the new model presented a significant improvement when compared to the original model. The original model achieved an RMSE of 0.217, while the redesigned model (with the period of occupation data incorporated) improved this value significantly to 0.066.

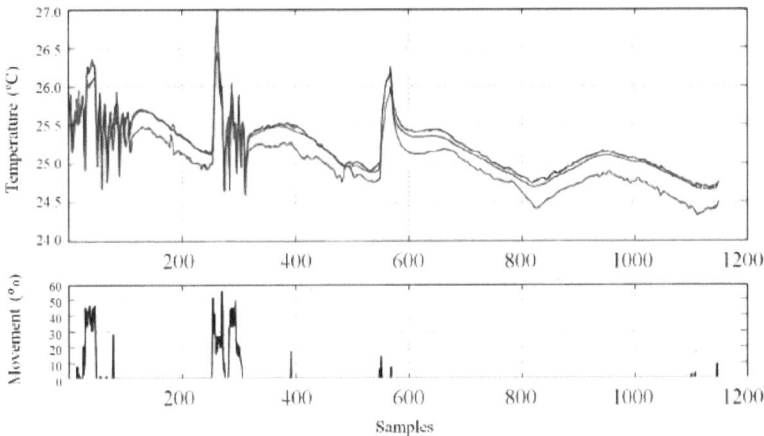

Fig. (6). Top: measured (red); predicted with the original model (magenta) and predicted with the new model (green). Bottom: movement signal [16].

The use of the redesigned model also translated in a better PMV computation. Fig. (7) shows the evolution of PMV, either computed with eq. (2-6), or obtained with the PMV model, using for that the original model for T_{ai}, and the redesigned one. As can be seen, the estimated PMV obtained with the new model is very good, and always better than using the original model. More information about these experiences can be found in the study [29].

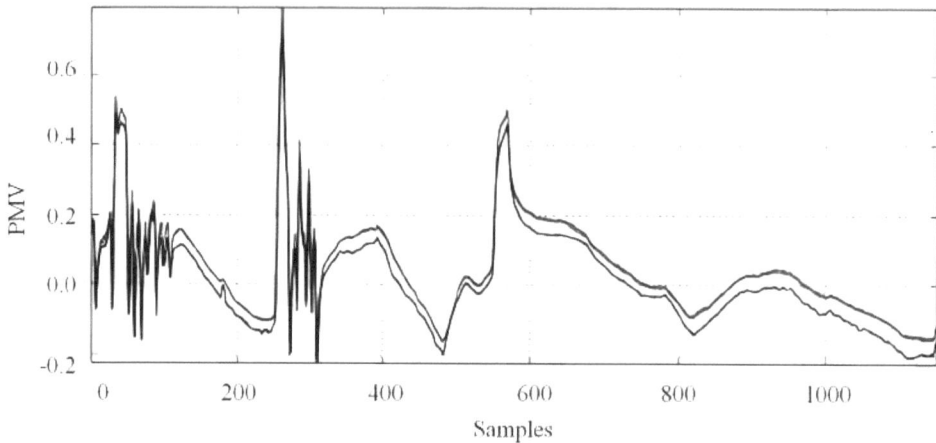

Fig. (7). PMV measured (red), estimated with the original model (black) and with the new model (green) [29].

When Model-Based Predictive Control (MBPC) is employed with the aim of controlling the HVAC systems in buildings, models must be used to forecast whether terms in the objective function and/or in the restrictions employed within a Prediction Horizon (PH). Again, there is not a single solution, but different approaches are available in the literature (see, for instance [30 - 34]).

The research group has proposed an intelligent MBPC HVAC system. In short, seven forecasting models were designed, three for atmospheric variables (air temperature – TA, Global Solar Radiation – SR, Relative Humidity – RH), and four interior variables (TA$_i$, RH^i, and M). These models supply the forecasts of those variables over PH (large arrows in Fig. (8)). These forecasts are supplied to a static PVM model, described earlier that outputs the forecasting PMV values to the MBPC algorithm. The PMV forecast is supplied to the MBPC algorithm, which uses it to ensure that the PMV lies in thermal comfort throughout PH, while minimising the economic cost corresponding to the HVAC energy consumption. Finally, as this scheme supports schedules, the HVAC is operated in such a way that it maintains thermal comfort only on the schedules supplied, minimising further the operational, economic cost. For more information on this control approach, please see [22].

As it is possible to see in Fig. (**8**), the forecasts of the motion signal are necessary to compute the forecasts of TA_i and. Forecasting occupancy is a complicated task, being among the most challenging problems in energy management systems. The occupant behaviour does not usually have a standard pattern, and it is the main reason for the forecasting errors resulting from the difference between predictions of simulation software and the actual situation.

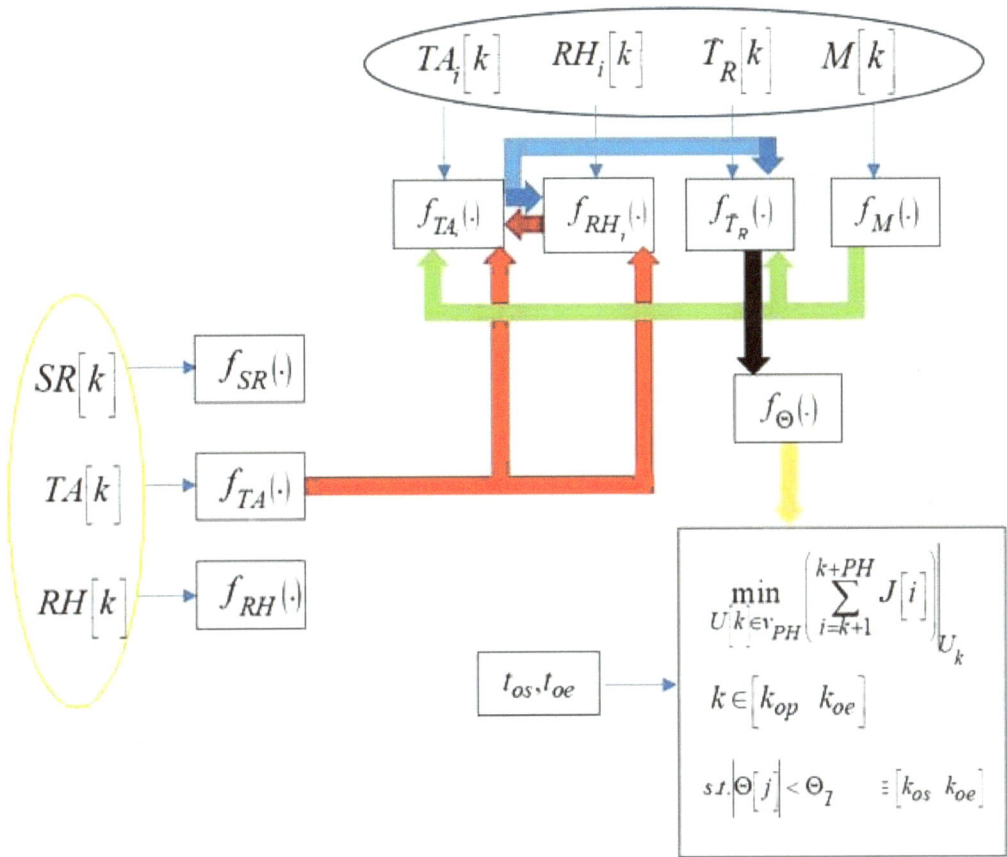

Fig. (8). Schematic diagram of the IMBPC HVAC approach [24].

The stochastic model assumes the presence of occupants as changing randomly and predicts at the next point in time the location of an occupant by analysing the location of the occupant at a previous point in time. In this sense, the most used models that have been efficiently implemented for predictions of the number of occupants in a room are the Markov models [35]. To predict the occupancy rate Dong and Lam [36] employed a data flow of a multi-sensor system coupled with a semi-Markov model. Sun *et al.* [37] also employed a stochastic model and showed that the arrival and departure time of occupants are exponentially distributed, and the number of occupants in a room of the building considered presented a quadratic distribution. To simulate the motion of an occupant in a building, Wang *et al.* [38] used a first-order homogeneous Markov model and achieved a quantitative relationship between the amount of time people move in and out of a building and the occupancy rate. Ding and co-workers [39] used a Markov model and a probability function, aiming to describe the electricity consumption caused by the randomness of occupancy in buildings.

The probabilistic models defined earlier need a large amount of data, and the result is not very accurate. The authors have previously proposed a more straightforward approach, presented in [22]. The forecast of occupant motion throughout the prediction horizon at each instant, assuming the existence of schedules, will be a constant value equal to the exponentially weighted average value of the movement signal, from the start of the schedule until that sample.

Denoting this average at instant k by s[k], the forecast of the M signal, throughout a prediction horizon PH, for an occupation period, eq. (8) gives:

$$\hat{M}[k+i]\Big|_{k} = \begin{cases} s[k], i \in PH \wedge k+i \in t_{oc} \\ 0, i \in PH \wedge k+i \notin t_{oc} \end{cases} \tag{8}$$

An exponentially weighted average is used for:

$$s[k] = (1-\alpha)M[k] + \alpha s[k-1] \tag{9}$$

where $\alpha \in \{0...1\}$ is a forgetting factor, which weights the current value of movement, against the previous average. A typical value is 0.8.

This method, unfortunately, as well as the others, does not present a good accuracy. To illustrate the performance obtained, Fig. (9) shows the evolution of different variables for one experiment in one lecture room in the University of Algarve. In that summer, three exams were scheduled to take place in that room: the first between 9h to 11h, the second from 11h to 13h, and the third between 16h to 18h. Accordingly, two scheduling periods were assumed: from 9h to 13h05m, and from 16h to 18h05m.

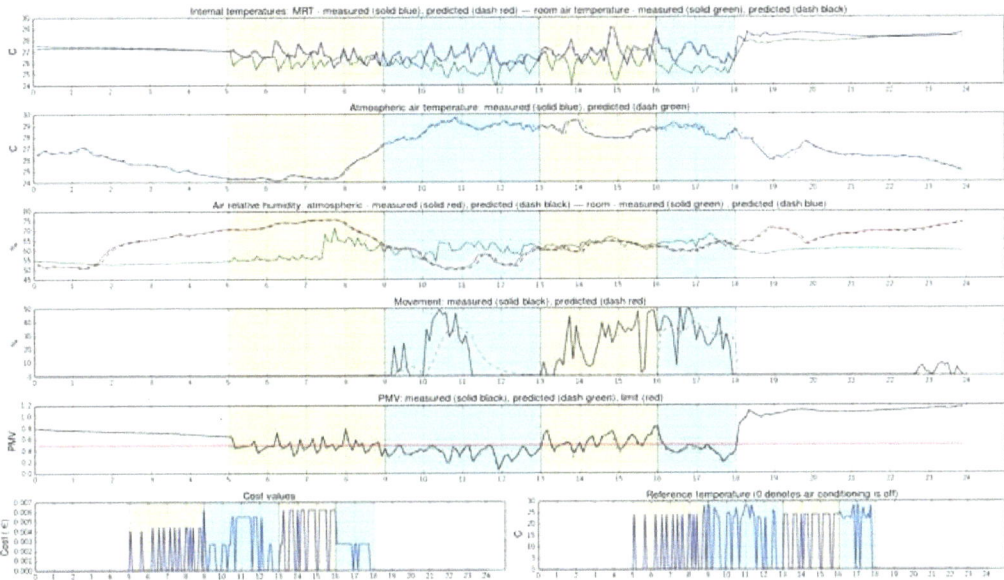

Fig. (9). Evolution of room variables [22].

In Fig. (**9**), the first graph presents how the room temperature evolves during the day, as well as the mean radiant temperature. The second presents the atmospheric air temperature. The third shows the room and atmospheric air relative humidity. The fourth graph and fifth graph present the occupant's motion and PMV signals, respectively. The sixth graph shows the evolution of the HVAC systems' reference temperature, where the 0 value indicates the scenario in which the unit is off.

Excellent one-step-ahead forecasts were obtained, except for the movement signal. Although the prediction method is simple, not requiring previous computation, the predictions only follow the trend of the signal. For more detail on this experience and the methods, please consult [22].

Eletric Consumption and Energy Efficiency

Following, the impact of occupation on the electric consumption and energy efficiency is discussed. One form to segment the electricity consumption is between non-residential and residential buildings' energy consumption.

In the former case, Mehreen and co-workers [40] analysed the existent connection among electric energy demand profiles and occupant's activities for a university service building. The effect of occupancy and building usage on energy demand in Finnish daycare and school buildings was discussed in [41].

To increase energy efficiency in the residential sector, López *et al.* [42] analysed the impact of active occupancy in the daily residential electricity demand profile in several regions of Spain. Muroni and co-workers [43], and Causone and co-workers [44] integrated into building performance simulation models profiles of occupancy and occupancy-related load profiles. Appliance load profiles are also employed in [45] for simulated Home Energy Management Systems (HEMS).

In contrast with previously presented works, the research developed here is focused in:

a. finding the impact on the number of occupants in electricity consumption of a residential house, and to determine which are the most affected appliances;
b. determining the impact of occupation in electricity consumption forecasting models.

The data that will be used is collected from a case study household, located in Gambelas, Algarve, in the south of Portugal. It is a single-family building, composed by two floors and with twenty different spaces. The building accounts with a PV installation composed by 20 Sharp NU-AK panels [46], each panel with 300 Wp. The inverter characteristics are described in [47], which also controls a battery with 11.5 kWh of storage capacity, described in [48]. The electric panel existent in the building is a Schneider panel consisting of 16 monophasic circuit breakers, plus a triphasic one. The house also has available a few Self-Powered Wireless Sensors (SPWS) for measuring room climate variables, and activity [24], and a few TP-Link HS100 Wi-Fi Smart Plugs (SP) [49], as well as one Intelligent Weather Station (IWS) [50].

This data is collected in the framework of a Portuguese FCT-funded project, NILM-forIHEM, which is currently under development. The objectives of this project are to improve appliance detection usage employing Non-Intrusive Load Monitoring (NILM) methods [51], and to use this disaggregated data, together with forecasting models, to improve the performance of HEMS.

Data Acquisition System

The building has a data acquisition system implemented to monitor many electric variables. The data that will be used for NILM identification is supplied by a Carlo Gavazzi (EM340) 3 phase energy meter [52]. EM340 supplies 45 distinct electric variables, sampled at 1 Hz. Additional electric variables are measured for every circuit breaker to provide approximate values for the NILM identification, as proposed by the project in which this activity is inserted. The devices making these measures are Circutor Wibeees (WBs) [53], which are plug and play wireless devices to acquire electric consumption values. Measurements of current, voltage, frequency, power factor, active inductive reactive, active reactive and apparent power, and capacitive reactive energy are obtained every second for the 16 monophasic circuit breakers, the same number for each phase of the triphasic one, together with totalised values. Every second, 198 variables are sampled by the WBs.

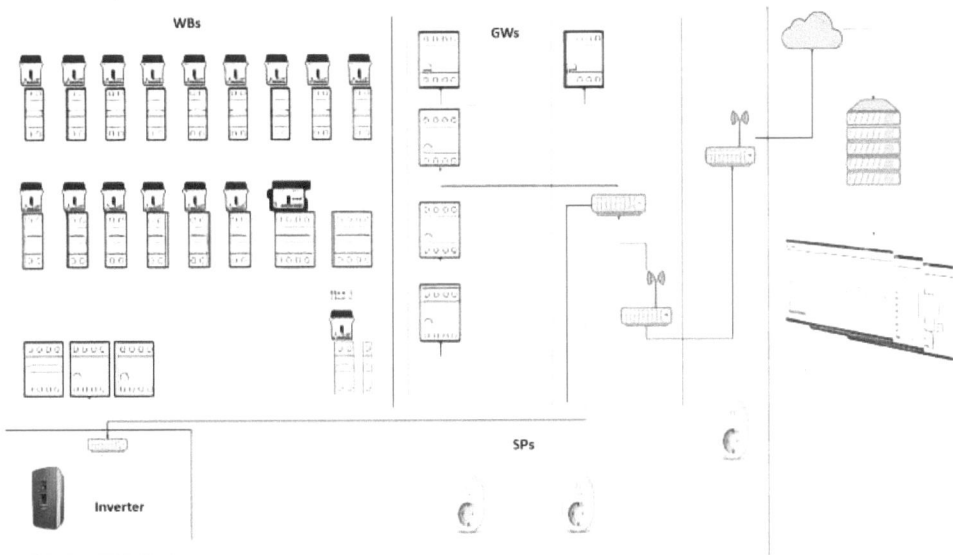

Fig. (10). Data acquisition diagram [54].

Several gateways and a technical wireless network are responsible for the data transmission from/to the measurement devices. For a description of the system implemented, the reader is invited to consult [54]. A diagram of the acquisition system is presented in Fig. (**10**), in which GW stands for Gateways and SP for smart plugins.

Electric Consumption and Occupation

To develop the models, the data used ranged from January to July 2020. In eleven days (13 and 18 Jan, 10 and 11 Feb, 01, 02, 03, 04, 07, 15 and 19 May) there was lack of some data and therefore those days were not considered. One second sampling interval acquired data was translated into daily and monthly data. Fig. (**11**) shows the monthly house electric consumption throughout the period considered. Please note that the data considered for May did not include seven days of that month. As expected, the highest electric consumption occurred in January.

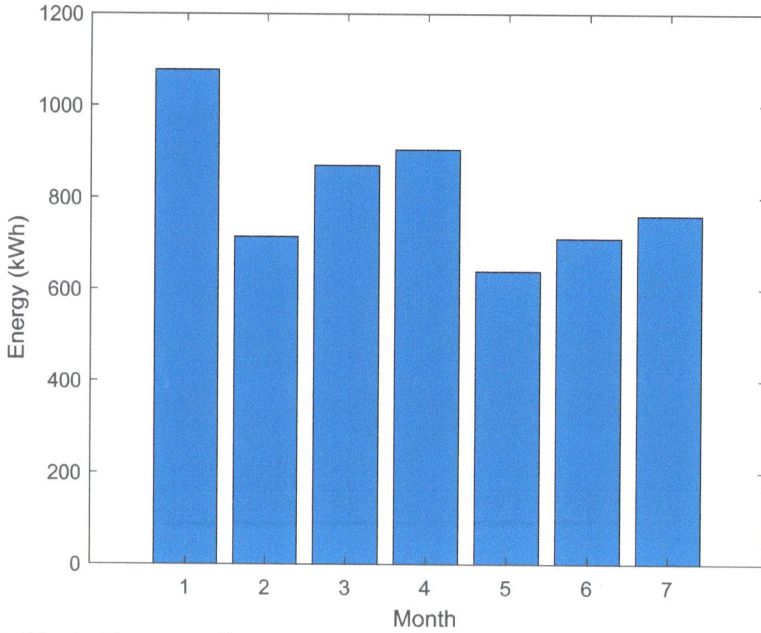

Fig. (11). Monthly electric consumption.

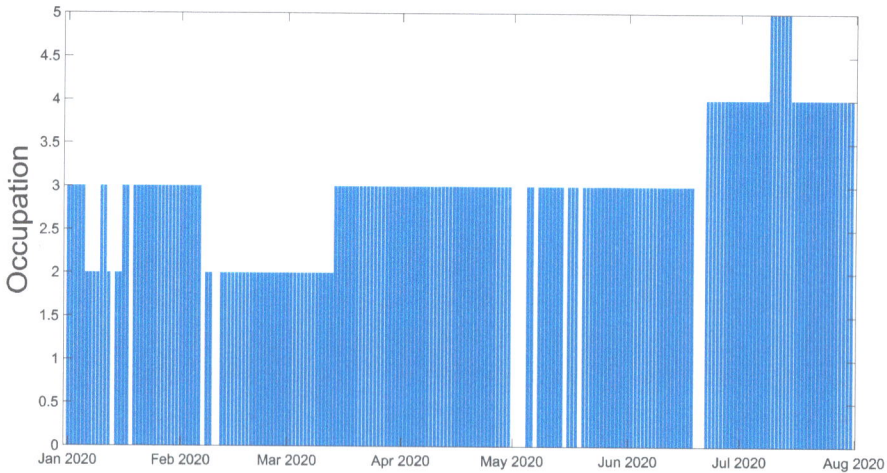

Fig. (12). Daily occupation between January and July.

Fig. (**12**) shows the daily occupation of the house, manually registered, for the period considered. If the Pearson correlation coefficient is computed between the daily electric consumption and the occupancy, it can be concluded there is no valid statistical correlation (0.03, with a p-value of 0.68).

Table **2** shows the disaggregation of the energy by the circuit breakers.

Table 2. Distribution of appliances through circuit breakers.

Breaker #	Appliances	Total Consumption (kWh)
1	Alarm	0
2	Swimming pool pump, external lightning	982
3	Lightning, Floor 1	1
4	Lightning Floor 0, iron, hairdryer, bathroom towels heater	289
5	Air conditioner (study)	31
6	Motors (garage, gate)	3
7	Air conditioners (bedrooms 2 and 3)	75
8	Electric Water Heater, washing machine	1,173
9	Dishwasher, microwave fryer, kettle	67
10	Air conditioner (bedroom 4)	160
11	Other kitchen appliances and illumination	282
12	Lounge lamp and other equipment (TV, audio, TV box, …)	125
13	Computer equipment, TV, aquarium, lamps, hall electric air heater and towels bathroom heater (Floor 1).	1,077
14	Air conditioner (Lounge)	43
15	Lightning, Floor 0	127
16	Stove nozzles	107
17	Oven	252
18	Stove nozzles	146
19	Data acquisition system	59

If appliance classes are roughly assigned to the different circuits, HVAC:4+5+7+10+13+14; Pump: 2; Light: 3+15; Water Heater: 8; Equipment: 12; Kitchen: 9+11; Oven: 16+17+18, the consumption distribution obtained is the one shown in Fig. (**13**).

Please note that this a rough classification (for example #4 consumption is assigned to HVAC class, but this circuit powers other appliances than the towers electric heater). Lightning consumption is insignificant (since the house is equipped with LED lamps), as well as other equipment. Kitchen appliances and oven consume around 17%, followed by the swimming pool water pump (20%). The electric water heater (and the washing machine) follow, with a percentage of 24%, and the highest consumption is assigned to the HVAC equipment (34%).

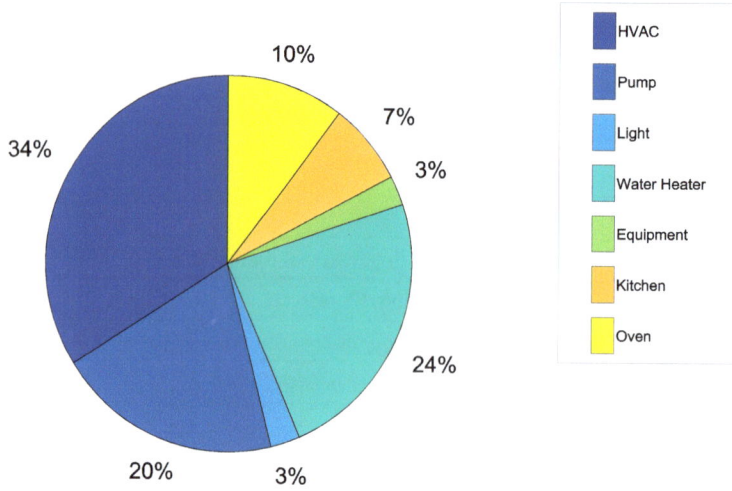

Fig. (13). Distribution of electricity consumption.

In contrast with the previously results regarding the Pearson correlation coefficient computed between the total electric consumption and the occupancy, now the correlations between the seven different classes and the occupation obtains statistically significant values: water pump, with (correlation of 0.16, a p-value of 0.02), other equipment (-0.17, 0.02) and kitchen (0.64, 0).

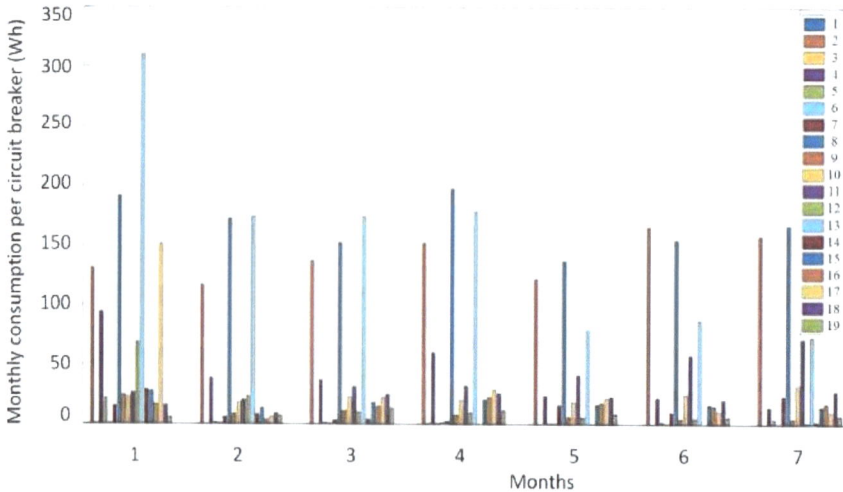

Fig. (14). Monthly consumption per circuit breaker.

Fig. (**14**) shows the monthly electric consumption, further disaggregated by the individual circuit breakers. The first noticeable fact is that monthly consumption per breaker is not constant. For instance, breaker #13, which supplies the hall electric air heater and towels bathroom heater, has the largest consumption in January, as expected. If the correlation of the consumption of each circuit with occupancy is determined, more statistically significant results are obtained, which are shown in Table **3**.

Table 3. Correlation between circuits consumption and occupation.

Circuit #	2	7	10	11	12	13	16	18
Correlation	0.16	0.27	0.32	0.66	-0.17	-0.10	0.25	0.36
P-value	0.02	0.00	0.00	0.00	0.02	0.02	0.00	0.00

Besides breaker #2, it can be observed that:

- a positive correlation of circuits #7 and 10 exists. This makes sense since, in principle, more people increases the usage of the air conditioners;
- the most considerable positive correlation, 0.66, is obtained in the kitchen appliances, circuit # 11. This is also expected;
- small negative correlations are found for circuits #12 (lounge) and 13 (computers, electric air heater and towels bathroom heater). These negative values might have to do with the largest consumption in January;
- a significant positive correlation is obtained for the stove nozzles, which makes sense. Notice that this is not the case of the use of the oven, which seems unaffected by occupancy. As the oven consumption is significant in relation to the nozzles, the combined consumption of the oven and the stove was not affected by occupancy, statistically speaking.

It is also possible to speculate that further disaggregating the energy up to the individual appliances would result in more significant correlations. This, however, will only be possible when the NILM methods are developed and applied to this data – constituting a future step in the research.

Collected data can also be analysed by the home occupants so that energy efficiency measures can take place. One is the change of the period of operation of the appliances, which occurred here.

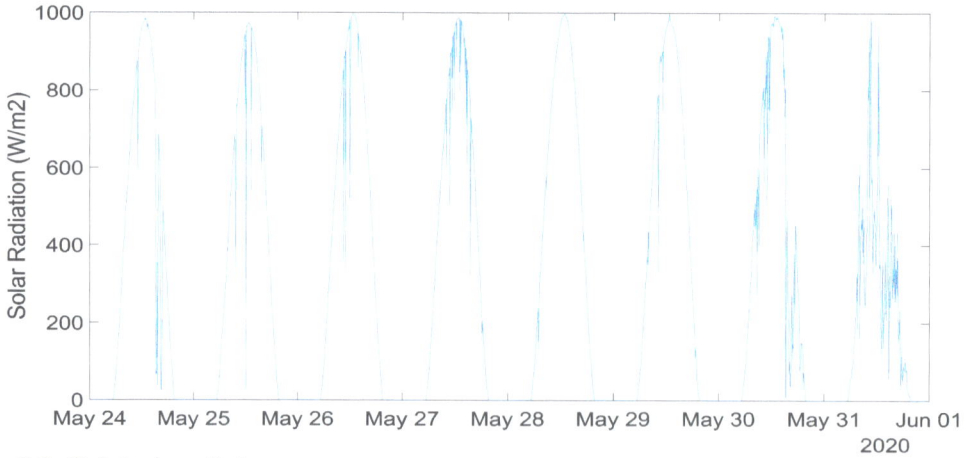

Fig. (15). Global solar radiation.

When solar radiation increased, PV power generated obviously increased. This means that more electricity can be exported to the grid and less can be obtained from the grid, assuming a constant pattern of daily consumption. When storage exists, however, more savings can be obtained by changing the time of usage of high consumption appliances. This is the case of the swimming water pump of the house, which typically must operate 7 hours a day and is responsible for 20% of the total consumption. Between January and mid-May the pump operated from 23:00 to 6:00, as the electricity tariff is lower at night. However, as solar radiation increased, PV production intensified, and the electricity exported to the mains grew. Fig. (**15**) shows the solar radiation in the last week of May. Please notice that data is now averaged in 5 minutes. The DC PV power generated is shown in Fig. (**16**).

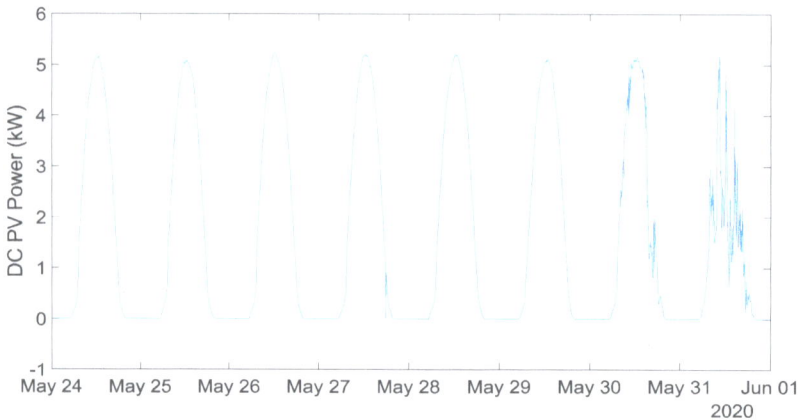

Fig. (16). DC PV power generated.

Fig. (**17**) shows the power associated with circuit breaker #2. The main consumption is related to the 7 hours operation of the water pump.

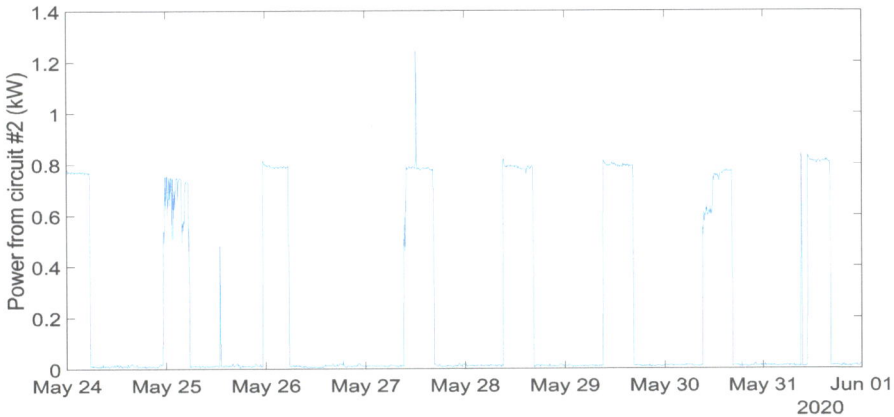

Fig. (17). Electric power associated with breaker #2.

As it can be seen, on 27th May, the period of operation changed from 23:00 - 6:00 to 9:00 -14:00. Typically, the energy was obtained from the mains at night, when there was no PV power generated, and the battery discharged until the minimum state of charge (SOC) of 5%. Fig. (**18**) illustrates the evolution of SOC throughout the period considered.

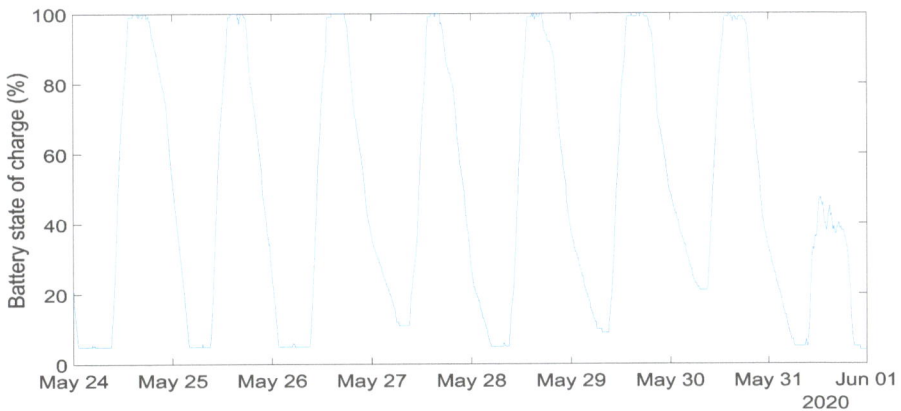

Fig. (18). Battery State of Charge.

As can be seen, the periods where the battery could not supply energy to the residence diminished when the period of operation of the water pump changed.

This was translated into a decrease in the power obtained from the grid, shown in Fig. (**19**).

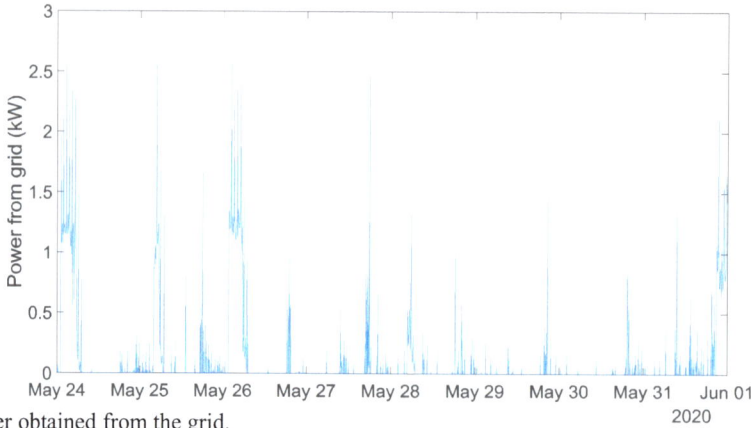

Fig. (19). Power obtained from the grid.

The decrease in the power obtained from the grid was achieved at the expense of a decrease in the power exported to the grid, as it can be seen in Fig. (**20**). Nevertheless, this is economically advantageous, as the price received for energy exported is a small fraction of the one related to energy obtained from the grid.

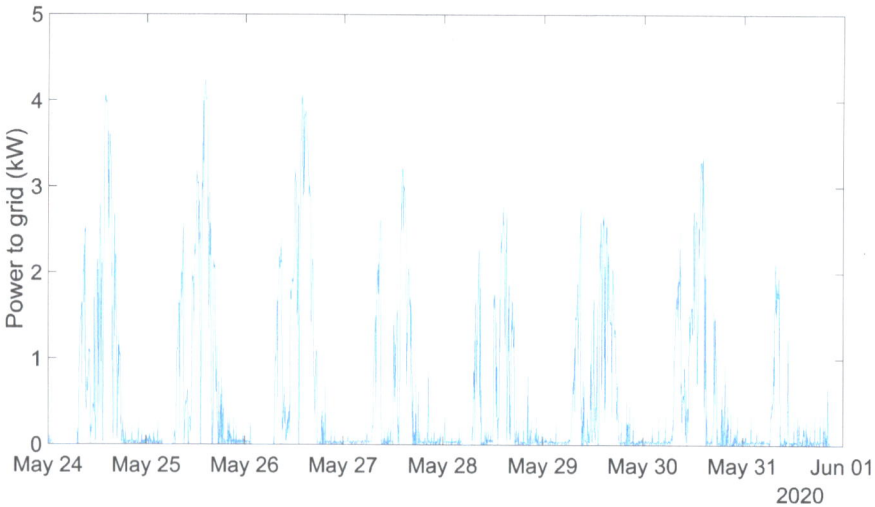

Fig. (20). Power exported to the grid.

The economic saving described above is just a small example of how electric information offers the possibility to be employed to (manually) improve the energy efficiency when operated by the occupant. These changes of the operation of schedulable devices, together with a better schedule of the inverter, can, however, be automatically achieved by the use of a HEMS. Preliminary results related to the use of this data for Model-Based Predictive Control HEMS can be found in [55].

Energy Consumption Forecasting Models and Occupancy

When MBPC algorithms are employed for the control of the inverter, forecasting models of electric consumption (and PV power generation) are needed, as discussed in the previous section. These models are employed for supplying predicted values over a specified Prediction Horizon (PH).

This section deals on the use of daily occupancy values on the design of forecasting models for energy consumption. Several approaches exist for predicting energy consumption in buildings. The reader can find excellent surveys in [56, 57].

As previously mentioned, Radial-Basis Function (RBF) Neural Networks (NN) are typically employed for predictive models. The hidden neurons employ a radial type of function, typically a Gaussian, their outputs being linear combined afterwards. Thus, the output of an RBF model is defined by eq. (10):

$$y[k] = w_{l+1} + \sum_{j}^{l} w_j e^{-\frac{\|\mathbf{i}[k]-\mathbf{C}(j)\|_2^2}{2\sigma_j^2}} \tag{10}$$

In (10), y[k] refers to the output, at instant k, i_j[k] is the jth input at that instant, w refers to the vector of linear weights, C(j) is the vector (extracted from a C matrix) of the centres associated with hidden neuron j, σ_j is its spread, and $\| \|_2$ represents the Euclidean distance.

As an RBF is a static model, dynamics are here introduced *via* external feedback. This way, the RBFs models described in (9) are employed as NAR (Nonlinear AutoRegressive) models or NAR models with eXogenous inputs (NARX). In (11), y is the variable being modelled, and considering only one exogenous input (v), for the sake of simplicity, the predicted values (), at instant k, is:

$$\hat{y}[k] = f\left(y\left[k-d_{o_1}\right], \ldots, y\left[k-d_{o_n}\right], v\left[k-d_{i_1}\right], \ldots, v\left[k-d_{i_n}\right]\right) \tag{11}$$

In (1), f(*k-d*) represents a RBF model (10), meaning that its arguments (the delays of y and v) represent the network input vector.

As the objective is to determine the evolution of the forecasts over a prediction horizon, (1) is iterated over that horizon. For k+1:

$$\hat{y}[k+1] = f\left(y\left[k+1-d_{o_1}\right], \ldots, y\left[k+1-d_{o_n}\right], v\left[k+1-d_{i_1}\right], \ldots, v\left[k+1-d_{i_n}\right]\right) \tag{12}$$

For the steps within the prediction horizon, it is possible not to have measured values for one or more terms in the argument of (11), depending on the indices of the delays. These must be obtained using previous predictions. As so, the computation of the predictions over a prediction horizon PH may require PH executions of the model (11).

To design a "good" data-driven model, three points must be addressed. First, from the available data, samples must be chosen and distributed between the design sets. This is called data selection. Once these sets are available, the features that will be supplied to the model must be determined. This is called feature selection and, in this case, besides the selection of the exogenous variables, the lags to be employed must also be chosen. This can be done together with topology selection, which in the case of this work, translates into the number of hidden neurons. Given the design sets and the feature and topology selected, the last step is to estimate the model parameters.

The ApproxHull algorithm proposed in [58] is used in this work. to select the data for training, testing and validation data from the set of available design data. As is described in [58], the ApproxHull is an incremental randomised approximate Convex Hull (CH) algorithm that selects the points involving the whole data. It is suitable to high dimension data, dealing with the complexity of memory and computational time efficiently. These convex hull vertices obtained are compulsorily inserted in the training set, ensuring the coverage of the whole operational range.

For feature and topology selection, a Multi-Objective Genetic Algorithm (MOGA) is employed in this work. The model design under study is considered as multi-objective optimisation, with the possibility to assign restrictions and priorities to each objective. The evolutionary algorithm searches the admissible space of the number of neurons and inputs - lags for the modelled and exogenous variables - for the RBF models. For a detailed explanation of MOGA [59], can be consulted.

A Levenberg-Marquardt algorithm [60, 61] is employed to determine the parameters of each model, with the aims of minimising an error criterion that exploits the linear-nonlinear relationship of the RBF NN model parameters, before being evaluated in MOGA [62, 63]. The initial values of the non-linear parameters (C and) are chosen randomly, or with the use of a clustering algorithm, w is determined as a linear least-squares solution. The early-stopping approach is the employed procedure to the termination [64], within a maximum number of iterations.

The models were developed, in the scope of the MOGA design, with a set of data divided into three sets: a training data set, employed to estimate the model parameters, a testing data set, for terminating the training, and a validation data set, used for performance comparison of the models obtained by employing MOGA (considering that it uses multiple objective formulations, its result is not a singular solution, but instead it is a set of non-dominated solutions). The minimisation objectives employed here are the RMSEs of the training (ε_{tr}) and the testing (ε_{te}) data sets, the complexity of the model ($O(\mu)$) and the forecasting error (ε_p). This forecasting error is obtained by summing the RMSEs along with the considered PH (presented in eq. (13)), with p data points, where D is a time-series, and E (presented in eq. (14)) is an error matrix (14):

$$\varepsilon(D, PH) = \sum_{i=1}^{PH} RMSE(E(D, PH), i) \tag{13}$$

$$E(D, PH) = \begin{bmatrix} e[1,1] & e[1,2] & \cdots & e[1, PH] \\ e[2,1] & e[2,2] & \cdots & e[2, PH] \\ \vdots & \vdots & \ddots & \vdots \\ e[p-PH,1] & e[p-PH,2] & \cdots & e[p-PH, PH] \end{bmatrix} \tag{14}$$

MOGA is set to be executed with 100 generations, with a population size of 100, a proportion of random emigrants of 0.10 and a crossover rate of 0.70. The admissible range of neurons varied from 2 to 20, while the admissible number of inputs vary from 2 to 20.

Data averaged in 15 minutes periods is used here. The goal is to model the total energy demand and to obtain small RMSE values over a prediction horizon of 48 steps-ahead (12 hours) to be employed, in future works, in MBPC HEMS. The lags associated for the modelled variable and the exogeneous variables belong to three different periods. Period 1 (P1) refers to lags immediately before the current sample; Period 2 (P2) refers lags centred one day before and Period 3 (P3) refers to the lags centred one week before the current sample.

Three models for load demand were designed using the Gambelas house data. The first design considers as exogeneous variables the daily number of occupants (v_1)

and the day encode (v_2) representing the position of the day within the week. For a description of the codification used, please see [65]. For this first model, the lag periods considered are: P1 [20, 1, 1]; P2 [4, 0, 0]; P3 [4, 0, 0]. This problem deals with the whole period of valid acquired data, *i.e.,* between January and July 2020.

The second design employs the same exogeneous variables as model 1, as well as the same lag periods; the difference lies on a smaller number of samples used, between May and July.

The third design employees an additional exogenous variable, the ambient temperature (v_3). For this third model, the lag periods are: P1 [20, 1, 1, 1]; P2 [4, 0, 0, 0]; P3 [4, 0, 0, 0]. The addition of the ambient temperature, in relation with the first model, aims to assess if the use of a weather-related variable, together with the other exogeneous variables, can provide a better prediction performance. The same period, May to July is considered. Please note that weather-data was only available from May on, which was the reason why this period was considered for models 2 and 3.

The use of the three models aims to provide two insights: i) between model 1 and models 2 and 3 the influence of the number of samples on the accuracy of the forecasting; ii) between models 2 and 3, the impact on the forecasting accuracy of the addition of the ambient temperature.

The ApproxHull results are shown in Table **4**. As it can be seen, a much larger number of samples was used in the first design.

Table 4. ApproxHull results for the three designs.

Design	Nº Total Samples	Nº Features	Nº Vertices	Nº Samples Training	Nº Samples Testing	Nº Samples Validation
1	15215	41	1985	9129	3043	3043
2	6310	41	1544	3786	1262	1262
3	6310	70	1622	3786	1262	1262

After MOGA execution, the number of non-dominated models for the three designs was 301, 351 and 422, respectively. Among those models, occupation (v_1) was present in 48, 65 and 165, and the day encodes (v_2) was selected in 68, 64 and 63 models.

Given the models within the non-dominated set, one best model was selected for each one of the problems, and the results for these models are presented in Table **5**.

In this table, $\|w\|_2$ denotes the 2-norm of the linear parameters, which is related to the model condition. The forecasting error, shown in the last column, is computed in two weeks of data, from 10th July to 24th July 2020.

Table 5. Selected models results.

Model	Features	Neurons	O(μ)	$\|w\|_2$	ε_{tr}	ε_{te}	ε_{va}	ε_p
1	16	20	340	641	0.176	0.168	0.178	9.98
2	15	17	272	1509	0.193	0.204	0.191	11.52
3	20	18	378	2173	0.190	0.207	0.202	10.87

As it can be seen, model 1 obtains better results than models 2 and 3, which shows that the larger number of design samples is beneficial both to one-ste--ahead forecasts (ε_{tr}, ε_{te} and ε_{va}) and to the whole PH (ε_p). Comparing models 2 and 3, it is evident that the use of an additional weather variable, the ambient temperature, is beneficial to the forecasting performance.

Equations (15), (16) and (17) present the selected models for designs 1, 2 and 3, respectively. Analysing the three models, only model 3 employed the number of occupants as an exogenous variable. Lags of the modelled variable were chosen from the three periods, a larger number of lags belonging to the first period, P1. In the last model (design 3) the ambient temperature also was considerably used.

$$y(k) = f \begin{pmatrix} y(k-1), y(k-2), y(k-3), y(k-6), y(k-9), y(k-10), \\ y(k-16), y(k-20), y(k-92), y(k-94), y(k-95), \\ y(k-99), y(k-100), y(k-669), y(k-671) \end{pmatrix} \quad \textbf{(15)}$$

$$y(k) = f \begin{pmatrix} y(k-1), y(k-6), y(k-7), y(k-9), y(k-11), y(k-17), \\ y(k-19), y(k-93), y(k-96), y(k-699), y(k-670), \\ y(k-671), y(k-672), y(k-673), y(k-674) \end{pmatrix} \quad \textbf{(16)}$$

$$y(k) = f \begin{pmatrix} y(k-1), y(k-2), y(k-4), y(k-9), y(k-16), y(k-18), \\ y(k-92), y(k-95), y(k-97), y(k-100), y(k-671), \\ , y(k-676), v1(k-1), v2(k-1), v3(k-1), v3(k-2), \\ v3(k-4), v3(k-10), v3(k-11), v3(k-14) \end{pmatrix} \quad \textbf{(17)}$$

The forecasting results for 7 days in July are shown in Figs. (**21** to **23**), for the three designs. The top plot illustrates the real values of the load demand, and the one-step-ahead predicted values; the bottom plot shows the evolution of the scaled prediction error over the prediction horizon.

As it can be seen, one-step-ahead predictions are very similar between the three models. From the analysis of the b) plots of Figs. (**21** to **23**), it is evident that the error is increasing over the prediction horizon, as expected. But the errors obtained 48-steps-ahead (12 hours ahead) are still small, obtaining values of 0.21, 0.25 and 0.235, for each one of the three models.

The smallest 48-step-ahead RMSE is achieved by model 1, obtained with a date between January and July. Regarding models 2 and 3, the use of the atmospheric temperature was beneficial for the forecasting performance.

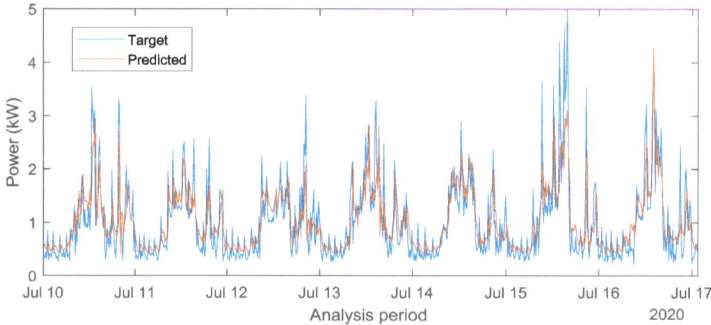

(a)

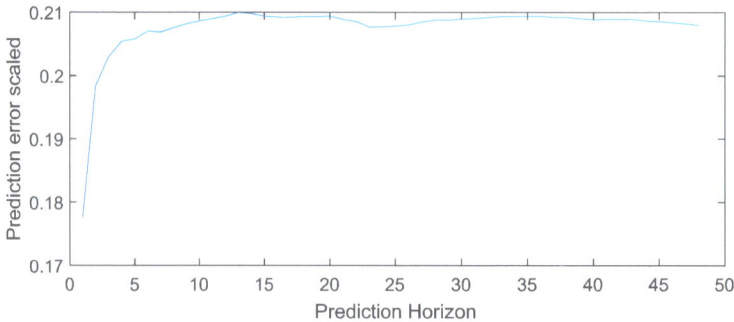

(b)

Fig. (21). Design 1 prediction results: **(a)** target and predicted values; **(b)** scaled RMSE values over the prediction horizon.

(a)

(b)

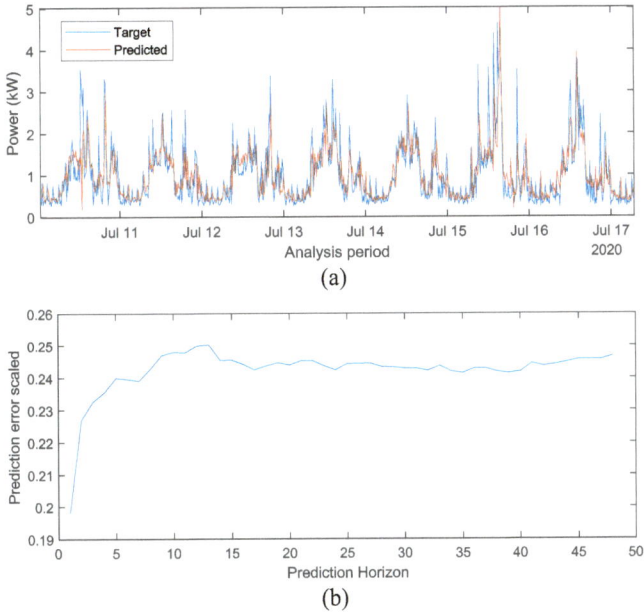

Fig. (22). Design 2 prediction results: **(a)** target and predicted values; **(b)** scaled RMSE over the prediction horizon.

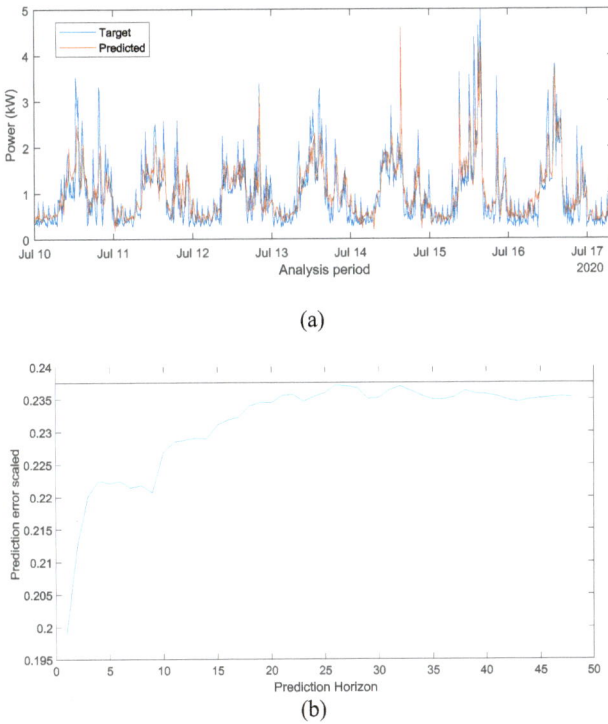

(a)

(b)

Fig. (23). Design 3 prediction results: **(a)** target and predicted values; **(b)** scaled RMSE values over the prediction horizon.

CONCLUDING REMARKS

It could be seen in the brief background information presented that given the percentage of energy consumption of building sector in the total energy consumption, it is of fundamental importance to decrease energy usage in buildings and, simultaneously, maintain the thermal comfort of its occupants. This can be obtained by integrating efficient passive solutions in the design and construction of buildings, the employment of renewable energy, and the use of efficient Home Energy Management Systems (HEMS), with a special emphasis on the control of existing HVAC systems. A good compromise between decreasing energy consumption in buildings while maintaining the thermal comfort of the occupants is fundamental in everyday's life. One of the methods to achieve this balance is the use of efficient HEMS, especially taking into account the control HVAC systems. Despite the recent advances in smart grids, IoT, and advances control algorithms, the use of the suboptimal solution is frequent. With a view of improving the current situation, this chapter discussed the impact of occupants on the building thermal comfort, HVAC control and home energy consumption forecasting.

The work started by presenting brief background information concerning the challenges of the building sector, considering the reduction of energy use. Then, it introduces the fundamental concepts of thermal comfort metrics, with focus on the Predicted Mean Vote (PMV). Subsequently, the PMV computation was discussed. As it involves iterative computations, that may consume a considerable amount of time in the cases where hundreds to thousands of iterations are needed within a single sampling period. Computational learning methods were used by the authors, namely Radial Basis Function Neural Networks, to address this issue. It was shown that excellent approximations were obtained with savings of 55 times regarding the standard iterative computation.

Following, the work presented results concerning the occupancy estimation and detection, and its impact on thermal comfort was also discussed. The authors essentially compared the predictive control of a case study in two situations: i) one model designed with data with no occupancy in the room, and ii) one model designed with data obtained during periods in which the room was occupied, in which the movement signal was incorporated. As it could be seen in the results, the second model presented a significant improvement concerning the first model in terms of accuracy and estimated PMV, highlighting the importance of the use of occupation data in model design.

These models were employed in an intelligent Model-Based Predictive Control for HVAC system, which minimises economic costs associated with the HVAC operation while maintaining the rooms in thermal comfort, for user-defined periods.

It is also needed to highlight that the economic saving, a very important factor for the occupants of a building, is just a small example of how electric information can be used to promote efficient energy use in a building even when manually operated by the occupant. As was shown in this work, these changes of the operation of schedulable devices, together with a better schedule of the inverter, can, however, be automatically achieved by the use of a HEMS. The automation of the systems reduces the interaction of the occupants with the systems - reducing the need of a high level of awareness of how the system should be best operated and also saving occupants' time.

The impact of the occupation on electric consumption was also an object of study, using data acquired from a residential building located in Gambelas, Faro, Portugal, ranging from January to July 2020. It was found out that no valid statistical correlation was obtained between total daily electric consumption and occupancy. However, when the correlation is computed between the disaggregated electric consumption of the building's appliances and occupancy data, statistical significance was obtained. It is also possible to speculate that further disaggregating the energy up to the individual appliances would result in more significant correlations. This, however, will only be possible to answer when the NILM methods, the focus of the project in which this work in inserted, are developed and applied to this data – constituting a future step in the research. The work also presented energy management measures to obtain better efficiency on the usage of the solar energy system existent in the house.

The last part of the work evaluated the energy consumption forecasting models having as exogenous variable the daily occupancy data. Three models were simulated. The first (called Design 1) has as modelling variable the total power demand, and as exogenous variables, the daily number of occupants and a day codification – that represents the different days of the week and considers holidays as well. Design 2 has the same characteristics but employs a smaller number of samples, also used in Design 3. The latter includes the ambient temperature on the data set. The problems aim to provide two comparisons: i) between Designs 1 and the other two, the influence of the number of design samples on the accuracy of the forecasting; ii) between Designs 2 and 3, the impact on the accuracy of the forecasting of adding the ambient temperature to the dataset.

Designs 1 and 2 considered 41 features, while the third one allowed 70 possible inputs, the number of occupants being one of those in all the three designs. The number of non-dominated models for the three designs was 301, 351 and 422, respectively. Among those models, the occupation was present in 48, 65 and 165 models, which shows that it is a relevant variable when designing forecasting load consumption models. The results showed that a larger number of design samples and the use of ambient temperature is beneficial for the forecasting accuracy.

CONSENT FOR PUBLICATION

Not Applicable.

CONFLICT OF INTEREST

The author confirms that this chapter contents have no conflict of interest.

ACKNOWLEDGEMENT

The authors would like to acknowledge the support of Programa Operacional Portugal 2020 and Operational Program CRESC Algarve 2020 grant 01/SAICT/2018. Antonio Ruano acknowledges the support of Fundação para a Ciência e Tecnologia, through IDMEC, under LAETA, grant UIDB/50022/2020.

REFERENCES

[1]   P. Nejat, F. Jomehzadeh, M.M. Taheri, M. Gohari, and M.Z. Abd, "Majid, "A global review of energy consumption, CO2 emissions and policy in the residential sector (with an overview of the top ten CO2 emitting countries)", *Renew. Sustain. Energy Rev.,* vol. 43, pp. 843-862, 2015. [http://dx.doi.org/10.1016/j.rser.2014.11.066]

[2]   DOE, "Quadrennial Technology Review: An assessment of Energy Technologies and Research Opportunities – Chapter 1: Energy Challenges", *Department of Energy*. USA2015.

[3]   G. Pau, M. Collotta, A. Ruano, and J. Qin, "Smart Home Energy Management", *Energies,* vol. 10, p. 382, 2017. [http://dx.doi.org/10.3390/en10030382]

[4]   P. H. Shaikh, N. B. M. Nor, P. Nallagownden, I. Elamvazuthi, and T. Ibrahim, "A review on optimised control systems for building energy and comfort management of smart sustainable buildings", *Renewable and Sustainable Energy Reviews,* vol. 34, pp. 409-429, 2014.

[5]   A.I. Dounis, and C. Caraiscos, "Advanced control systems engineering for energy and comfort management in a building environment-A review", *Renew. Sustain. Energy Rev.,* vol. 13, pp. 1246-1261, 2009. [http://dx.doi.org/10.1016/j.rser.2008.09.015]

[6]   T.A. Nguyen, and M. Aiello, "Energy intelligent buildings based on user activity: A survey", *Energy and buildings,* vol. 56, pp. 244-257, 2013. [http://dx.doi.org/10.1016/j.enbuild.2012.09.005]

[7]   S.B. Sadineni, S. Madala, and R.F. Boehm, "Passive building energy savings: A review of building envelope components", *Renew. Sustain. Energy Rev.,* vol. 15, pp. 3617-3631, 2011. [http://dx.doi.org/10.1016/j.rser.2011.07.014]

[8]   R. Pacheco, J. Ordóñez, and G. Martínez, "Energy efficient design of building: A review", *Renew. Sustain. Energy Rev.,* vol. 16, pp. 3559-3573, 2012.

[http://dx.doi.org/10.1016/j.rser.2012.03.045]

[9]   K. M. and K. M., "Ten questions concerning model predictive control for energy efficient buildings", *Build. Environ.,* vol. 105, pp. 403-412, 2016.
[http://dx.doi.org/10.1016/j.buildenv.2016.05.034]

[10]   P.F. Pereira, N.M.M. Ramos, R.M.S.F. Almeida, and M.L. Simões, "Methodology for detection of occupant actions in residential buildings using indoor environment monitoring systems", *Building and Environment,* vol. 146, pp. 107-118, 2018.
[http://dx.doi.org/10.1016/j.buildenv.2018.09.047]

[11]   ASHRAE, *Thermal Environmental Conditions for Human Occupancy,* 2004.

[12]   P.O. Fanger, *Thermal comfort: analysis and applications in environmental engineering.* McGraw-Hill: New York, 1972.

[13]   S. Zhang, and Z. Lin, "Standard effective temperature based adaptive-rational thermal comfort model", *Appl. Energy,* vol. 264, p. 9, 2020.
[http://dx.doi.org/10.1016/j.apenergy.2020.114723]

[14]   A.E. Ruano, and P.M. Ferreira, "Neural Network based HVAC Predictive Control", *IFAC Proceedings Volumes (IFAC-PapersOnline),* vol. 19, 2014 pp. 3617-3622 24-29 AUG.
[http://dx.doi.org/10.3182/20140824-6-ZA-1003.01051]

[15]   S. Atthajariyakul, and T. Leephakpreeda, "Neural computing thermal comfort index for HVAC systems", *Energy Convers. Manage.,* vol. 46, pp. 2553-2565, 2005.
[http://dx.doi.org/10.1016/j.enconman.2004.12.007]

[16]   M. Kumar, and I.N. Kar, "Non-linear HVAC computations using least square support vector machines", *Energy Convers. Manage.,* vol. 50, pp. 1411-1418, 2009.
[http://dx.doi.org/10.1016/j.enconman.2009.03.009]

[17]   J. Yao, and J. Xu, "Research on the BPNN in the Prediction of PMV", *Applied Mechanics and Mechanical Engineering,* vol. 29-32, pp. 2804-2808, 2010.
[http://dx.doi.org/10.4028/www.scientific.net/AMM.29-32.2804]

[18]   Q. Chai, H.Q. Wang, Y.C. Zhai, and L. Yang, "Using machine learning algorithms to predict occupants' thermal comfort in naturally ventilated residential buildings", *Energy Build.,* vol. 217, p. 13, 2020.
[http://dx.doi.org/10.1016/j.enbuild.2020.109937]

[19]   P.M. Ferreira, A.E.B. Ruano, and E.A. Faria, "Design and Implementation of a real-time data acquisition system for the identification of dynamic temperature models in a hydroponic greenhouse", *Acta Hortic.,* no. 519, pp. 191-199, 2000.
[http://dx.doi.org/10.17660/ActaHortic.2000.519.19]

[20]   P.M. Ferreira, S.M. Silva, A.E. Ruano, A.T. Negrier, and E.Z.E. Conceicao, "Neural Network PMV Estimation for Model-Based Predictive Control of HVAC Systems", *2012 IEEE International Joint Conference on Neural Networks (IJCNN),* 2012 pp. 15-22 Brisbane, Australia
[http://dx.doi.org/10.1109/IJCNN.2012.6252365]

[21]   P.M. Ferreira, A.E. Ruano, S. Silva, and E.Z.E. Conceicao, "Neural Networks based predictive control for thermal comfort and energy savings in public buildings", *Energy Build.,* vol. 55, pp. 238-251, 2012.
[http://dx.doi.org/10.1016/j.enbuild.2012.08.002]

[22]   A.E. Ruano, S. Pesteh, S. Silva, H. Duarte, G. Mestre, and P.M. Ferreira, "The IMBPC HVAC system: A complete MBPC solution for existing HVAC systems", *Energy Build.,* vol. 120, pp. 145-158, 2016.
[http://dx.doi.org/10.1016/j.enbuild.2016.03.043]

[23]   M. Corporation, *Tmote Sky.,* 1998. http://www.eecs.harvard.edu/~konrad/projects/shimmer/references/tmote-sky-datasheet.pdf

[24]   A. Ruano, S. Silva, H. Duarte, and P.M. Ferreira, "Wireless Sensors and IoT Platform for Intelligent HVAC Control", *Appl. Sci. (Basel),* vol. 8, p. 370, 2018.
[http://dx.doi.org/10.3390/app8030370]

[25]   S. Meyn, A. Surana, Y. Lin, S. Oggianu, S. Narayanan, and T. Frewen, "A Sensor-Utility-Network Method for Estimation of Occupancy in Buildings", In: *48th IEEE Conf. CDC/CCC.* China, 2009, pp. 1494-1500.
[http://dx.doi.org/10.1109/CDC.2009.5400442]

[26]   N. Cao, J. Ting, S. Sen, and A. Raychowdhury, "Smart Sensing for HVAC Control: Collaborative Intelligence in Optical and IR Cameras", *IEEE Trans. Ind. Electron.,* vol. 65, pp. 9785-9794, 2018.
[http://dx.doi.org/10.1109/TIE.2018.2818665]

[27]   W. Wang, T. Hong, N. Li, R.Q. Wang, and J. Chen, "Linking energy-cyber-physical systems with occupancy prediction and interpretation through WiFi probe-based ensemble classification", *Applied Energy,* vol. 236, pp. 55-69, 2019.
[http://dx.doi.org/10.1016/j.apenergy.2018.11.079]

[28]   P.F. Pereira, and N.M.M. Ramos, "Detection of occupant actions in buildings through change point analysis of in-situ measurements", *Energy and Buildings,* vol. 173, pp. 365-377, 2018.
[http://dx.doi.org/10.1016/j.enbuild.2018.05.050]

[29]   A.E. Ruano, S. Silva, S. Pesteh, P.M. Ferreira, H. Duarte, and G. Mestre, "Improving a neural networks based HVAC predictive control approach", *9th IEEE International Symposium on Intelligent Signal Processing (WISP 2015),* 2015pp. 90-95 Siena, Italy
[http://dx.doi.org/10.1109/WISP.2015.7139168]

[30]   R. Godina, E.M.G. Rodrigues, E. Pouresmaeil, J.C.O. Matias, and J.P.S. Catalão, "Model Predictive Control Home Energy Management and Optimization Strategy with Demand Response", *Appl. Sci. (Basel),* vol. 8, p. 408, 2018.
[http://dx.doi.org/10.3390/app8030408]

[31]   M. Razmara, M. Maasoumy, M. Shahbakhti, and R.D. Robinett, "Optimal exergy control of building HVAC system", *Appl. Energy,* vol. 156, pp. 555-565, 2015.
[http://dx.doi.org/10.1016/j.apenergy.2015.07.051]

[32]   O. Alrumayh, and K. Bhattacharya, *"Model predictive control based home energy management system in smart grid,"* in *2015 IEEE Electrical Power and Energy Conference.* EPEC, 2015, pp. 152-157.

[33]   M. Castilla, J.D. Alvarez, J.E. Normey-Rico, and F. Rodriguez, "Thermal comfort control using a non-linear MPC strategy: A real case of study in a bioclimatic building", *J. Process Contr.,* vol. 24, pp. 703-713, 2014.
[http://dx.doi.org/10.1016/j.jprocont.2013.08.009]

[34]   Y.D. Ma, A. Kelman, A. Daly, and F. Borrelli, "Predictive Control for Energy Efficient Buildings with Thermal Storage", *IEEE Contr. Syst. Mag.,* vol. 32, pp. 44-64, 2012.
[http://dx.doi.org/10.1109/MCS.2011.2172532]

[35]   B.F. Balvedi, E. Ghisi, and R. Lamberts, "A review of occupant behaviour in residential buildings", *Energy and Buildings,* vol. 174, pp. 495-505, 2018.
[http://dx.doi.org/10.1016/j.enbuild.2018.06.049]

[36]   B. Dong, and K.P. Lam, "A real-time model predictive control for building heating and cooling systems based on the occupancy behavior pattern detection and local weather forecasting", *Building Simulation,* vol. 7, pp. 89-106, 2014.
[http://dx.doi.org/10.1007/s12273-013-0142-7]

[37]   K. Sun, D. Yan, T. Hong, and S. Guo, "Stochastic modeling of overtime occupancy and its application in building energy simulation and calibration", *Building and Environment,* vol. 79, pp. 1-12, 2014.
[http://dx.doi.org/10.1016/j.buildenv.2014.04.030]

[38]   C. Wang, D. Yan, and Y. Jiang, "A novel approach for building occupancy simulation", *Building*

Simulation, vol. 4, pp. 149-167, 2011.
[http://dx.doi.org/10.1007/s12273-011-0044-5]

[39] Y. Ding, Q. Wang, Z. Wang, S. Han, and N. Zhu, "An occupancy-based model for building electricity consumption prediction: A case study of three campus buildings in Tianjin", *Energy and Buildings,* vol. 202, p. 109412, 2019.
[http://dx.doi.org/10.1016/j.enbuild.2019.109412]

[40] M.S. Gul, and S. Patidar, "Understanding the energy consumption and occupancy of a multi-purpose academic building", *Energy and Buildings,* vol. 87, pp. 155-165, 2015.
[http://dx.doi.org/10.1016/j.enbuild.2014.11.027]

[41] T. Sekki, M. Airaksinen, and A. Saari, "Impact of building usage and occupancy on energy consumption in Finnish daycare and school buildings", *Energy and Buildings,* vol. 105, pp. 247-57, 2015.
[http://dx.doi.org/10.1016/j.enbuild.2015.07.036]

[42] M.A. López, I. Santiago, F.J. Bellido-Outeiriño, A. Moreno-Munoz, and D. Trillo-Montero, "Active occupation profiles in the residential sector in Spain as an indicator of energy consumption", *2012 IEEE Second International Conference on Consumer Electronics - Berlin (ICCE-Berlin),* 2012pp. 1-5
[http://dx.doi.org/10.1109/ICCE-Berlin.2012.6336472]

[43] A. Muroni, I. Gaetani, P-J. Hoes, and J.L.M. Hensen, "Occupant behavior in identical residential buildings: A case study for occupancy profiles extraction and application to building performance simulation", *Building Simulation,* vol. 12, pp. 1047-1061, 2019.

[44] F. Causone, S. Carlucci, M. Ferrando, A. Marchenko, and S. Erba, "A data-driven procedure to model occupancy and occupant-related electric load profiles in residential buildings for energy simulation", *Energy and Buildings,* vol. 202, p. 109342, 2019.
[http://dx.doi.org/10.1016/j.enbuild.2019.109342]

[45] J-N. Louis, A. Caló, K. Leiviskä, and E. Pongrácz, "Modelling home electricity management for sustainability: The impact of response levels, technological deployment & occupancy", *Energy and Buildings,* vol. 119, pp. 218-232, 2016.

[46] *Sharp NU-AK PV panels.,* 2020. https://www.sharp.co.uk/cps/rde/xchg/gb/hs.xsl/-/html/produc--details-solar-modules-2189.htm?product=NUAK300B

[47] *Kostal Plenticore Plus Inverter.,* 2020.https://www.kostal-solar-electric.com/en-gb/products/hybid-inverters/plenticore-plus

[48] *BYD Battery Box HV.,* 2020. https://www.eft-systems.de/en/The%20B-BOX/product/Battery%20 Box%20HV/3

[49] *TP-Link WiFi Smart Plugs.,* 2020 . https://www.tp-link.com/pt/home-networking/smart-plug/hs100/

[50] G. Mestre, A. Ruano, H. Duarte, S. Silva, H. Khosravani, S. Pesteh, P.M. Ferreira, and R. Horta, "An Intelligent Weather Station", *Sensors (Basel),* vol. 15, no. 12, pp. 31005-31022, 2015.
[http://dx.doi.org/10.3390/s151229841] [PMID: 26690433]

[51] A. Ruano, A. Hernandez, J. Ureña, M. Ruano, and J. Garcia, "NILM Techniques for Intelligent Home Energy Management and Ambient Assisted Living: A Review", *Energies,* vol. 12, p. 2203, 2019.
[http://dx.doi.org/10.3390/en12112203]

[52] *Carlo   Gavazzi   EM340,*   2020.https://www.carlogavazzi.co.uk/blog/carlo-gavazzi-enery-solutions/em340-utilises-touchscreen-technology

[53] *Wibeee   Consumption   Analyzers.,*   2020.   http://circutor.com/en/products/measurement-and   -control/fixed-power-analyzers/consumption-analyzers

[54] A. Ruano, K. Bot, and M.G. Ruano, "Home Energy Management System in an Algarve residence. First results", *CONTROLO 2020: Proceedings of the 14th APCA International Conference on Automatic Control and Soft Computing,* vol. 695, Lecture Notes in Electrical Engineering, 695, G. J.

A, B.-C. M., and C. J.P., Eds., ed Bragança, Portugal: Springer Science and Business Media Deutschland GmbH, pp. 332-341, 2021.

[55]   A. Ruano, H. Qassemi, H. Inoussa, M. Marzouq, H.E. Fadili, and S. Bennani, "A Model-based Predictive Control approach for Home Energy Management Systems. First results", *The 6th International Conference on Wireless Technologies, Embedded and Intelligent Systems (WITS-2020),* 2020 Fez, Morocco

[56]   K.P. Amber, R. Ahmad, M.W. Aslam, A. Kousar, M. Usman, and M.S. Khan, "Intelligent techniques for forecasting electricity consumption of buildings", *Energy,* vol. 157, pp. 886-893, 2018.
[http://dx.doi.org/10.1016/j.energy.2018.05.155]

[57]   Y. Wei, X. Zhang, Y. Shi, L. Xia, S. Pan, and J. Wu, "A review of data-driven approaches for prediction and classification of building energy consumption", *Renewable and Sustainable Energy Reviews,* vol. 82, pp. 1027-1047, 2018.
[http://dx.doi.org/10.1016/j.rser.2017.09.108]

[58]   H.R. Khosravani, A.E. Ruano, and P.M. Ferreira, "A convex hull-based data selection method for data driven models", *Applied Soft Computing,* vol. 47, pp. 515-533, 2016.
[http://dx.doi.org/10.1016/j.asoc.2016.06.014]

[59]   P. Ferreira, and A. Ruano, "Evolutionary Multiobjective Neural Network Models Identification: Evolving Task-Optimised Models", *New Advances in Intelligent Signal Processing,* vol. 372, A. Ruano and A. Várkonyi-Kóczy, Eds., ed: Springer Berlin / Heidelberg, pp. 21-53, 2011.
[http://dx.doi.org/10.1007/978-3-642-11739-8_2]

[60]   M. Levenberg, *A method for the solution of certain non-linear problems in least squares,* 1964.

[61]   D.W. Marquardt, "An algorithm for least-squares estimation of nonlinear parameters", *Journal of the Society for Industrial and Applied Mathematics,* vol. 11, pp. 431-441, 1963.
[http://dx.doi.org/10.1137/0111030]

[62]   A.E.B. Ruano, D.I. Jones, and P.J. Fleming, "A New Formulation of the Learning Problem for a Neural Network Controller", *30th IEEE Conference on Decision and Control Brighton, UK,* 1991pp. 865-866
[http://dx.doi.org/10.1109/CDC.1991.261439]

[63]   P.M. Ferreira, and A.E. Ruano, "Exploiting the separability of linear and nonlinear parameters in radial basis function networks", *Adaptive Systems for Signal Processing, Communications, and Control Symposium (AS-SPCC) Lake Louise, Canada,* pp. 321-326, 2000.
[http://dx.doi.org/10.1109/ASSPCC.2000.882493]

[64]   S. Haykin, *Neural Networks: A Comprehensive Foundation.,* 2nd edPrentice Hall, 1999.

[65]   P.M. Ferreira, R. Pestana, and A.E. Ruano, "Improving the Identification of RBF Predictive Models to Forecast the Portuguese Electricity Consumption", *IFAC Proceedings Volumes (IFAC-PapersOnline),* vol. 1, pp. 208-213, 2010.
[http://dx.doi.org/10.3182/20100329-3-PT-3006.00039]

CHAPTER 5

Detecting Occupant Actions in Buildings and the Drivers of Their Behaviour

Pedro F. Pereira[1,*] and **Nuno M. M. Ramos**[1]

[1] *CONSTRUCT – LFC, Faculty of Engineering (FEUP), University of Porto, Rua Dr. Roberto Frias s/n, 4200-465 Porto, Portugal*

Abstract: The new directive 2018/844/EU on Energy Performance of Buildings and Energy Efficiency, supports the change of building into smarter, more energy-efficient and include the perspective of the occupants' needs. The knowledge of occupant behaviour is the centre of the balance between the buildings energy efficiency and its indoor environmental quality. This chapter presents a state-of-the-art of the buildings occupant behaviour, presenting the new developments and future trends. It summarises research in which a series of methodologies were developed to supply relevant data to the building management systems (BMS). These methodologies use a monitoring system based on environmental sensors, namely relative humidity, temperature, and carbon dioxide. New methods to detect the occupant actions in the operation of building systems were summarised and compared. The methodologies were based on statistical tools and machine learning techniques. They can be applied to different case studies since they can adapt to the local environment under a self-learning strategy. The drivers of behaviour for the operation of those building systems were also analysed. Two methodologies that allowed to predict the occupant actions taking into account the parameters that influenced the occupant behaviour were described. It was also possible establishing the seasonality of drivers of behaviour. The overall results highlight that the actions, motivations, and impacts of a specific set of occupants performed in building systems can significantly vary depending on the room and on environmental parameters.

Keywords: Action detection, Drivers of behavior, Intelligent buildings, *In situ* data acquisition, Occupant behaviour.

* **Corresponding author Pedro F. Pereira:** CONSTRUCT – LFC, Faculty of Engineering (FEUP), University of Porto, Rua Dr. Roberto Frias s/n, 4200-465 Porto, Portugal; Tel: +351225082257; E-mail: fpfp@fe.up.pt

Enedir Ghisi, Ricardo Forgiarini Rupp and Pedro Fernandes Pereira (Eds.)

INTRODUCTION

Scope

Nowadays, energy efficiency in buildings is a recurring subject and one way to reduce CO_2 emissions. The 2018/844/EU [1] directive, recently approved to update the Directive 2010/31/EU [2] and the Directive 2012/27/EU [3], introduced a new concept in the balance between energy efficiency and the indoor environmental quality: the occupants. This new concept turns the focus of the energy efficiency studies to occupant centric.

Occupant Behaviour

Solar passive methods were the first studies in the energy efficiency of buildings [4]. These initial studies focused on systems and building components, drawing attention away from the study of the building occupants [5]. However, in recent years, it can be observed an increase in studies related to occupant behaviour in buildings [6].

One of the main factors contributing to the increase of the studies in occupant behaviours relies on the fact that occupancy is considered a predominant factor in the building performance and one of the main variables for the exiting gap between the simulated and the post-occupancy monitoring data [7 - 10]. The literature studied this effect from different perspectives. The occupant behaviour studies could be divided into two distinct research fields: the energy consumption impacts of occupants; and the indoor environmental quality impacts [11, 12].

This research field contains several frameworks, ontologies and reviews. One of the most cited ontology was written by Hong *et al.* [13]. According to the authors, there was a need to create an ontology to organise the research of energy-related occupant behaviour in buildings. Therefore, the authors divided the studies into four categories: drivers, needs, actions and systems (DNAS) (Fig. **1**). The first category groups the works that studied the environmental factors that motivate the occupants to perform an action that fulfills a need; the second category groups the studies where the occupants' requirements of the indoor environment of their dwelling; the third category compiles the studies on the actions performed by the occupants with the systems and the fourth category groups the buildings active and passive components that the occupants operate to fulfil the occupants' needs.

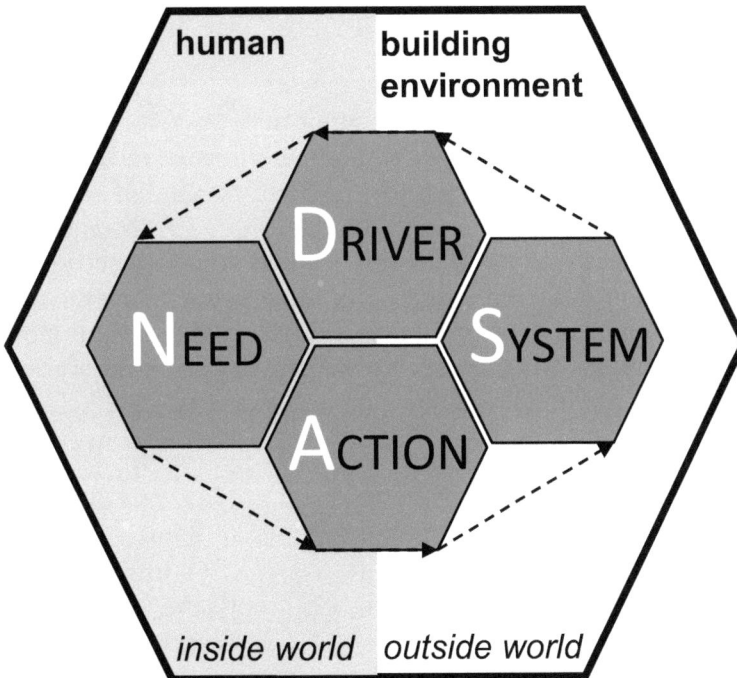

Fig. (1). DNAS ontology, extracted from [13].

In order to reduce the gap between the building simulations and the post-occupancy monitoring data, researchers have adopted two different approaches: the macro studies and the room-level. The former has the utility for city-level management and creates databases of the occupant behaviour divided into clusters to provide designers with reliable information for simulation [10]. However, the latter is more indicated to be used in the BMS that have to consider the occupants' specificities to fulfil their needs. If the former databases were employed in the dwelling BMS without the necessary adaptations to the specificities of the occupants, the occupants' needs would not be fulfilled, and the balance between occupants satisfaction, energy efficiency and indoor environmental quality will not be met. Otherwise, if the drivers of occupants' behaviours were studied, the BMS could anticipate the occupants' needs and better deal with the occupants' impacts on energy consumptions and indoor environmental quality [14 - 16]. Literature also pointed to a great diversity of drivers for the same occupants' action [17, 18], which highlights the need to focus on occupants' specificities from each dwelling [10] or room [19, 20].

State-of-the-Art

Occupant Behaviour Data-acquisition

The study of the occupants' behaviour could be performed in three different ways: monitoring, surveys, and laboratory tests [7]. The first kind of studies are the most common and accepted by the research community [7, 16]. Monitoring the indoor built environment is generally a non-intrusive and non-destructive technique that allows data collection with reliability. The other types of methods to study the occupants' behaviour tend to produce distorted and conditioned data. The laboratory tests may lead occupants to behave in a non-natural way, and the surveys can be or short and reliable or long and inaccurate [16]. Thus the surveys can be used complementarily to the monitoring systems. However, surveys use alone is not recommendable [21].

As stated before, the study of occupant behaviour focused essentially on the occupants impacts on energy consumption and IEQ impacts. Therefore, the sensors measuring energy consumption [22] and the sensors to study the parameters of IEQ are often used [23, 24]. To assess IEQ, three types of parameters are used: physical; chemical and microbiological [25]. However, in residential environments the studies tend to be more simple and focus on carbon dioxide (CO_2), particulate matter (PM) and volatile organic compounds (VOC) or the total volatile organic compounds (TVOC) [26, 27]. These parameters are the most common for the assessment of IAQ [28]. Furthermore, because of the capacity attributed to CO_2 of providing adequate estimations of all other pollutants related to human bio-effluents [29 - 31], CO_2 monitoring is in many cases conducted without any association of other IAQ parameter [32, 33]. Temperature (T) is one of the most used sensors to monitor the indoor environment because it is closely related to the human comfort level [34]. In the same field of studies and, according to Brager and De Dear [35], different levels of accuracy can be achieved in physical measurements linked to comfort parameters. The least demanding category for these studies include at least one sensor of temperature and relative humidity (RH) placed above the floor. Temperature and relative humidity sensors can also be used to detect occupant behaviours [6]. Temperature, relative humidity and CO_2 sensors are also good indicators of occupancy and occupant action [6, 36 - 38]. However, specific sensors to monitor occupancy are becoming more frequent such as passive infrared sensor (PIR), microwave, ultrasound, video cameras, infra-red cameras, and wearable sensors [21, 39 - 41].

Intelligent Buildings

The magnitude of monitoring systems, and consequently, the big data generated by these sensors, imply the existence of building management systems (BMS).

This BMS encourages the use of information and communication technologies (ICT), that is also being driven by the directive 2018/844/EU [1]. Thus, conditions were created to improve the buildings abilities, providing them with the intelligence of perception of the interior and exterior environmental parameters, taking into account the occupants and considering the requirements of indoor environmental quality (IEQ) to optimise the energy consumption (Fig. **2**). To these improvements take place, the BMS must be gifted with artificial intelligence (AI). The definition of AI varies among the research community, and according to Winston [42], artificial intelligence is the computational study that makes perception, reasoning and action possible.

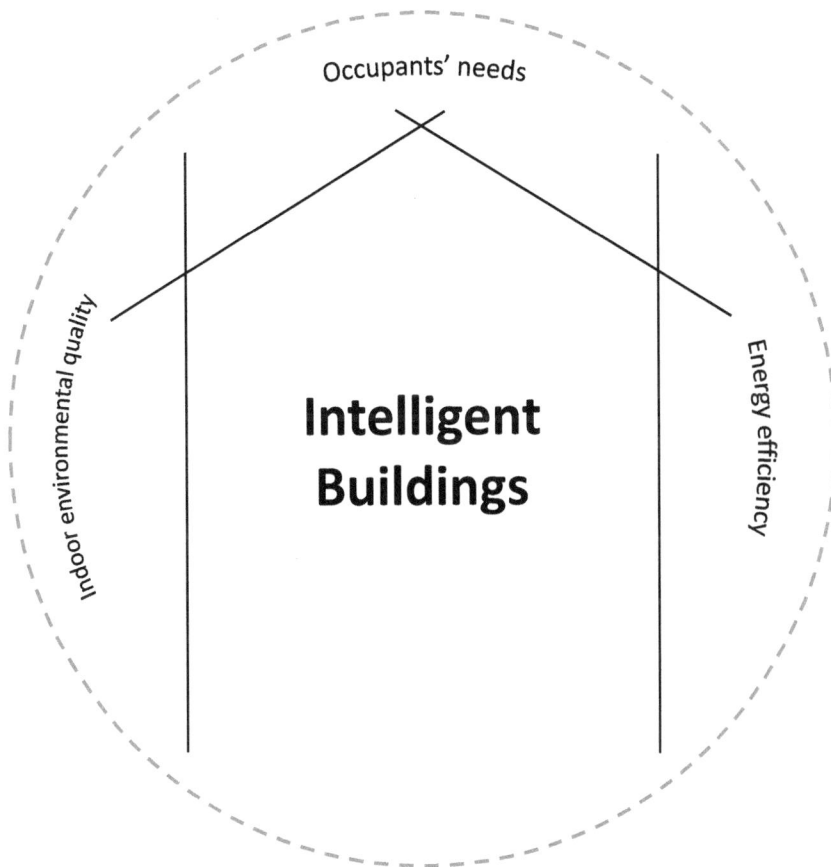

Fig. (2). Intelligent Buildings, focused on occupant behaviour, adapted from [20].

The area of studies of intelligent buildings (IB), although current, is not new. The beginning of discussions on this topic, at least theoretical, is considered to date back to the 1980s. The opening of the first world intelligent building was reported to be in the beginning of 1984 in Connecticut [43]. It was an office building called "Cityplace". The building had a useful area of about 0.12 km². In this first case, the building's intelligence comprised many functions that a single computer could control. This building had a fibre optic network connecting the equipment of the following systems: lighting, fire protection, transportation, security, heating, ventilation, and, above all, telecommunications and electronic office services [44]. A few months later, the concept of the IB evolves to artificial intelligence (AI). This new concept of the future IB described buildings with the capacity of thinking for themselves. On May 13, 1984, the same newspaper wrote an article describing the concept of a new generation of buildings capable of thinking for themselves. This future building would have the capacity of controlling some systems accordingly to the habits learned from the occupants [45]. The definition and field of application of the EI followed this initial vision, and constant evolution is happening. This new kind of building is linked with the ICT which are in constant revolution. Thus, the definition of an intelligent building is not unique and transversal to all research groups. In a review of 2002, the authors Wigginton and Harris [46], found nearly 30 definitions for IB. The evolution of the IB evolves from the simple inclusion of technology in the buildings to the inclusion of AI and the need to interact and fulfil the occupants' requirements Wong, *et al.* [47]. Several researchers are currently criticising the use of technology, *per se*. The IB concept in office and commercial buildings is currently well disseminated; however, its application is only beginning in residential buildings. This difference relies on the fact that office and commercial buildings, in general, are constituted by large open spaces with many occupants/users that have heterogeneous requirements not able to be fulfilled entirely. Thus it is necessary to control the indoor environmental quality centrally. Furthermore, in office and commercial buildings, the occupancy is well defined in time and number. At home, the IB need to fulfil the specificities of the occupants and learn with them. Only the recent technology and big data analysis, and machine learning techniques enable these features.

Summarising

The concept of IB has to be linked with technology and automation in buildings. However, the automation of the IB needs to be driven by the occupants' requirements in the most efficient way. The concept of an IB is also linked with the use of "intelligence" is the other stages of the building, starting with an intelligent design, focused on passive strategies to achieve energy efficiency. Thus, the concept of an IB is associated with the saving of resources without neglecting the IEQ.

Digital Twins

Linked with the thematic of the intelligent buildings, the recent introduction of the fourth revolution (widely known as "Industry 4.0") to the industry of architecture, engineering and construction (AEC), brought the term "digital twin" (DT). According to the scientific literature, the term "digital twin" was used in 2010 by NASA. Digital twins are platforms capable of merging physical, digital and biological worlds (Sebastian *et al.* 2018). The general concept of a DT is the creation of a digital replica of the real building. To that be accomplished, real-time data is used in order to assure the constant update of the real building in a digital way. In AEC industry, this concept is typically connected with Building Information Modelling (BIM), monitoring systems, building simulation, knowledge discovery from databases (KDD), and the Internet of Things (IoT) [48 - 50]. The deployment of the DT can then encompass numerous tools and technologies, and in literature, there are different frameworks for the creations of a DT [51 - 55]. One of the most recent reviews in this research field considers the existence of three technology architectures of the DT: data related, high-fidelity modelling, and model-based simulation [51]. This work considers the former perspective as the base technologies of the DT, the second the core and the latter an important technology (Fig. **3**).

Other comprehensive data-driven approaches were adopted by Blume, *et al.* [55], which uses the Cross Industry Standard Process for Data Mining (CRISP-DM) concept. CRISP-DM approach comprises three main steps: defining requirements; creating the digital twin, and deploying the digital twin (Fig. **4**). The second step comprises probably the most challenging part of the DT and comprises the KDD. This second phase included handling the big databases created by the monitoring systems massively applied in the buildings. There is a need to treat and understand the data, and model them to predict and anticipate future events. In the third phase, DM results are comparatively evaluated, and visual analysis is submitted to the managing entity take actions.

As said before, of the most complex phases of a digital twin comprises the knowledge discovery from databases (KDD). The main base of the KDD is known as data mining (DM), which consists in the analysis of an extensive database, in order to be able to observe non-obvious relationships between the data, thus giving unique, interpretable and potentially useful information. For the task of DM, the knowledge to be extracted can be obtained in two ways: one in which the analyst helps the system to build the model by defining the classes and the examples in each category; another in which the inductive learning of pattern extraction is not supervised, being carried out based on observation and discovery, classes are not defined, having the DM system to observe and recognise the patterns by itself. It is usual to refer to DM as having two types of analysis: forecast and description [56].

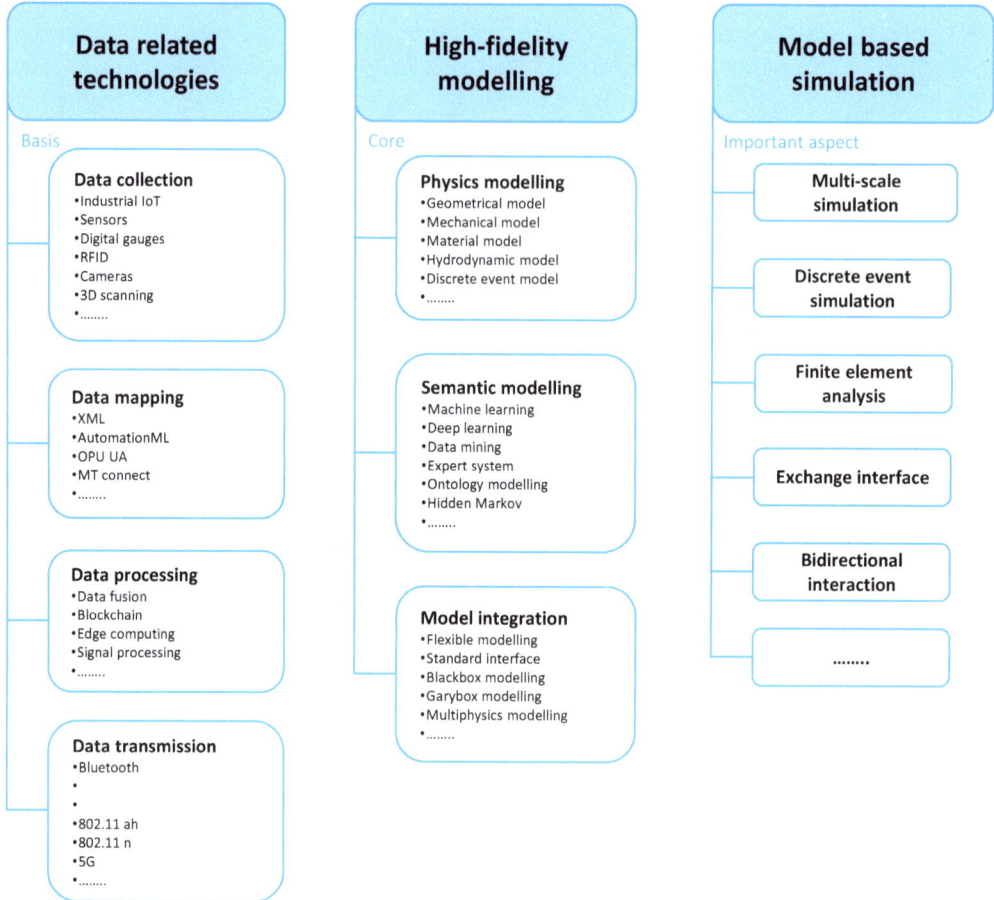

Data related technologies	High-fidelity modelling	Model based simulation
Basis	Core	Important aspect
Data collection •Industrial IoT •Sensors •Digital gauges •RFID •Cameras •3D scanning •……..	**Physics modelling** •Geometrical model •Mechanical model •Material model •Hydrodynamic model •Discrete event model •……..	**Multi-scale simulation**
Data mapping •XML •AutomationML •OPU UA •MT connect •……..	**Semantic modelling** •Machine learning •Deep learning •Data mining •Expert system •Ontology modelling •Hidden Markov •……..	**Discrete event simulation** **Finite element analysis**
Data processing •Data fusion •Blockchain •Edge computing •Signal processing •……..	**Model integration** •Flexible modelling •Standard interface •Blackbox modelling •Garybox modelling •Multiphysics modelling •……..	**Exchange interface** **Bidirectional interaction**
Data transmission •Bluetooth • •802.11 ah •802.11 n •5G •……..		………

Fig. (3). Technology architecture for digital twin, extracted from Liu, *et al.* [51].

The description methods characterise the general properties of the constituent data of the database, pointing out the interesting characteristics, while in the forecasting methods, inferences are made about the available data, as well as predictions about other variables or the behaviour of new data sets. In data mining, machine learning (ML) techniques and artificial intelligence (AI) are commonly used to create numerical models. In the present chapter, the ML and AI models are related to occupant behaviour.

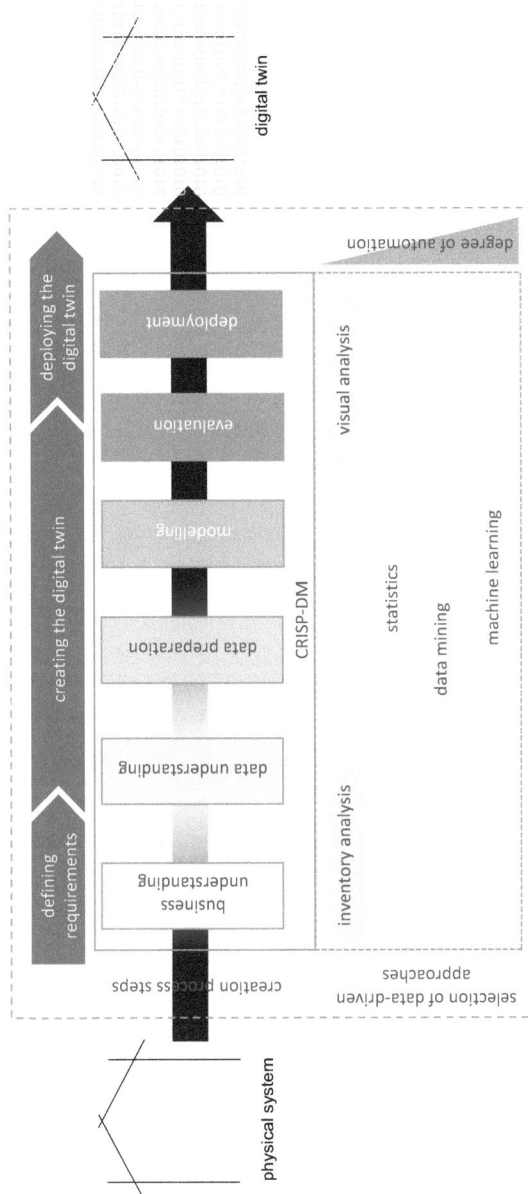

Fig. (4). Workflow based on the CRISP-DM approach extracted from Blume, *et al.* [55].

Machine Learning Techniques Used in Occupant Behaviour

In buildings, machine learning techniques can be used to model occupant behaviour. In the present chapter, it will be used to recognise human activities and drivers of behaviour. In buildings, machine learning techniques start with the recognition of human activities. The techniques can follow different models, including normally: generic algorithms [57 - 59], linear and logistic regressions [20, 60, 61], hidden Markov models (HMM) [37, 62 - 66], and decision trees [67 - 69]. However, machine learning has many techniques to predict occupant behaviour [6, 60, 70 - 77].

The change point analysis (CPA) is a statistical tool that detects abrupt changes in time series. This tool can detect the shifts in time and in number. The applicability of CPA is wide, being applied to many scientific fields such as medicine, geology, meteorology, literature, traffic, finance, genetics, finance, buildings, among others [78 - 81]. The CPA is explained in statistics as a change of a previously followed distribution. According to Chen and Gupta [80], the CPA has two challenges: first, the tools have to prove the hypothesis of having a change in the distribution of the time series; and secondly, to find that change in time, if any change exist. In CPA literature, this is known as the "change point detection problem" [78]. The CPA can be performed under two different circumstances:

- Online – when the time series has not stopped to be filled and are continuously being incremented. The reason why this type of analysis is also known as incremental. In this way, the new data as to be analysed as a potential cause of the abrupt shift from the previously followed distribution. This kind of analysis is frequent in many fields such as signal processing, traffic, quality control and indoor built environment [71, 82, 83],
- Offline – when the time series are not growing anymore, and the entire dataset is at the disposable to be mined in a retrospective way.

The CPA starts with the test of a model that the operator chooses to fit the time series distribution. In CPA, there are different models to be used in the adjustment of the time series. This research field considers the most common models in CPA as being the: linear regression change, variance change, and mean change (or the last two at the same time) [80, 84] (Fig. **5**). However, one of the difficulties of the process is to have a time series distribution with no hypothesis of change. The CPA is also considered too rigid, which makes it not always compatible with the task of finding changes in time series [79]. The best model selected to fit each time series has been approached by some researchers that formulated their methodologies [79, 80, 84]. However, the rules established by these authors are not considered generically applicable because they depend on some factors that are not known prior to the CPA [85]. Therefore, it is the operator's task to choose the model to fit the dataset and calibrate it based on the graphical analysis or comparing the change of the data series with previous changes where there is knowledge of the occurrence of an event [70]. Although these limitations are pointed to the method, Guralnik and Srivastava [78] defend that the CPA tool is an efficient, repeatable, and consistent way to detect abrupt shifts in time series.

The application of the CPA in buildings is not comprehensive, but some studies are already using it in the research field of occupant behaviour. Occupancy was detected by the authors' Szczurek and Maciejewska [86] using time series of CO_2 concentrations, and the authors, Li and Dong [81], used the CPA tool to detect changes in an occupancy presence distribution given by motion sensors. The study of [70] differs from the two above because the authors use the CPA tool to find occupant actions instead of occupancy.

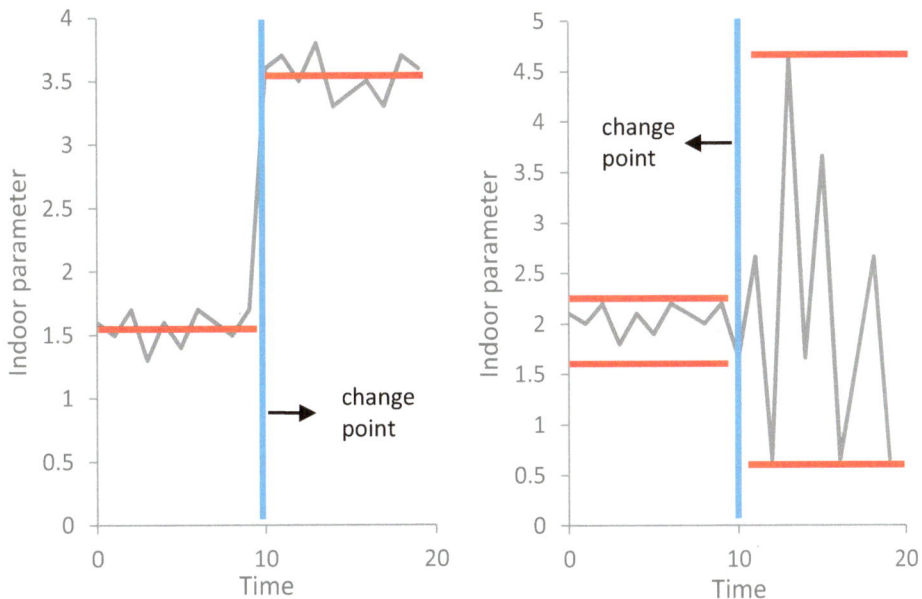

Fig. (5). Examples of a change point in the mean (left) and the variance (right), extracted from [70].

Other techniques could be used to detect abrupt changes in a dataset. In the industry, these techniques are used to predict and anticipate failures based on some abnormal behaviours. These techniques to detect early failure are commonly known as an anomaly, novelty, and outlier detection [75]. Statistical analysis is among the most simple and also effective technique to detect abrupt changes [6]. The analysis is based on the fact that abnormal behaviours trying to be discovered correspond to a data set outliers. Therefore, assuming a Gaussian distribution, this method identifies the values with low probability as outliers. The method of the boxplots stands out to determine the outliers of a data set because of its graphical representation [6]. This method was used by Tukey [86]. In this work, the author found that 99.3% of the data were found within a defined interval. This interval was defined between Q1−1.5 IQR and Q3+1.5 IQR. IQR was set as the interquartile range between quartile 1 (Q1) and quartile 3 Q3 (Q3-Q1). Therefore, the values out this interval (above and bellow) are considered extreme values - outliers.

The self-learning algorithm proposed by Pereira, *et al.* [6] was specially developed to detect occupant actions in buildings. This algorithm is capable of detecting the most appropriated parameter to detect occupant actions in buildings. It relies on the condition that the occupant actions that are set to be detected cause the outliers (extreme values) of the monitored data series. Therefore, the extreme values of that parameter correspond to the exact moment that the occupant performed an action.

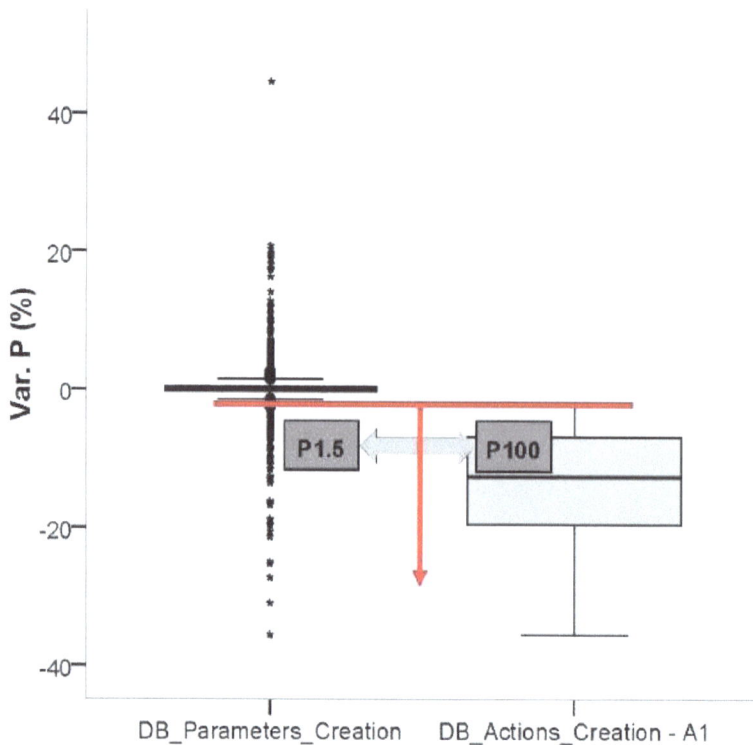

Fig. (6). Box plots representing the methodology do detect actions from data series, extracted from [6].

Fig. (**6**) represents the basis of this methodology graphically. The box plot on the left corresponds to the monitored data relative to the variation of the parameter P (water vapour pressure) between two sequential monitoring instants (10 min). The box plot on the right corresponds to the data obtained after an action being performed by the occupants. As it can be seen in Fig. (**6**), in this case, the maximum value (percentile 100 − P_{100}) of the parameter of the right box plot corresponds to the percentile 1.5 ($P_{1}.5$) of the total monitored dataset. The methodology includes a complex flowchart used to determine the best parameters to detect each occupant actions. This methodology relies on indexes following the confusion matrix to validate the parameters as drivers to detect occupant actions.

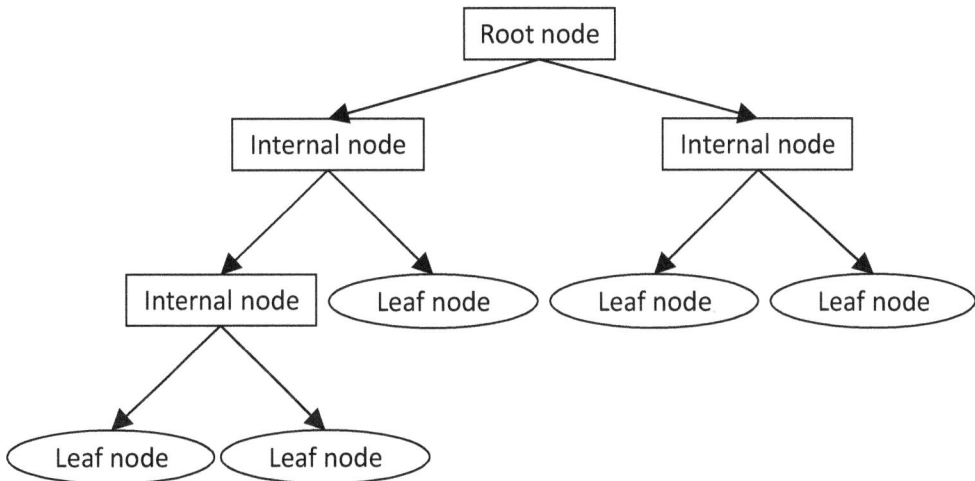

Fig. (7). Simple structure of a decision tree, extracted from [91].

Another tool available in data mining is the decision tree [87]. Decision trees are flowcharts used to classify models graphically in supervised approaches (Fig. **7**). The representation is simple and could metaphorically be compared to a tree [67, 76, 87 - 90]. A decision tree comprises nodes and paths that can be metaphorically described as one root, branches and leaves [88, 89]. Between the branches/paths are internal nodes that correspond to the predictor attributes. The number of branches following a node corresponds to the possible values of the predictor attribute. Therefore, the branches divide the sample in different ways (classifications). The division promoted by the branches from the root and the subsequent internal nodes to the final leaves assures the classification of the sample. Therefore, a decision tree is drawn from the root downwards. The root is the most determinant predictor attribute. Constructing a decision tree model could be divided into a two-step process (Fig. **8**): the learning step; and the classification [89]. A data set is used in the learning process, where the attributes and actions are known [67]. The data set used to train the model could have some incoherencies, and thus, the relevance of each attribute for the resulting action is taken into account. This process is controlled by a training subset and a test subset. Then, the decision tree algorithm generates a decision tree that can classify other data sets. Thus, the goal of a decision tree model is to attribute labels from a data set. In this sense, a decision tree model is useful to predict the value of a target attribute (label attribute), taking into account a data set of input attributes (predictor attribute).

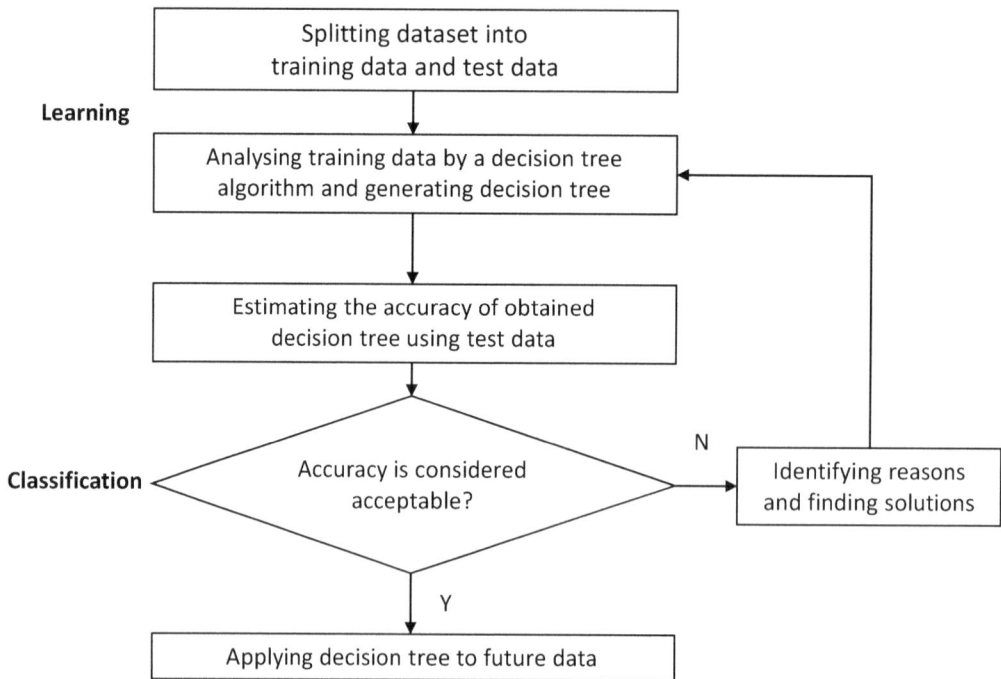

Fig. (8). Procedure of decision tree generation, extracted from [89].

The classification model is based on a machine-learning algorithm. The first two algorithms are known as ID3 and CART. These algorithms were developed independently by different authors and constituted the fundamental basis of further studies in this research field. The ID3 algorithm was further developed into the C4.5 algorithm by the same author [92]. Between the three decision tree algorithms, the C4.5 has two advantages: compared to CART, it works better with continuous attributes, and compared with ID3, it does not have the over-fitted problem of the ID3 algorithm [91]. One of the differences between the C4.5 and ID3 algorithms relies on using the information gain ratio as the standard of attributes selection rather than the information gain used by the ID3 algorithm [91]. The information gain ratio is calculated for the selection of every node. This operation guarantees the selection of the attribute of the data that most effectively splits its set of samples into subsets which occurs with the attribute corresponding to the maximum information gain ratio. According to the literature [91, 92], the draw of a decision tree using the C4.5 algorithm can be divided into four parts:

1 – Calculation of the information carried by the distribution (Entropy of sample P)

$$Entropy(P) = -\sum_{i=1}^{n} p_i \times log_2(p_i) \qquad (5)$$

where p_i is the nonzero probability that an arbitrary attribute in P, and m is the number of classes of the sample.

2 – Calculation of the split Entropy of sample data set P

The amount of information still needed after splitting is given by:

$$Entropy_A(P) = -\sum_{i=1}^{n} \frac{|P_i|}{|P|} \times Entropy(P_i) \qquad (6)$$

where acts as the weight of the ith subset of the sample P, and Entropy (Pi) is the information required to classify the attribute of the sample P based on the partitioning caused by the attribute A.

3 – Obtaining the information gain of attribute A

The information gain of attribute A is determined as a way to test if the selected attribute A can reduce overall Entropy:

$$Gain(A) = Entropy(P) - Entropy_A(P) \qquad (7)$$

In practice, the objective of the algorithm is to split the sample with the attribute (A) that would best classify the sample, and therefore, the amount of information still needed to finish the classification of the sample P is reduced (lower). Thus, the higher better the attribute and lower the entropy.

4 – Calculate the information gain ratio

This step is the main difference between C4.5 and ID3 algorithms. C4.5 introduces a split information value:

$$SplitEntropy_A(P) = -\sum_{i=1}^{n} \frac{|P_i|}{|P|} \times log_2\left(\frac{|P_i|}{|P|}\right) \qquad (8)$$

This value represents the potential information generated by splitting the training data set, P, into n partitions, corresponding to the n outcomes of a test on attribute A.

Then the gain ratio of the attribute A is defined as:

$$GainRatio(A) = \frac{Gain(A)}{SplitEntropy_A(P)} \qquad (9)$$

The finding of each node of the decision tree is based on this procedure. The attribute that maximises the gain ratio is designated to the considered node to split the sample. This procedure's continuous execution divides the sample P into several subsets until no more information is needed to classify the sample.

Literature attributes good accuracy to the classifications of a sample given by a decision tree [92]. However, depending on the dataset, some paths could have faulty values due to noise or outliers of the sample and training data.

In the buildings sector, the decision tree algorithm has been used to: discover occupancy patterns in office spaces [87]; predict residential energy demand [93]; model activities of daily living [69]; analyse and rank the relationships between human factors and environmental conditions, and IEQ satisfaction [94]; analyse the human and spatial factors of the clusters (residential rooms) formed following IEQ attributes [19].

Detecting occupant actions is important in the way that a BMS needs to know the occupants' behaviour, but the main purpose of this knowledge is to predict their actions and needs. It is important to discover the occupants' drivers to behave to anticipate the occupants' impact on energy consumption and IEQ [14 - 16]. The literature pointed to a great diversity of drivers for the same occupants' action [17, 18]. Nevertheless, only a few research works had a small scale approach [10]. Furthermore, if it is intended to have BMS capable of managing the homes accordingly to their occupants' requirements [18]. Therefore, the focus of the studies should be on the specificities of the occupants from each dwelling [10] or room [19, 20].

Some studies used linear and logistic regressions to analyse the relationship between the actions and their drivers [95]. The authors Stazi, *et al*. [95] highlighted the use of logistic regressions when there is more than one potential driver to be correlated with an action. The study developed by Huchuk, *et al*. [60], compared different machine learning techniques used to predict residential buildings occupancy. The results highlighted logistic regressions as being among the simplest and more accurate ML techniques. One of the first authors to use logistic regressions to correlate occupant actions with potential drivers was the author Herkel, *et al*. [96]. The author used the "logit-function" previously developed by Nicol [97] to link the action of windows opening with the outdoor temperature. More recent studies [41] used logistic regression to associate the windows operations with the occupants' drivers to perform that action. The logistic regressions were also used by Park and Choi [98] to model and predict the behaviour of occupants. The predictions were used in the BMS, optimising it [99]. In work performed by Pereira, *et al*. [20], logistic regressions were used to model some occupant action in the residential environment. The results highlighted that one action was correlated to different drivers depending on the season. The same action had different drivers on each room from the studied flat. Therefore, the actions logistic regression model created in one room did not predict with accuracy the same action in a different room from the same flat.

The result of a logistic regression represents the probability of a given event occurs. Therefore, the regressions assume results within the range 0 to 1. One of the advantages of the logistic regressions compared to other ML techniques is the easiness of analysing the results, mainly the extremes (0 and 1) [60]. Another advantage of this ML technique compared to linear regressions relies on how the logistic regressions could be performed on datasets not normally distributed and constituted by binary series [100]. The "logit-function", indicated in equation (10), developed by Nicol [97], can be used to calculate the probability distribution of a particular event.

$$P(A_i) = \frac{e^{\alpha + \beta_1 X_1 + \dots + \beta_n X_n}}{1 + e^{\alpha + \beta_1 X_1 + \dots + \beta_n X_n}} \tag{10}$$

Where $P(A_i)$ is the probability of occurrence of action i (A_i), α is an intercept-related function constant, β is a slope-related function constant, and X_i (i=1,...,n) are the independent variables of the function.

Other techniques can be used to correlate occupant actions and the potential drivers of behaviour. The Pearson correlation coefficient (PCC), also known as Pearson's ρ, is a statistical tool that uses linear correlations between two variables. The PCC range between -1 and 1. The positive part of the range indicates positive linear correlations, and the negative part indicates negative linear correlations. While the use of PCC is discouraged if the variables are categorical, Spearman's rank correlation coefficient (SRCC) can be used. Furthermore, if the dataset does not pass the normality test of Kolmogorov–Smirnov and Shapiro–Wilk, the use of PCC is not recommended, but the SRCC can be used [101]. In statistics, SPCC or Spearman's ρ, is a nonparametric measure of rank correlation. It assesses how well the relationship between two variables can be described using a monotonic function. The values of the Spearman correlation coefficient take the same values as the Pearson correlation coefficient. However, while SPCC evaluates monotonic relationships (that could be linear or not), PCC test linear relationships.

The value of the Spearman coefficient (ρ) is given by equation (11), accordingly to Hollander, *et al.* [102].

$$\rho = 1 - \frac{6 \sum_{i=1}^{n} d_i^2}{n^3 - n} \tag{11}$$

Where: n – is the number of pairs (xi, yi), and di the difference between the posts xi and yi.

While the interpretation of the extreme values of Spearman and Pearson correlation coefficients is easy, the qualitative analysis of the values between those extremes is not so obvious. Therefore, Hinkle *et al.* [103] created a classification of the coefficients that can be found in Table **1**.

Table 1. Quantitative classification of Spearman and Pearson correlation coefficients, adapted from [104].

Correlation Coefficient	Qualitative Classification
± (0.9 to 1.0)	Very high correlation
± (0.7 to 0.9)	High correlation
± (0.5 to 0.7)	Moderate correlation
± (0.3 to 0.5)	Low correlation
0.0 to ± 0.3	Negligible correlation

The correlations coefficients of Pearson and Spearman were used in the built environment mainly to correlate spatial and human factors with IAQ [104, 105], energy consumptions [106, 107], health conditions [108, 109] or productivity (service buildings) [110, 111]. In literature, few studies of occupant behaviour were found using the correlations coefficients of Pearson or Spearman. One of the few studies found correlates the household characteristics and occupants' behaviour with the impact on energy consumption [112]. Another study used the PCC to find the drivers of the occupants' behaviour in a green-certified office building relative to the blind operation and the frequency of those operations [113]. Pereira and Ramos [18] used the Spearman rank correlation coefficient to find the drivers for occupants operation of windows and roller shutters in different compartments of the same flat.

Gap and Objectives

The digital twins (DT) constitute a tool with great potential for application in future buildings as the main platform of the BMS. The perspective is that the DT will be the future IB brain, reducing the energy consumption in buildings and respecting the occupant needs and the thresholds defined by the literature to the IEQ. Machine-learning techniques should be developed and adapted to buildings to endow the DT with AI. The literature has compared machine-learning techniques [60,67,73], but some techniques were not thoroughly studied or need to be compared to others to test their accuracy [79, 114]. In this chapter, three different machine-learning techniques to detect occupant actions in the same dwelling are summarised and compared to each other. The drivers to occupant behaviour is also introduced with two different approaches.

MATERIALS AND METHODS

An *in-situ* data acquisition campaign needs to be prepared in advance to prevent problems from occurring. The indoor built environment research field has identified the most common errors in data acquisition campaigns in inhabited buildings [115]. The authors Hnat *et al.* [115] compiled a list of the main problems encountered during ten different studies, involving about twenty buildings: power supply; wireless network access; availability of systems for home use with a large number of sensors; and various problems linked with the occupants (changes in the layout of the monitoring systems, the collaboration of the occupants concerning surveys, permission to carry out visits and restrictions on the placement of sensors due to aesthetic changes). The authors proposed a series of checks to avoid the successive occurrence of errors [115]:

- Verify the connections between each system/subsystem,
- Verify the connections with the data storage,
- Verify the data extraction at a defined time step,
- Stipulate a minimum communication frequency for all sensors,
- Verify the extreme and mean measured values when extracted,

• Ensure uniformity in the sensor data and time,
• Ensure that there is sufficient storage space.

The main *in-situ* data acquisition campaigns are based on two main types: monitoring systems and interviews or surveys. In general, the use of sensors to measure the occupancy of occupants and their actions in the operation of housing and interactions with the building envelope is considered preferable [116]. The indoor environment parameters and the outdoor parameters from meteorological stations are also used in studies of user behaviour [21].

In the following sections, the main resources used in this work will be presented, namely:

• Case study,
• Monitoring system,
• Surveys carried out,
• Software used.

Case Study

The case study used in this work corresponds to a flat in a residential building located in the northwest of Portugal. This flat is located about 6 km from the Atlantic Ocean (Fig. **9**). The building has five floors above ground, with the study apartment located on the fourth floor. The studied flat has superior and inferiorly similar occupied flats. The flat has glass windows facing north, south and east. The building envelope consists of a double masonry wall with 2 cm of thermal insulation and an airbox between the masonry panels. The facade cladding is made of natural stone, which is part of a ventilated facade. The exterior glazing consists of double frames with roller shutters inside. The building airtightness was measured by means of a Blowerdoor model EU-1000 from Retrotec, and the ACH_{50} is 3.17 h^{-1}. The flat had a centralised mechanical extract ventilation (MEV) with air grills in bathrooms and the kitchen. The air is admitted mainly due to windows permeability because the are no specific air inlets in the façades. The users can manually regulate the air vent grills located in the bathrooms an in the kitchen. The flat general characteristics can be found in Table **2**.

Table 2. Dwellings' general characteristics.

Compartment	Floor Area	Windows Area	Volume	Windows Orientation	Type of Occupancy
	(m^2)	(m^2)	(m^3)	-	-
Dining room (DR)	37.95	11.5	97.53	S	Occasional

(Table 2) cont.....

Compartment	Floor Area	Windows Area	Volume	Windows Orientation	Type of Occupancy
Living Room (LR)	13.94	2.1	35.83	N	Daily
Kitchen (K)	14.95	4.8	36.63	S	Daily
Main Room WC (W1)	4.92	0.5	12.64	E	Daily
WC 2 (W2)	4.96	0.5	12.74	E	Occasional
WC 3 (W3)	-	-	-	-	Occasional
Main Room (R1)	18.27	10.1	46.95	S and E	Daily
Room2 (R2)	14.59	2.1	37.50	N	Rarely

The flat was studied for more than two years in normal occupancy conditions.

Data Acquisition Strategy

In this study, two different methods were used for data acquisition: Monitoring system and surveys. A monitoring system was installed inside the apartment with T, RH and CO_2 sensors and data from a weather station located close to the apartment under study ($\approx$ 400 m) was used. The occupants registered in a daily journal their actions and the occupancy.

Monitoring System

The rooms R1, R2, K, W1, W2, DR, and LR were monitored. Temperature and relative humidity sensors have been distributed to cover transversal and longitudinal profiles of the rooms, according to the recommendations of ISO-7726 [117]. Sensors were placed to avoid abnormal values caused by the direct sunlight within the four seasons. The CO_2 sensors were placed in the breathing zone (0.75 and 1.80 above the floor) of the case study, according to ASHRAE-62.1 [118]. The recommendation of putting the sensors 0.6 m away from walls was not possible in all the cases. The precipitation, solar radiation, wind direction and velocity, temperature and relative humidity were also obtained from a local weather station. The weather station's equipment is from the manufacture Davis, model Vantage Pro 2 Plus, and is located at approximately 400 m from the case study [119]. In the north façade and south façade balconies, T and RH sensors were also placed to produce redundant information. The sensors specifications can be consulted in Table 3. The temperature and relative humidity sensors were programmed to acquire data every 10 minutes, while the CO_2 sensors were programmed to record data every 30 seconds. The monitoring campaign lasted two years.

Fig. (9). Case study, images extracted from Google maps.

Table 3. Main characteristics of used equipment.

Sensor Model	Parameter	Precision	Range	Time Response	Resolution
HOBO UX100	Temperature	±0.21°C – 0°C to 50°C	-20°C a 70°C	4 min – 90%	0.024°C at 25°C
	Relative humidity	±2.5% – 10% to 90%	1% a 95%	11 seg – 90%	0.05% at 25°C
Telaire 7001	CO_2	±50 ppm or 5% measured (the biggest)	0 a 5000ppm (cable adaptation)	1 min – 90%	1 ppm

Surveys

Surveys were used to complement the monitoring campaign. The occupants were requested to fill surveys in a daily base registering the data and time of the following actions/states were registered:

- State of occupancy,
- Window operation,
- Operation of roller shutters,
- Showering,
- Cooking,
- Use of heating.

In the first year, it was decided to ask the occupants to complete a daily filling survey throughout the year. However, it was found that daily journals tend to be the same, and no great changes were found within the year. The reason was that the occupants were displeased with the excessive workload that these inquiries entailed and failed to fill them accurately. Thus, in the second year, it was requested to be filled a full month representative of summer and winter. Between the first year of surveys and the second, there was a spacing of 5 months.

Software Used

To program and download the data from the Monitoring System, the specific programs of the equipment were used.

Microsoft Excel was used as the fundamental base of work for the database, and IBM's SPSS was also used for statistical treatment in addition to Microsoft Excel.

To identify occupant action in section "Detection of Occupant Actions", an R package from the software RStudio was used. The R package used is called "changepoint".

To construct the decision tree of section "Detection of Occupant Actions" the open-source data mining software RapidMiner was used.

Detection of Occupant Actions

In this section, three machine learning techniques applied to the case study will be compared. All the techniques were used to detect occupant action based on a dataset composed of time series. In the first part of this section, a change point analysis (CPA) is described. In the second part, an outlier method is shown, and at last, a decision tree method will be used.

Change Point Analysis (CPA)

In work detailed in [70], the authors proposed a methodology based on the statistical tool CPA to detect occupant action in buildings based on the time series of the indoor environmental parameter. Five steps comprised the methodology. The first step corresponded to the data collection of monitored indoor environmental parameters: T, RH and CO_2. The second step was based on the use of the R package to detect abrupt changes. The third step was to compare the software change points with the real action occurrence based on the surveys and reed switch sensors. This comparison was made in only 8% of the total monitored time. The fourth step comprised the interactive calibration process of the CPA model, in the software, based on the two indexed results. The last step is a refinement criterion that was only applied to the CPA models that were validated in the step before. This criterion reduced the detections of false-trues actions. The software detected the same occupants' action consecutively. Thus, the authors applied a minimum distance between change points in order to the software the same occupants' action has many change points. An example of the result of the application of the methodology can be seen in Fig. (**10**). As can be seen, although there are five false-negatives (red dashed line), only one action was not detected. Four of the false-negatives have in the proximities a false-true. This corresponds to the same action of windows opening that was too early detected (considering the accuracy proposed for the methodology of ± four-time steps that correspond to 2 minutes with the CO_2 parameter) from its register in the survey filled by the occupants. These errors could be made in the surveys and not in the CPA model. The only action not detected by the CPA was the one near midnight on day 22.

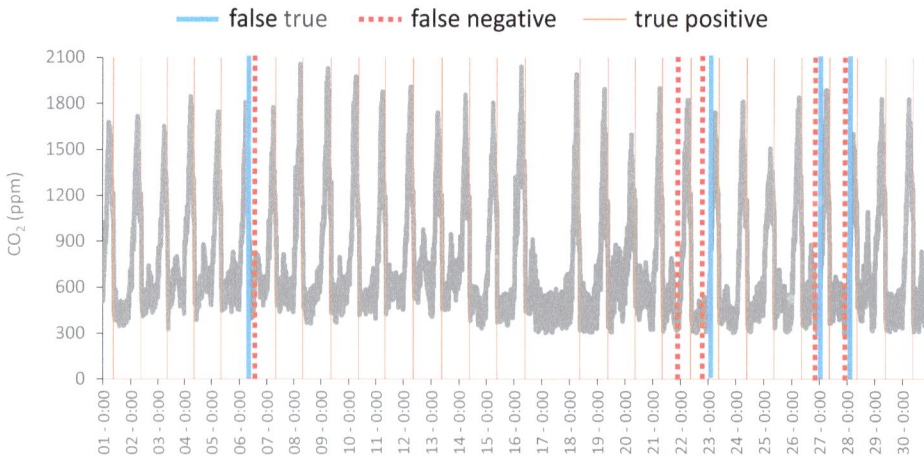

Fig. (10). Result of the CPA methodology to detect windows opening in one month, adapted from [70].

The methodology proposed was applied to detect the action of windows opening in two rooms, showering, cooking and heating. The methodology lead to high accuracies; however, the methodology had two possible could have two types of errors: the false-negatives and the false-positives. The two indexes developed by the authors, index A and index B, detect these errors. Running the CPA software in all the data available (one year), the change point analysis was able to detect the occupant actions of windows opening with an accuracy of 97%, showering with an accuracy of 97%, heating with an accuracy of 100% and cooking with an accuracy of 85%.

Fig. (11). Flowchart of the methodology proposed in [6], extracted from [6].

New Technique Based on the Outlier Detection

In work detailed in [6], the authors developed a methodology to detect occupant actions in buildings based on an adaptation of the standard outlier detection from time series. This method is based on the principle that occupants' actions promote considerable changes in the indoor environmental parameters. These changes correspond to the extreme values of the monitored indoor environmental parameters distribution. Three steps compose the methodology. The first is for the creation of the databases needed for the model from the monitoring data. The second step is the iterative process of running the algorithm created (Fig. **11**). In this step, the parameters that could be triggers for the action to be detected are set (or not if there are no parameters available that could be used to detect, with accuracy, the intended actions). The third step is to run the algorithm to the entire data. The authors Pereira *et al.* [6] applied the method to detect the actions of windows opening in two different rooms of the same flat, showering, cooking and heating. The accuracy of this methodology, calculated accordingly to the confusion matrix [120], is very high (> 99.7%).

Decision Tree

The decision tree was used to discover the parameters that could be predictors of the actions intended to be discovered. In this case, only the actions of windows opening were targeted for detection.

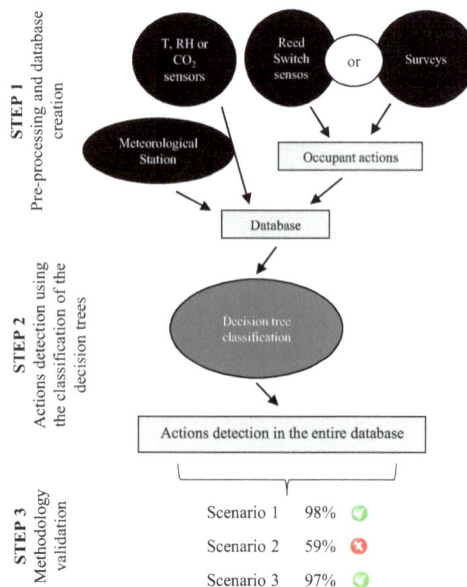

Fig. (12). Methodology used to detect actions using decision trees.

The methodology to have the decision tree is based on three steps (Fig. **12**). The first step comprises the pre-processing operations used to delete the faulty values and create the database. The second step is a very intuitive procedure as the software Rapidminer is creating nodes according to the restrictions placed on the software and the parameters introduced in the database. In the present study, the "gain ratio" criterion was used (C4.5 algorithm). The potential for obtaining reliable results increases with the placement of a low "confidence" and without restricting the number of nodes ("maximal depth" equal to -1). In this step, the model is trained with a few percentages of data were the actions are known. One year of data was used in the present case, and one month (approximately 8%) was used to train the model. The idea is to classify the model based on one month and then apply the decision tree rules to the entire data (one year). The decision tree created for the operation of the window of room R1 and room LR is presented in Figs. (**13** and **14**), respectively. In the third step, the Rapid miner's decision tables were used to classify the accuracy of the model.

For room R1, the accuracy was 87.7%, and for room LR, the accuracy was 96.7%.

Comparison Between Methodologies of Occupant Actions Detection

There are many ways of calculating the accuracy of a model, which difficult the comparison between them. The accuracy calculation accordingly to the confusion matrix reference [120], the true-negatives are considered. In a time series, with data logging intervals of 10 minutes and action occurring once or twice a day, if a methodology does not have too many problems identifying the true-negatives, the accuracy is automatically very high even if no true-positive is found. Because of that, the author of [70] and [6] stated the indexes A and B. Therefore, comparing the mean values of the two indexes for the same action detections from the three methodologies, the CPA had the best performance, followed by the new methodologies adapted from the outlier technique and, at last, the decision tree.. However, the methodology using the decision tree had high accuracies with values of 87% in one room and 97% in another.

Detection of the Drivers of Occupant Behaviour

Spearman's Rank Correlation Coefficient

The work developed in detail in the study [18] developed a methodology to find the motivation behind residential buildings occupants. Spearman's rank correlation coefficient was used to hierarchise the motivations by importance and influence on the occupant actions. The authors applied the methodology in the actions of windows opening (in five rooms from the same flat) and in roller shutters (three rooms of the same flat). The authors found different behaviours between the seasons of the year, and the rank of the motivations was different accordingly to the room and season (Table **4**). The rank was possible to construct accordingly to the absolute value of the SRCC.

Table 4. Ranking the motivations by influence on the occupant windows operation, extracted from [18].

System	Compartment	1st			2nd			3rd		
		S	I	Trans.	S	I	Trans.	S	I	Trans.
Windows	R1	OC	DP	OC	DP	OC	DP	*	*	*
	R2	SR	-	*	DP	-	*	Te	-	*
	LR	T	-	T	DP	-	*	O	-	*
	W1	DP	B/S	B/S	Te	DP	DP	B/S	*	*
	W2	DP	DP	DP	T	B/S	B/S	*	*	*
Solar protections	R1	O	DP	DP	DP	O	O	T	T	T
	R2	DP	DP	DP	*	*	*	*	*	*
	LR	T	DP	DP	DP	T	O	O	O	*
* no Sherman correlation coefficient was found above 0.30										
- this action has not occurred in this period										

Logistic Regression

The work delved deep in [20], used logistic regressions to predict occupant actions based on the motivations correlated with them. The rank of the independent variables considered as drivers of windows opening was made using the absolute value of the β for each logistic equation/windows model (Table **5**). It was also possible to prove that the windows model constructed in one room does not fit well with the windows in other room of the same type in the same flat (Fig. **15**).

Table 5. Ranking the motivations by influence on the occupant windows operation, extracted from [20].

Window room	Drivers			
	1st	2nd	3rd	4th
R1	D	OC	CO_2	-
R2	Te	D	Wv	-
R3	O and D	Te	Ti	Pi
R4	B	D	Te	-
R5	B	O	Te	-

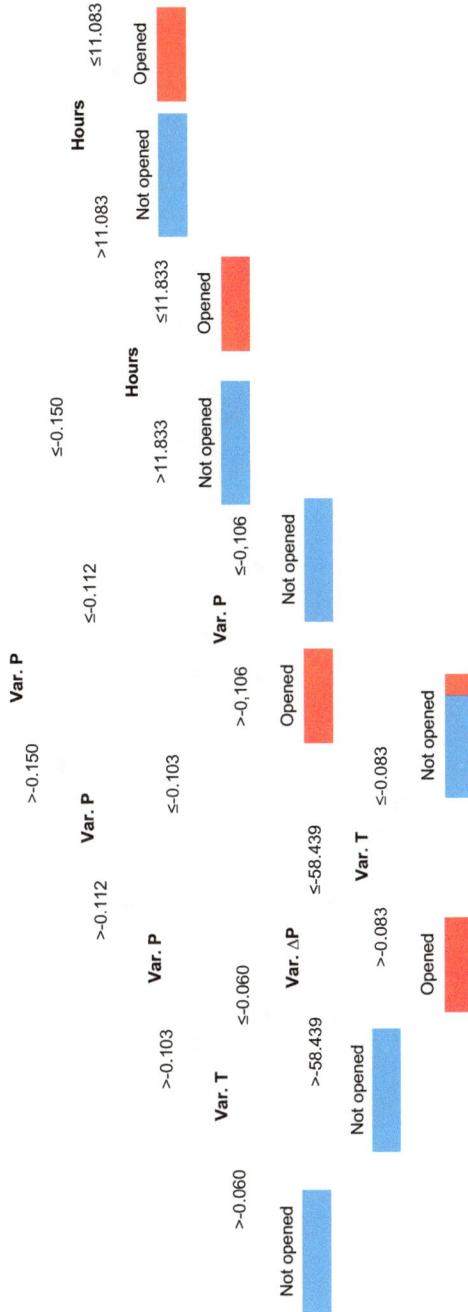

Fig. (13). Decision tree created for the windows opening of the room R1.

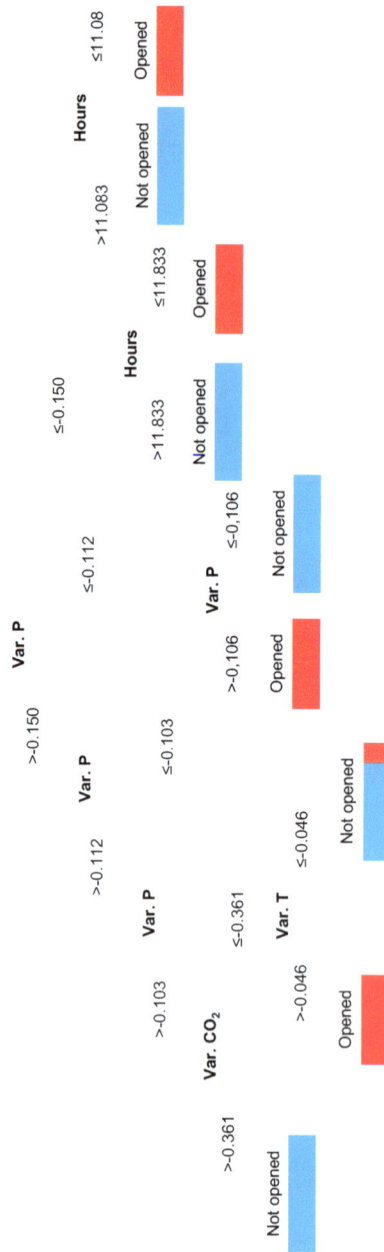

Fig. (14). Decision tree created for the windows opening of the room LR.

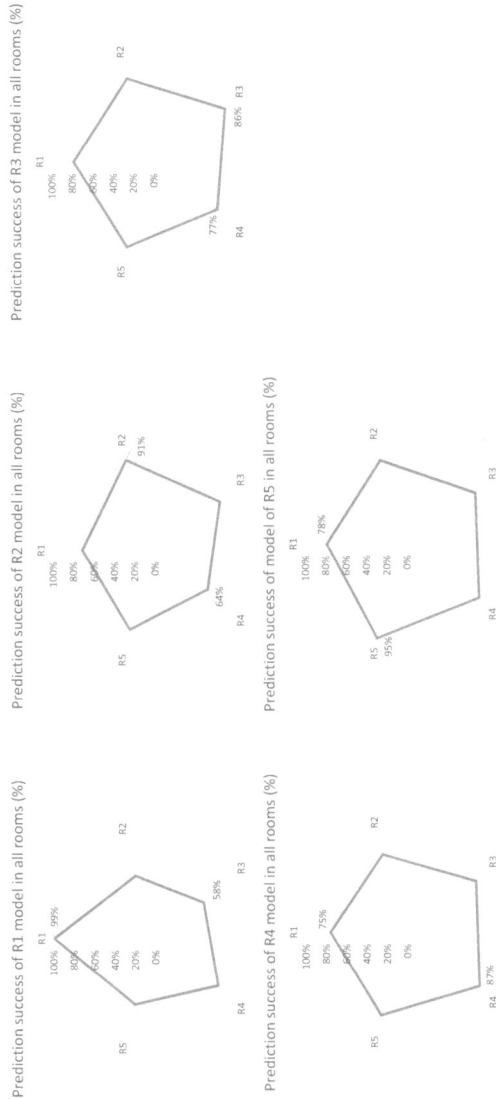

Fig. (15). Decision tree created for the windows opening of the room LR, extracted from [20].

Comparison Between Methodologies of Occupant Actions Drivers Detection

The application of Spearman's rank correlation coefficient and the logistic regression to rang the driver to occupants operation of windows lead to different conclusions. Considering room R1, the drivers of the "day period" and "state of occupancy" were considered in the two techniques but in a changed hierarchy. On R2, the "exterior temperature" and the "day period" was considered but in different orders and with the SRCC, the "solar radiation" was the main predictor, and the logistic regression found the "wind direction" as the third drivers of behaviour. In the R3/LR, the "state of occupancy", the "day period", and the "indoor temperature" were considered as drivers to occupants' windows opening, but the techniques put the drivers in a different order. The same happened in the R4/W1; the same drivers were found but ranked in a different way. In R5/W2, some bigger differences were found, "state of occupancy" and "outdoor temperature" was considered in the logistic regression model but not in the SRCC. The use of the SRCC finds the "day period" and the "indoor temperature" as a driver. The Previous occurrence of a bath was the only coincident predictor.

CONCLUSION

The monitoring system installed and the surveys carried out produce a quantity of data whose treatment requires the use of advanced tools. Surveys are presently still an important way to data acquisition, but their accuracy can be questionable. With the increase in the size of the databases, there is a need to use advanced tools to discover knowledge inside databases. Machine-learning techniques have to be adaptable and used in buildings to improve the knowledge of occupants. The intelligence of the future buildings will have to consider the occupants as the centre of their function. The current trend is to use digital twins to manage the facilities in all the building stages, using BIM as a midterm software or the dashboard. The knowledge of occupant actions and the possibility of predicting them, considering their drivers, is essential to endow a building with real intelligence. With this knowledge, the home management system can inform the occupants or automatically correct a system if actions are required, considering the energy efficiency, the IEQ thresholds and the occupants' patterns mined.

The application of three techniques to detect the same action in the same room leads to the conclusion that the CPA is the best technique. However, the other two methodologies, one based on the decision tree and the other in an adaptation of the outlier technique, also showed high accuracies, above 90%.

The application of two machine-learning techniques to detect occupants behaviour drivers emphasised the need to study and model occupants at a room-scale. One room's logistic models did not fit the same action in another room of the same flat. Furthermore, different methodologies can rank drivers of the same action in a different way. The seasonality of motivations is an important factor. The drivers of occupants' actions vary within the season.

CONSENT FOR PUBLICATION

Not Applicable.

CONFLICT OF INTEREST

The author confirms that this chapter contents have no conflict of interest.

ACKNOWLEDGEMENT

This work was financially supported by Base Funding - UIDB/04708/2020 of the CONSTRUCT - Instituto de I&D em Estruturas e Construções funded by national funds through the Fundação para a Ciência e a Tecnologia FCT/MCTES (PIDDAC).

The authors would like to thank the Municipal Department of Civil Protection, Porto for the outdoor weather data.

REFERENCES

[1]   A. González-Briones, J. Prieto, F. De La Prieta, E. Herrera-Viedma, and J.M. Corchado, "Energy Optimization Using a Case-Based Reasoning Strategy", *Sensors (Basel),* vol. 18, no. 3, p. 865, 2018. [http://dx.doi.org/10.3390/s18030865] [PMID: 29543729]

[2]   A. González-Briones, J. Prieto, F. De La Prieta, E. Herrera-Viedma, and J.M. Corchado, "Energy Optimization Using a Case-Based Reasoning Strategy", *Sensors (Basel),* vol. 18, no. 3, p. 865, 2018. [http://dx.doi.org/10.3390/s18030865] [PMID: 29543729]

[3]   "Energy Efficiency Directive (EEA), 2012",

[4]   E. Maldonado, *Sistemas solares passivos (passive solar systems - in portuguese),* 1985.

[5]   D. Yan, and T. Hong, *EBC Annex 66 - Definition and Simulation of Occupant Behavior in Buildings,* 2014.

[6]   P.F. Pereira, N.M.M. Ramos, R.M.S.F. Almeida, and M.L. Simões, "Methodology for detection of occupant actions in residential buildings using indoor environment monitoring systems", *Building and Environment,* vol. 146, pp. 107-118, 2018. [http://dx.doi.org/10.1016/j.buildenv.2018.09.047]

[7]   D. Yan, "Occupant behavior modeling for building performance simulation: Current state and future challenges", *Energy and Buildings, Article,* vol. 107, pp. 264-278, 2015.

[8]   K. Gram-Hanssen, "Residential heat comfort practices: Understanding users", *Build. Res. Inform,* vol. 38, no. 2, pp. 175-186, 2010. [http://dx.doi.org/10.1080/09613210903541527]

[9]   D. Calì, T. Osterhage, R. Streblow, and D. Müller, "Energy performance gap in refurbished German dwellings: Lesson learned from a field test," Energy and Buildings", *Article,* vol. 127, pp. 1146-1158, 2016. [http://dx.doi.org/10.1016/j.enbuild.2016.05.020]

[10]   S. Gilani, and W. O'Brien, "Review of current methods, opportunities, and challenges for in-situ monitoring to support occupant modelling in office spaces", *Journal of Building Performance Simulation,* vol. 10, no. 5-6, p. 444p. 470, 2017. [http://dx.doi.org/10.1080/19401493.2016.1255258]

[11] S. D'Oca, T. Hong, and J. Langevin, "The human dimensions of energy use in buildings: A review", *Renewable and Sustainable Energy Reviews,* vol. 81, pp. 731-742, 2018.
[http://dx.doi.org/10.1016/j.rser.2017.08.019]

[12] P. Šujanová, M. Rychtáriková, T. Sotto Mayor, and A. Hyder, "A Healthy, Energy-Efficient and Comfortable Indoor Environment, a Review", *Energies,* vol. 12, no. 8, p. 1414, 2019.
[http://dx.doi.org/10.3390/en12081414]

[13] T. Hong, S. D'Oca, W.J.N. Turner, and S.C. Taylor-Lange, ""An ontology to represent energy-related occupant behavior in buildings. Part I: Introduction to the DNAs framework," Building and Environment", *Article,* vol. 92, pp. 764-777, 2015.
[http://dx.doi.org/10.1016/j.buildenv.2015.02.019]

[14] V. Fabi, R.V. Andersen, S. Corgnati, and B.W. Olesen, ""Occupants' window opening behaviour: A literature review of factors influencing occupant behaviour and models," (in English), Building and Environment", *RE:view,* vol. 58, pp. 188-198, 2012.

[15] F. Stazi, F. Naspi, and M. D'Orazio, ""A literature review on driving factors and contextual events influencing occupants' behaviours in buildings," Building and Environment", *RE:view,* vol. 118, pp. 40-66, 2017.

[16] T.A. Nguyen, and M. Aiello, "Energy intelligent buildings based on user activity: A survey", *Energy Build.,* vol. 56, pp. 244-257, 2013.
[http://dx.doi.org/10.1016/j.enbuild.2012.09.005]

[17] P.F. Pereira, N.M.M. Ramos, and J.M.P.Q. Delgado, *Intelligent Residential Buildings and the Behaviour of the Occupants - State of the Art.* 1st ed.. Springer, 2019.

[18] P.F. Pereira, and N.M.M. Ramos, "Occupant behaviour motivations in the residential context – An investigation of variation patterns and seasonality effect", *Building and Environment,* vol. 148, pp. 535-546, 2019.
[http://dx.doi.org/10.1016/j.buildenv.2018.10.053]

[19] P. F. Pereira, N. M. M. Ramos, and A. Ferreira, "Room-scale analysis of spatial and human factors affecting indoor environmental quality in Porto residential flats", *Building and Environment,* vol. 186, p. 107376, 2020.

[20] P.F. Pereira, N.M.M. Ramos, and M.L. Simões, "Data-driven occupant actions prediction to achieve an intelligent building", *Building Research & Information,* vol. 48, no. 5, pp. 485-500, 2020.
[http://dx.doi.org/10.1080/09613218.2019.1692648]

[21] M. Jia, R.S. Srinivasan, and A.A. Raheem, ""From occupancy to occupant behavior: An analytical survey of data acquisition technologies, modeling methodologies and simulation coupling mechanisms for building energy efficiency," Renewable and Sustainable Energy Reviews", *RE:view,* vol. 68, pp. 525-540, 2017.

[22] A. González-Briones, J. Prieto, F. De La Prieta, E. Herrera-Viedma, and J.M. Corchado, "Energy Optimization Using a Case-Based Reasoning Strategy", *Sensors (Basel),* vol. 18, no. 3, p. 865, 2018.
[http://dx.doi.org/10.3390/s18030865] [PMID: 29543729]

[23] G. Marques, and R. Pitarma, "mHealth: Indoor Environmental Quality Measuring System for Enhanced Health and Well-Being Based on Internet of Things", *Journal of Sensor and Actuator Networks,* vol. 8, no. 3, p. 43, 2019.
[http://dx.doi.org/10.3390/jsan8030043]

[24] T. Parkinson, A. Parkinson, and R. de Dear, "Continuous IEQ monitoring system: Context and development", *Building and Environment,* vol. 149, pp. 15-25, 2019.

[25] F. Cappitelli, "Chemical–physical and Microbiological Measurements for Indoor Air Quality Assessment at the Ca' Granda Historical Archive, Milan (Italy)", *Water, Air, and Soil Pollution, journal article,* vol. 201, no. 1, pp. 109-120, 2009.
[http://dx.doi.org/10.1007/s11270-008-9931-5]

[26] J. Kim, T. Hong, M. Lee, and K. Jeong, "Analysing the real-time indoor environmental quality factors considering the influence of the building occupants' behaviors and the ventilation", *Building and Environment,* vol. 156, pp. 99-109, 2019.

[27] R.E. Militello-Hourigan, and S.L. Miller, "The impacts of cooking and an assessment of indoor air quality in Colorado passive and tightly constructed homes", *Building and Environment,* vol. 144, pp. 573-582, 2018.
[http://dx.doi.org/10.1016/j.buildenv.2018.08.044]

[28] L. Cony Renaud Salis, M. Abadie, P. Wargocki, and C. Rode, "Towards the definition of indicators for assessment of indoor air quality and energy performance in low-energy residential buildings", *Energy and Buildings,* vol. 152, pp. 492-502, 2017.
[http://dx.doi.org/10.1016/j.enbuild.2017.07.054]

[29] D. Bienfait, "Guidelines for ventilation requirements in buildings", *Commission of the European Communities, Directorate General for Science, Research and Development, Brussels,* vol. 11, 1992.

[30] A.K. Persily, "The relationship between indoor air quality and carbon dioxide", *Presented at the Proceedings 7th Indoor Air Quality and Climate, Nagoya, Japan,* 1996. 21-26 July, 21-26 July.

[31] W. Ye, X. Zhang, J. Gao, G. Cao, X. Zhou, and X. Su, "Indoor air pollutants, ventilation rate determinants and potential control strategies in Chinese dwellings: A literature review", *Sci. Total Environ.,* vol. 586, pp. 696-729, 2017.
[http://dx.doi.org/10.1016/j.scitotenv.2017.02.047] [PMID: 28215812]

[32] M. Laskari, S. Karatasou, and M. Santamouris, "A methodology for the determination of indoor environmental quality in residential buildings through the monitoring of fundamental environmental parameters: A proposed Dwelling Environmental Quality Index", *Indoor Built Environ.,* vol. 26, no. 6, pp. 813-827, 2017.
[http://dx.doi.org/10.1177/1420326X16660175]

[33] J. Fernández-Agüera, S. Domínguez-Amarillo, C. Alonso, and F. Martín-Consuegra, "Thermal comfort and indoor air quality in low-income housing in Spain: The influence of airtightness and occupant behaviour", *Energy and Buildings,* vol. 199, pp. 102-114, 2019.

[34] P. O. Fanger, *Thermal Comfort: New York.,* 1972.

[35] G.S. Brager, and R.J. De Dear, "Thermal adaptation in the built environment: A literature review", *Energy Build,* vol. 27, no. 1, pp. 83-96, 1998.
[http://dx.doi.org/10.1016/S0378-7788(97)00053-4]

[36] S. Wolf, J.K. Møller, M.A. Bitsch, J. Krogstie, and H. Madsen, "A Markov-Switching model for building occupant activity estimation", *Energy and Buildings,* vol. 183, pp. 672-683, 2019.
[http://dx.doi.org/10.1016/j.enbuild.2018.11.041]

[37] L.M. Candanedo, V. Feldheim, and D. Deramaix, ""A methodology based on Hidden Markov Models for occupancy detection and a case study in a low energy residential building," Energy and Buildings", *Article,* vol. 148, pp. 327-341, 2017.
[http://dx.doi.org/10.1016/j.enbuild.2017.05.031]

[38] F. Haldi, and D. Robinson, "On the behaviour and adaptation of office occupants", *Build. Environ,* vol. 43, no. 12, pp. 2163-2177, 2008.
[http://dx.doi.org/10.1016/j.buildenv.2008.01.003]

[39] T. Weng, and Y. Agarwal, "From buildings to smart buildings-sensing and actuation to improve energy efficiency", *IEEE Design and Test of Computers, Article,* vol. 29, no. 4, pp. 36-44, 2012.
[http://dx.doi.org/10.1109/MDT.2012.2211855]

[40] Z. Yang, B. Becerik-Gerber, N. Li, and M. Orosz, ""A systematic approach to occupancy modeling in ambient sensor-rich buildings," SIMULATION", *Article,* vol. 90, no. 8, pp. 960-977, 2014.

[41] D. Calì, R.K. Andersen, D. Müller, and B.W. Olesen, "Analysis of occupants' behavior related to the

use of windows in German households," (in English), Building and Environment", *Article,* vol. 103, pp. 54-69, 2016.

[42]   P.H. Winston, *Artificial Intelligence,* 1993.

[43]   S.J. Marcus, *The 'Intelligent' Buildings,* 1983 .http://www.nytimes.com/1983/12/01/business/the-intelligent-buildings.html

[44]   A. T.-p. So and W. L. Chan, *Intelligent Building Systems,* Kluwer Academic Publishers, 1999.

[45]   J. Sinopoli, *Smart Building Systems for Architects, Owners, and Builders,* Elsevier Inc, 2010.

[46]   M. Wigginton, and J. Harris, *Intelligent skins,* 2002.

[47]   J.K.W. Wong, H. Li, and S.W. Wang, "Intelligent building research: A review", *Autom. Construct.,* vol. 14, no. 1, pp. 143-159, 2005.
[http://dx.doi.org/10.1016/j.autcon.2004.06.001]

[48]   A. Khalil, S. Stravoravdis, and D. Backes, "Categorisation of building data in the digital documentation of heritage buildings", *Applied Geomatics.*

[49]   J. Ridley, "Connecting the buildings of the past to the cities of the future with digital twins", *50 Forward 50 Back: The Recent History and Essential Future of Sustainable Cities - Proceedings of the CTBUH 10th World Congress,* pp. 221-227, 2019.

[50]   S. Y. Teng, M. Touš, W. D. Leong, B. S. How, H. L. Lam, and V. Máša, "Recent advances on industrial data-driven energy savings: Digital twins and infrastructures", *Renewable and Sustainable Energy Reviews,* vol. 135, p. 110208, 2021.
[http://dx.doi.org/10.1016/j.rser.2020.110208]

[51]   M. Liu, S. Fang, H. Dong, and C. Xu, "Review of digital twin about concepts, technologies, and industrial applications", *Journal of Manufacturing Systems,* 2020.
[http://dx.doi.org/10.1016/j.jmsy.2020.06.017]

[52]   Ž. Turk, and R. Klinc, "A social–product–process framework for construction", *Building Research & Information,* vol. 48, no. 7, pp. 747-762, 2020.

[53]   J. Autiosalo, J. Vepsäläinen, R. Viitala, and K. Tammi, "A Feature-Based Framework for Structuring Industrial Digital Twins", *IEEE Access,* vol. 8, pp. 1193-1208, 2020.
[http://dx.doi.org/10.1109/ACCESS.2019.2950507]

[54]   P. Jouan, and P. Hallot, "Digital Twin: Research Framework to Support Preventive Conservation Policies", *ISPRS Int. J. Geoinf.,* vol. 9, no. 4, p. 228, 2020.
[http://dx.doi.org/10.3390/ijgi9040228]

[55]   C. Blume, S. Blume, S. Thiede, and C. Herrmann, "Data-Driven Digital Twins for Technical Building Services Operation in Factories: A Cooling Tower Case Study", *Journal of Manufacturing and Materials Processing,* vol. 4, no. 4, p. 97, 2020.
[http://dx.doi.org/10.3390/jmmp4040097]

[56]   J. Han, and M. Kamber, *Data Mining: Concepts and Techniques.* The Morgan Kaufmann, 2006.

[57]   U. Rutishauser, J. Joller, and R. Douglas, "Control and learning of ambience by an intelligent building", *IEEE Trans. Syst. Man Cybern. A Syst. Hum.,* vol. 35, no. 1, pp. 121-132, 2005.
[http://dx.doi.org/10.1109/TSMCA.2004.838459]

[58]   D. Kolokotsa, D. Rovas, E. Kosmatopoulos, and K. Kalaitzakis, "A roadmap towards intelligent net zero- and positive-energy buildings", *Sol. Energy,* vol. 85, no. 12, pp. 3067-3084, 2011.
[http://dx.doi.org/10.1016/j.solener.2010.09.001]

[59]   Ó. Pérez, M. Piccardi, J. García, M.A. Patricio, and J.M. Molina, "Comparison between genetic algorithms and the Baum-Welch algorithm in learning HMMs for human activity classification", *Lecture Notes in Computer Science (including subseries Lecture Notes in Artificial Intelligence and Lecture Notes in Bioinformatics),LNCS,* vol. 4448, pp. 399-406, 2007.

[http://dx.doi.org/10.1007/978-3-540-71805-5_44]

[60] B. Huchuk, S. Sanner, and W. O'Brien, "Comparison of machine learning models for occupancy prediction in residential buildings using connected thermostat data", *Building and Environment,* vol. 160, p. 106177, 2019.
[http://dx.doi.org/10.1016/j.buildenv.2019.106177]

[61] F. Haldi, and D. Robinson, "Modelling occupants' personal characteristics for thermal comfort prediction", *Int. J. Biometeorol.,* vol. 55, no. 5, pp. 681-694, 2011.
[http://dx.doi.org/10.1007/s00484-010-0383-4] [PMID: 21347586]

[62] B. Dong, K.P. Lam, and C.P. Neuman, "Integrated building control based on occupant behavior pattern detection and local weather forecasting", *Proceedings of Building Simulation,* pp. 193-200, 2011 .

[63] Z.B. Zhao, W.S. Xu, and D.Z. Cheng, "User behavior detection framework based on NBP for energy efficiency", *Autom. Construct.,* vol. 26, pp. 69-76, 2012.
[http://dx.doi.org/10.1016/j.autcon.2012.04.001]

[64] H.S. Ahmed, B.M. Faouzi, and J. Caelen, "Detection and classification of the behavior of people in an intelligent building by camera", *Int. J. Smart Sensing Intell. Syst.,* vol. 6, no. 4, pp. 1317-1342, 2013.
[http://dx.doi.org/10.21307/ijssis-2017-592]

[65] B. Dong, and K.P. Lam, "A real-time model predictive control for building heating and cooling systems based on the occupancy behavior pattern detection and local weather forecasting", *Build. Simul.,* vol. 7, no. 1, pp. 89-106, 2014.
[http://dx.doi.org/10.1007/s12273-013-0142-7]

[66] J. Liisberg, J.K. Møller, H. Bloem, J. Cipriano, G. Mor, and H. Madsen, "Hidden Markov Models for indirect classification of occupant behaviour," Sustainable Cities and Society", *Article,* vol. 27, pp. 83-98, 2016.

[67] J. De Boeck, K. Verpoorten, K. Luyten, and K. Coninx, "A comparison between decision trees and markov models to support proactive interfaces", *Proceedings - International Workshop on Database and Expert Systems Applications, DEXA,* pp. 94-98, 2007.
[http://dx.doi.org/10.1109/DEXA.2007.94]

[68] K. Bao, F. Allerding, and H. Schmeck, "User behavior prediction for energy management in smart homes", *Proceedings - 2011 8th International Conference on Fuzzy Systems and Knowledge Discovery, FSKD,* vol. 2, pp. 1335-1339, 2011.
[http://dx.doi.org/10.1109/FSKD.2011.6019758]

[69] M. Prossegger, and A. Bouchachia, "Multi-resident activity recognition using incremental decision trees", *Lecture Notes in Computer Science (including subseries Lecture Notes in Artificial Intelligence and Lecture Notes in Bioinformatics), LNAI,* vol. 8779, pp. 182-191, 2014.
[http://dx.doi.org/10.1007/978-3-319-11298-5_19]

[70] P.F. Pereira, and N.M.M. Ramos, "Detection of occupant actions in buildings through change point analysis of in-situ measurements", *Energy and Buildings,* vol. 173, pp. 365-377, 2018.
[http://dx.doi.org/10.1016/j.enbuild.2018.05.050]

[71] S. Aminikhanghahi, and D.J. Cook, "A survey of methods for time series change point detection," Knowledge and Information Systems", *Article,* vol. 51, no. 2, pp. 339-367, 2017.

[72] R. K. Jain, K. M. Smith, P. J. Culligan, and J. E. Taylor, "Forecasting energy consumption of multi-family residential buildings using support vector regression: Investigating the impact of temporal and spatial monitoring granularity on performance accuracy", *Applied Energy,* vol. 123, no. 0, pp. 168-178, 2014.
[http://dx.doi.org/10.1016/j.apenergy.2014.02.057]

[73] Y. Peng, A. Rysanek, Z. Nagy, and A. Schlüter, "Using machine learning techniques for occupancy-prediction-based cooling control in office buildings", *Applied Energy,* vol. 211, pp. 1343-1358, 2018.

[http://dx.doi.org/10.1016/j.apenergy.2017.12.002]

[74] G.I. Webb, M.J. Pazzani, and D. Billsus, "Machine learning for user modeling," User Modeling and User-Adapted Interaction", *Article,* vol. 11, no. 1-2, pp. 19-29, 2001.

[75] R.P. Ribeiro, P. Pereira, and J. Gama, "Sequential anomalies: a study in the Railway Industry", *Machine Learning,* vol. 105, no. 1, pp. 127-153, 2016.

[76] S.H. Ryu, and H.J. Moon, "Development of an occupancy prediction model using indoor environmental data based on machine learning techniques," Building and Environment", *Article,* vol. 107, pp. 1-9, 2016.

[77] M.T. Mulia, S.H. Supangkat, and N. Hariyanto, "A review on building occupancy estimation methods", *2017 International Conference on ICT For Smart Society (ICISS),* pp. 1-7, 2017. [http://dx.doi.org/10.1109/ICTSS.2017.8288878]

[78] V. Guralnik, and J. Srivastava, "Event Detection from Time Series Data", *5th ACM SIGKDD international conference on Knowledge discovery and data mining,* 1999

[79] C. Beaulieu, J. Chen, and J.L. Sarmiento, "Change-point analysis as a tool to detect abrupt climate variations", *Philosophical Transactions of the Royal Society A. Mathematical, Physical and Engineering Sciences, Article,* vol. 370, no. 1962, pp. 1228-1249, 2012. [http://dx.doi.org/10.1098/rsta.2011.0383]

[80] J. Chen, and A. K. Gupta, "Parametric statistical change point analysis: With applications to genetics, medicine, and finance",

[81] Z. Li, and B. Dong, "A new modeling approach for short-term prediction of occupancy in residential buildings," Building and Environment", *Article,* vol. 121, pp. 277-290, 2017.

[82] Y. Mei, "Sequential change-point detection when unknown parameters are present in the pre-change distribution," Annals of Statistics", *Article,* vol. 34, no. 1, pp. 92-122, 2006.

[83] Q. Lu, X. Xie, A. K. Parlikad, and J. M. Schooling, "Digital twin-enabled anomaly detection for built asset monitoring in operation and maintenance", *Automation in Construction,* vol. 118, p. 103277, 2020.

[84] J. Bai, and P. Perron, "Estimating and testing linear models with multiple structural changes," Econometrica", *Article,* vol. 66, no. 1, pp. 47-78, 1998.

[85] R. Killick, and I.A. Eckley, "Changepoint: An R package for changepoint analysis," Journal of Statistical Software", *Article,* vol. 58, no. 3, pp. 1-19, 2014.

[86] J.W. Tukey, *Exploratory Data Analysis,* vol. 2, Addison-Wesley Publishing Company, 1977.

[87] S. D'Oca, and T. Hong, "Occupancy schedules learning process through a data mining framework," Energy and Buildings", *Article,* vol. 88, pp. 395-408, 2015.

[88] Y. Zhao, and Y. Zhang, "Comparison of decision tree methods for finding active objects", *Advances in Space Research,* vol. 41, no. 12, pp. 1955-1959, 2008. [http://dx.doi.org/10.1016/j.asr.2007.07.020]

[89] Z. Yu, F. Haghighat, B.C.M. Fung, and H. Yoshino, "A decision tree method for building energy demand modeling", *Energy and Buildings,* vol. 42, no. 10, pp. 1637-1646, 2010. [http://dx.doi.org/10.1016/j.enbuild.2010.04.006]

[90] X. Ren, D. Yan, and T. Hong, "Data mining of space heating system performance in affordable housing," Building and Environment", *Article,* vol. 89, pp. 1-13, 2015.

[91] X. Meng, P. Zhang, Y. Xu, and H. Xie, "Construction of decision tree based on C4.5 algorithm for online voltage stability assessment", *International Journal of Electrical Power & Energy Systems,* vol. 118, p. 105793, 2020. [http://dx.doi.org/10.1016/j.ijepes.2019.105793]

[92] J. Han, M. Kamber, and J. Pei, 8 - Classification: Basic Concepts.*Data Mining.,* J. Han, M. Kamber, J.

Pei, Eds., 3rd ed. Morgan Kaufmann: Boston, 2012, pp. 327-391.
[http://dx.doi.org/10.1016/B978-0-12-381479-1.00008-3]

[93]   J. Zhao, B. Lasternas, K. P. Lam, R. Yun, and V. Loftness, "Occupant behavior and schedule modeling for building energy simulation through office appliance power consumption data mining", *Energy and Buildings,* vol. 82, no. 0, pp. 341-355, 2014.
[http://dx.doi.org/10.1016/j.enbuild.2014.07.033]

[94]   J-H. Choi, and J. Moon, "Impacts of human and spatial factors on user satisfaction in office environments", *Building and Environment,* vol. 114, pp. 23-35, 2017.
[http://dx.doi.org/10.1016/j.buildenv.2016.12.003]

[95]   F. Stazi, and F. Naspi, "Modelling window status in school classrooms. Results from a case study in Italy", *Building and Environment, Article,* vol. 111, pp. 24-32, 2017.

[96]   S. Herkel, U. Knapp, and J. Pfafferott, "Towards a model of user behaviour regarding the manual control of windows in office buildings", *Building and Environment,,* vol. 43, no. 4, pp. 588-600, 2008.
[http://dx.doi.org/10.1016/j.buildenv.2006.06.031]

[97]   J.F. Nicol, "Characterising occupant behaviour in buildings: Towards a stochastic model of occupant use of windows, lights, blinds, heaters and fans", *Proceedings of the 7th International IBPSA Conference,* no. 7, pp. 1073-1078, 2001.

[98]   J. Park, and C-S. Choi, "Modeling occupant behavior of the manual control of windows in residential buildings", *Indoor Air,* vol. 29, no. 2, pp. 242-251, 2019.
[http://dx.doi.org/10.1111/ina.12522] [PMID: 30468527]

[99]   H. Kim, T. Hong, and J. Kim, "Automatic ventilation control algorithm considering the indoor environmental quality factors and occupant ventilation behavior using a logistic regression model", *Building and Environment,* vol. 153, pp. 46-59, 2019.
[http://dx.doi.org/10.1016/j.buildenv.2019.02.032]

[100] H. B. Gunay, W. O'Brien, and I. Beausoleil-Morrison, "A critical review of observation studies, modeling, and simulation of adaptive occupant behaviors in offices", *Building and Environment,* vol. 70, no. Supplement C, pp. 31-47, 2013.
[http://dx.doi.org/10.1016/j.buildenv.2013.07.020]

[101] A. Lehman, N. O'Rourke, L. Hatcher, and E. Stepanski, *JMP for Basic Univariate and Multivariate Statistics: Methods for Researchers and Social Scientists.* 2nd ed. SAS, 2013.

[102] M. Hollander, D.A. Wolfe, E. Chicken, J.W. Sons, Ed., *Nonparametric Statistical Methods.,* 2013.

[103] D.E. Hinkle, W. Wiersma, and S.G. Jurs, "Applied Statistics for the Behavioral Sciences", *Malawai Medical Journal,* 5th edBoston, 2003.

[104] S. Murtyas, A. Hagishima, and N. H. Kusumaningdyah, "On-site measurement and evaluations of indoor thermal environment in low-cost dwellings of urban Kampung district", *Building and Environment,* vol. 184, p. 107239, 2020.
[http://dx.doi.org/10.1016/j.buildenv.2020.107239]

[105] F. Villanueva, A. Tapia, M. Amo-Salas, A. Notario, B. Cabañas, and E. Martínez, "Levels and sources of volatile organic compounds including carbonyls in indoor air of homes of Puertollano, the most industrialised city in central Iberian Peninsula. Estimation of health risk", *International Journal of Hygiene and Environmental Health,* vol. 218, no. 6, pp. 522-534, 2015.

[106] Y. Ding, and X. Liu, "A comparative analysis of data-driven methods in building energy benchmarking", *Energy and Buildings,* vol. 209, p. 109711, 2020.
[http://dx.doi.org/10.1016/j.enbuild.2019.109711]

[107] C. Won, S. No, and Q. Alhadidi, "Factors Affecting Energy Performance of Large-Scale Office Buildings: Analysis of Benchmarking Data from New York City and Chicago", *Energies,* vol. 12, no. 24, p. 4783, 2019.
[http://dx.doi.org/10.3390/en12244783]

[108] G. R. Ana, A. S. Alli, D. C. Uhiara, and D. G. Shendell, "Indoor air quality and reported health symptoms among hair dressers in salons in Ibadan, Nigeria", *Journal of Chemical Health & Safety,* vol. 26, no. 1, pp. 23-30, 2019.
[http://dx.doi.org/10.1016/j.jchas.2018.09.004]

[109] M.A. Gopang, M. Nebhwani, A. Khatri, and H.B. Marri, "An assessment of occupational health and safety measures and performance of SMEs: An empirical investigation", *Safety Science,* vol. 93, pp. 127-133, 2017.

[110] H. Mallawaarachchi, L. De Silva, and R. Rameezdeen, "Modelling the relationship between green built environment and occupants' productivity", *Facilities,* vol. 35, no. 3/4, pp. 170-187, 2017.
[http://dx.doi.org/10.1108/F-07-2015-0052]

[111] M. A. Hiyassat, M. A. Hiyari, and G. J. Sweis, "Factors affecting construction labour productivity: a case study of Jordan", *International Journal of Construction Management,* vol. 16, no. 2, pp. 138-149, 2016.
[http://dx.doi.org/10.1080/15623599.2016.1142266]

[112] K. Desipri, N.Z. Legaki, and V. Assimakopoulos, "Determinants of domestic electricity consumption and energy behavior: A Greek case study", *IISA 2014, The 5th International Conference on Information, Intelligence, Systems and Applications,* pp. 144-149, 2014.
[http://dx.doi.org/10.1109/IISA.2014.6878760]

[113] S. M. Jubaer Alam, and Z. Shari, "Occupants Interaction with Window Blinds in A Green-Certified Office Building in Putrajaya, Malaysia", *Journal of Design and Built Environment,* vol. 19, no. 1, pp. 60-73, 2019.

[114] T. Ahmad, H. Chen, Y. Guo, and J. Wang, "A comprehensive overview on the data driven and large scale based approaches for forecasting of building energy demand: A review", *Energy and Buildings,* vol. 165, pp. 301-320, 2018.
[http://dx.doi.org/10.1016/j.enbuild.2018.01.017]

[115] T.W. Hnat, "The hitchhiker's guide to successful residential sensing deployments", *Presented at the 9th ACM Conference on Embedded Networked Sensor Systems, Seattle, WA, Conference Paper,* 2011. http://www.scopus.com/inward/record.url?eid=2-s2.0-83455176221&partnerID=40&md5=d534d005d c639f5ae71334bd94efff38
[http://dx.doi.org/10.1145/2070942.2070966]

[116] T. Hong, D. Yan, S. D'Oca, and C.F. Chen, "Ten questions concerning occupant behavior in buildings: The big picture," Building and Environment", *Article,* vol. 114, pp. 518-530, 2017.

[117] "Ergonomics of the thermal environment - Instruments for measuring physical quantities",

[118] *Ventilation for Acceptable Indoor Air Quality,* 2013.

[119] DMPC, *Meteorological Station of Porto,* 2018. https://www.wunderground.com/personal-weathe- -station/dashboard?ID=IPORTOPO9

[120] T. Fawcett, "An introduction to ROC analysis", *Pattern Recognition Letters,* vol. 27, no. 8, pp. 861- 874, 2006.
[http://dx.doi.org/10.1016/j.patrec.2005.10.010]

CHAPTER 6

The (Not So) Close Relationship Between Occupancy and Windows Operation

Aline Schaefer[1,*], **João Vitor Eccel**[1] and **Enedir Ghisi**[1]

[1] *Federal University of Santa Catarina, Energy Efficiency in Buildings Laboratory, Florianópolis, Brazil*

Abstract: Occupancy is one of the main factors to understand the operation and energy consumption of a building due to the variability of human behaviour. However, the variability of human behaviour is not taken into account in many thermal and energy performance studies, causing inconsistencies between simulation results and reality. One of the reasons for these inconsistencies also relies on adopting an opening availability schedule which is strictly limited to the occupancy schedule of a room, especially in residential buildings, which may not represent the reality. Thus, the aim of this study is to investigate the dependency relationship between the room's occupancy schedule and the operation of openings in residential buildings. Data on occupancy of rooms and openings operation were obtained through a database obtained by means of application of questionnaires in low-income houses in Florianópolis, southern Brazil. Descriptive analysis by means of association measures was performed in order to evaluate the level of relationship between occupancy and openings operation. In addition, cluster analysis was performed to identify different patterns of occupant behaviour in the sample. The main result has shown that the opening operation schedule often does not depend on whether the room is occupied or not and seems to rely more accordingly to a daily routine, such as the time one wakes up or goes to sleep, or leaving and coming back home.

Keywords: Association measures, Cluster analysis, Exploratory data analysis, Low-income houses, Occupancy, Occupant pattern behaviour, Occupant profile, Openings operation, Residential building, Representative profile.

INTRODUCTION

Buildings are great contributors to environmental impacts, being energy consumption one of the most significant factors [1]. Globally, 40% of energy consumption is attributed to buildings, and the residential sector accounts for approximately 30% of that [2]. In Europe, energy consumption inresidential

* **Corresponding author Aline Schaefer:** Federal University of Santa Catarina, Energy Efficiency in Buildings Laboratory, Florianópolis, Brazil; Tel: +55(48)99907.5485; E-mail: alineschaefer.au@gmail.com

Enedir Ghisi, Ricardo Forgiarini Rupp and Pedro Fernandes Pereira (Eds.)

buildings represents approximately 26% of the total energy consumption [3], while in the USA it represents 22.5% [4]. In Brazil, according to data from the National Energy Balance (BEN), in 2018, buildings were responsible for approximately 50% of the national energy consumption [5]. Also according to BEN, the energy consumption of residential buildings has relevant participation in this demand, being responsible for approximately 20% of the energy consumption in buildings [5]. Thus, due to the great impact that national buildings have on the demand for energy, alternatives aimed at making them energy efficient can have a very positive impact in reducing the national demand for such a resource.

Alternatives such as the use of more efficient appliances, the adoption of envelopes with high thermal performance and the use of natural lighting may reflect a significant reduction in energy consumption, without jeopardizing the comfort of building users. For example, Palacios-Garcia *et al.* [6] found an effective reduction in energy consumption by the artificial lighting system when exchanging ordinary lamps for LED lamps. Their study concluded that replacing 50% of standard bulbs with LED bulbs would reduce the annual energy consumption of the lighting system by 40%, and that an 80% replacement would reduce consumption by 65%. However, in order to be sure that the buildings will be more efficient after adopting such alternatives, there is a need to perform computer simulations, which make it possible to verify the impact of the strategies even before adopting them.

The performance of simulations can result in important data regarding the use and performance of buildings. However, in order to obtain results that are consistent with reality, it is necessary that the input data also be. The main factors influencing energy consumption in buildings can be classified as: (1) climate, (2) building envelope, (3) building energy services and systems, (4) building operation and maintenance, (5) user activities and behaviour and (6) quality of the internal environment [7]. Such influencing factors should be used in the simulations, and the more consistent with the reality of the building they are, the greater the likelihood that sound conclusions will be obtained from the simulations.

According to IEA [7], the first three factors mentioned above are linked to variables that influence building performance. They are usually calculated from standard values of the other three factors, which are linked to the functioning of the building. When using standardised data, it is assumed that all buildings analysed work within the same standard. This calculation methodology allows a coherent comparison between the energy performance of buildings, but does not allow to conclude about the real energy consumption of buildings. D'Oca and Hong [8] argue that differences between the energy consumption obtained through simulations and the actual energy consumption are the result of assumptions that are not based on practice. Thus, when the objective of the study is to understand the real energy consumption of the building, all the main influencing factors must be analysed, in order to be consistent with the reality of the buildings under study, such as, for example, the occupant behaviour [8].

The identification of occupant behaviour patterns can be based on several aspects that integrate the relationship between occupant and building, such as, for example, the pattern of turning on lamps, using appliances, using curtains, changing clothes, occupancy, windows operation, among others. Occupancy is one of the main points to understand how a building works. Page *et al.* [9] defend the idea that being in the building is the primary condition to be interaction between occupant and building and that, therefore, occupancy is, in a way, the basis for all other models. For example, when the occupants of a building stay in a room more frequently, there is a tendency for the energy consumption in such a room to be higher, since the appliances installed there will be used for a longer time, and the lamps will stay on for longer. In addition, there is a tendency that users will act in favour of maintaining a comfortable temperature, either by activating air-conditioning devices or operating windows. Thus, when using real occupancy data in thermal and energy simulations of buildings, a better understanding about the functioning of the building can be obtained, as well as conclusions consistent with reality.

There are several studies that show the importance of creating building occupancy patterns. Aerts *et al.* [10] identified seven occupancy models for households in Belgium, based on data from a national time use survey, in which 6400 people in 3455 households reported their activities and movements in a diary, starting at 4a.m. and ending at 3:50 in the morning of the next day. For the authors, occupancy patterns are of great value, since they incorporate the great variability of users' behaviours without the need to make complicated simulations. Erickson *et al.* [11] created an occupancy model from data extracted from sensors, in order to use it to program the operation of an air-conditioning system. After integrating the occupancy data with the air-conditioning system, the air-conditioning devices stopped working at maximum power unnecessarily and it was possible to reduce the annual energy consumption in the building by up to 42%.

As for the effects of the operation of the openings, in terms of the energy consumption of Brazilian residential buildings, Sorgato *et al.* [12] concluded that the operation of windows is an important alternative for maintaining thermal comfort in buildings. As Brazil is a country of abundant winds, the operation of windows can improve the thermal comfort of users without the need to use air-conditioning systems. Sorgato *et al.* [12] concluded that the occupant's behaviour in relation to the operation of windows is a factor that needs to be considered when analysing the energy performance of buildings. Wang and Greenberg [13] and Moghadam *et al.* [14], corroborate this conclusion, stating that the user's behaviour in relation to opening and closing windows has a relevant impact on the energy consumption of buildings.

Several studies highlight the dependence between occupancy and the pattern of operation of the windows, that is, they state that the occupancy of a room is a fundamental and mandatory condition for a window to be open. However, in practice, there is a strong tendency to have windows open in a certain room even when it is not occupied, especially in single-family residential buildings. In this sense, this study aimed to analyse the degree of association between occupancy and the pattern of opening windows in the same room in single-family residential buildings. For this, measures of association between data on occupancy and operation of windows collected in low-income housing in southern Brazil were applied. Additionally, a cluster analysis was performed in order to identify how many and which are the different occupant profiles that can be found in the sample used in this study.

Studies on Occupant Behaviour

Bonte *et al.* [15] claim that human behaviour is based on two action groups. The first concerns universal reactions, which occur by human nature. The second group concerns actions motivated by the past and life experiences. Thus, depending on the user, the behaviour can be very different. In the same sense, Papakostas and Stiropoulos [16] state that user's activities in a building can occur due to physiological needs, cultural and social norms, or the combination of these factors. As each user is different, even if they have the same cultural and social norms, their actions in the building are unlikely to be the same.

For Chen *et al.* [17], the lack of understanding about occupant behaviour in buildings is an obstacle to improving energy efficiency. Besides them, Silva and Ghisi [18] state that the uncertainties in the results of simulations arise from the uncertainties related to the input data, such as the number of occupants in a certain room and the occupancy in the building as a whole. Bonte *et al.* [15] maintain that the user behaviour currently used in computer simulation tools is passively modelled, assuming that users do not interfere in occupied rooms. However, in simulations in which users' behaviour is not assumed, standardised behaviours are adopted, which often do not correspond to the user's reality. Examples of this are: assuming that all users will sleep within a certain time interval, that the windows in all rooms are open at the same time, that the use of appliances is the same in all buildings, among others.

Haldi and Robinson [19] used real characteristics of users' behaviour to propose building use modelling techniques, related to opening windows. They analysed the building of the Laboratory of Solar and Physical Energy (LESO), located in Switzerland. Occupancy data was collected using infrared sensors and window operating data using micro switches installed on them. After the data collection period, windows operation models were created based on the building's occupancy pattern, which allowed results to be closer to reality when thermal and energy performance analyses were performed.

Erickson *et al.* [11] created an occupancy model from data extracted from sensors, in order to use it to program the operation of the air-conditioning system in a building. Initially, the air-conditioners did not consider the degree of occupancy of the rooms, and, many times, they operated at maximum power, with few occupants in the rooms. Therefore, the study proposed integration between the air-conditioning system and the model of occupancy in order to reduce energy consumption. To obtain occupancy data, sensors were used at border points (doors and corridors) between the rooms. After collecting occupancy data for five days, occupancy models were created, and the use of them in the simulations resulted in a reduction in energy consumption. The simulations indicated that, using the occupancy models, the air-conditioning system did not work at maximum power unnecessarily and that the reductions in annual energy consumption in the building reached 42% on average, taking into account the ASHRAE comfort limits.

Aerts *et al.* [10] used data from the time use survey and the family budget survey conducted in Belgium in 2005 to identify recurrent occupancy models, as well as to create a probabilistic occupancy model capable of reproducing occupancy sequences in a very similar way to actual occupancy sequences. The time use survey contained 12,800 occupancy sequences. Using the hierarchical data clustering technique, it was possible to identify seven occupancy models.

Andersen *et al.* [20] proposed a method to create patterns of user behaviour in relation to opening and closing windows. In this study, measurements were taken in fifteen households in Denmark for eight months. The houses were separated into four groups according to the type of ventilation (natural or mechanical) and the type of ownership (owner or tenant). For the study, the internal and external environmental conditions were monitored, in order to relate the users' behaviour with the environmental conditions. In addition, sensors were installed in the most frequently used windows to check the pattern of opening and closing them. The authors used logistic regression to calculate the probability of an operating event (opening or closing) of windows. With the equation and measurements (which indicated how the environmental data varied over time), the probability of opening or closing the windows was calculated. With the opening and closing probability data calculated every ten minutes, the model was then created, defining when the windows would be open and when they would be closed. Finally, the authors identified that the behaviour of individuals varied considerably, with different user behaviours leading to differences of up to 300% in energy consumption in identical buildings. Still, the authors proposed four reference user models for use in simulations, stating that the models would have the capacity to significantly increase the validity of the output data.

In another study, Andersen *et al.* [21] compared actual energy consumption results from five flats in Denmark with results from simulations using window operation and heating system operation models for the same five flats. The purpose of the study was to check for discrepancies between the results. For the creation of the model, measurements were made in the internal and external environments, in addition to the application of questionnaires. To perform the simulation, the stochastic models and the collected environmental data were used as input data. In the study, the authors created a stochastic model of window operation using an algorithm. As a result, the correlation between the average of the ten simulations and the average of the measurements in the five flats was weak. However, when comparing the average of the simulations with the flats individually, there were cases of strong correlation between the results. Because of this, it was stated that the models were able to make realistic predictions. In addition, the authors concluded that the fact that the models worked poorly for some flats and reasonably well for others indicated that applying only one model of user behaviour to several buildings is not an effective practice, as it will generate inconsistent results.

Such studies show the need to know the building occupancy data correctly. Reliable data about occupancy in buildings is a strong indication of where activities are most frequently carried out. In addition to collecting real data, it is necessary to adopt specific data analysis techniques, from which behaviour profiles can be determined. Another point observed is that most studies present a user profile model which imposes as a condition for a certain action to happen - such as opening windows - that the room has to be occupied, otherwise the action would not happen. The adoption of this condition, if mistaken, can lead to errors in thermal performance studies. For this reason, this study investigates if there is a relationship between occupancy in a given room and the operation of its openings.

Explanatory Data Analysis

Data analysis does not need necessarily to be complex and difficult to understand. Exploratory data analysis is an example. Promoted by the North American statistician John Tukey, exploratory data analysis refers to the study approach of a data set whose intention is to explore the trends, relationships between variables and patterns that may be implicit in a given data sample. Exploratory analysis techniques assist in the process of extracting relevant information from a data set. As pointed out by Ramos [22], in the exploratory analysis the average and discrepant behaviours are identified, compared and the interdependence between variables is investigated. In addition, exploratory data analysis helps in the process of investigating and identifying trends.

In this study, exploratory data analysis was used to understand the relationships between two variables of interest and the existence of one or more patterns. Thus, two exploratory analysis techniques were used: measures of association between variables and cluster analysis.

Association measures are coefficients used to quantify the degree of similarity between variables. As this study is a sample described by binary variables, the review of association measures for dichotomous variables was prioritized. As presented by Dias *et al.* [23], the use of binary variables for classification purposes had great repercussions in the years between 1960 and 1970, as a result of the evolution in computational calculations, responding to problems as, for example, in disciplines such as taxonomy. The binary variables described the presence or absence of a certain attribute, allowing the application of statistical techniques to assess the degree of proximity between the entities.

Several statistical measures have been developed over the years to allow the assessment of similarity or association between variables or entities. Bussab *et al.* [24], Finch and Huynh [25] and Azambuja [26] present several of them. The calculation used by these measures uses an association matrix of common characteristics (Table **1**). In this matrix, the pairs where there is the presence of a certain attribute in the two variables (represented by the letter "a"), the pairs where there is the presence of one of the attributes in one of the variables, but the absence of the same attribute in the other variable (letters "b" and "c") or absence of a certain attribute in both variables (represented by the letter "d"). The different measures of association reflect on the inclusion or exclusion of common absences in the numerator or denominator, and also on the weight given to the concordances and disagreements. Some of these measures are shown in Table **2**.

Table 1. Matrix of association of common characteristics.

Association Matrix		Variable 1	
		Yes (1)	**No (0)**
Variable 2	Yes (1)	a	b
	No (0)	c	d

Table 2. Association measures for binary variables.

Association Measure	Numerator Criteria	Denominator Criteria	Weights Criteria
Russel and Rao (RR)	Common absences excluded	All combinations included	Equal weight for concordances and disagreements
Simple Agreement (SM)	Common absences included	All combinations included	Equal weight for concordances and disagreements
Sokal and Sneath I (SS1)	Common absences included	All combinations included	Double weight for concordances
Rogers and Tanimoto (RT)	Common absences included	All combinations included	Double weight for disagreements
Jaccard (JACCARD)	Common absences excluded	Common absences excluded	Equal weight for concordances and disagreements
Dice (DICE)	Common absences excluded	Common absences excluded	Double weight for concordances
Sokal and Sneath II (SS2)	Common absences excluded	Common absences excluded	Double weight for disagreements
Kulczynski (K1)	Common absences excluded	All concordances excluded	Equal weight for concordances and disagreements
Sokal and Sneath III (SS3)	Common absences included	All concordances excluded	Equal weight for concordances and disagreements

As for cluster analysis, it is an exploratory, non-theoretical and non-inferential technique, which encompasses a variety of algorithms whose objective is to group similar objects [24]. It is an analytical technique to develop significant subgroups of individuals or objects, classifying a sample of entities (individuals or objects) in a smaller number of mutually exclusive groups, based on the similarities between the entities [27].

Cluster analysis performs an innate task for all human beings and is more present in our lives than we think. The simple fact of organizing a cabinet separating the pieces and the colours in different stacks is already being carried out a work of cluster analysis. However, when this activity requires differentiation into groups based on a large number of characteristics, this process becomes more complex, requiring the application of specific techniques to achieve this result.

Cluster analysis generally involves three steps: (1) determining a measure of similarity or association between objects (individuals), (2) applying some partition technique, which establishes the criteria for forming clusters, and (3) verification of the profile of each group based on the description of the characteristics of each group [28].

The measure of similarity or association is a mathematical criterion that represents the metric difference between two individuals, obtained by calculating the distances between two objects. The most used measure of dissimilarity is the Euclidean distance, also known as the straight line distance, obtained from the square root of the sum of the squares of the difference between each variable of two objects. The City-Block distance, Mahalanobis distance, Minkowski metric and Pearson correlation are other examples of similarity measures widely used. These and other measures of similarity or distance can be seen in Bussab *et al.* [24] and Mingoti [29].

Partition techniques are commonly classified into hierarchical and non-hierarchical and represent a set of procedures used as a criterion to separate groups from the distance between objects (similarity measure). Hierarchical techniques can be agglomerative or divisive and are characterised by the formation of a tree. This property allows the construction of a graph called a dendrogram, which indicates the level of similarity obtained with each new union of two groups. Fig. (**1**) exemplifies the construction of a dendogram. The horizontal axis shows the objects involved in the analysis, while the vertical axis, the level of similarity obtained with each new union. The cut line, also called the stop rule, indicates when the partitioning process will be interrupted, determining the number of clusters formed.

Some partition algorithms most used in hierarchical processes are the Simple Link Method, Complete Link Method, Mean Distance Method, Centroid Method and Ward Method. Each of these methods defines a different way of using the similarity measure to determine the distances between two objects.

For example, while in the Simple Link Method the similarity between two clusters is measured by the two closest elements (diagram "b" in Fig. **2**), in the Complete Link Method it is measured by the most distant objects (diagram "c") and in Average Distance Method this measure is given by the average of the distances between all pairs of elements possible to be formed between two clusters (diagram "a"). Details on these and other algorithms can be found in Bussab *et al.* [24], Mingoti [29] and Kaufman and Rousseeuw [31].

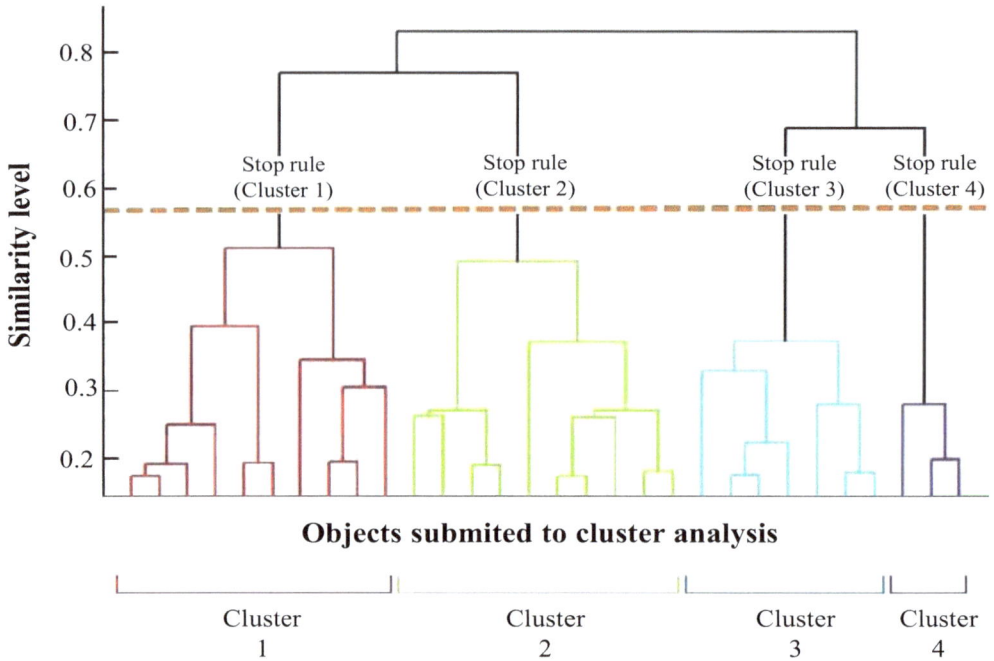

Fig. (1). Example of a dendrogram. Source: Based on Mathworks [30].

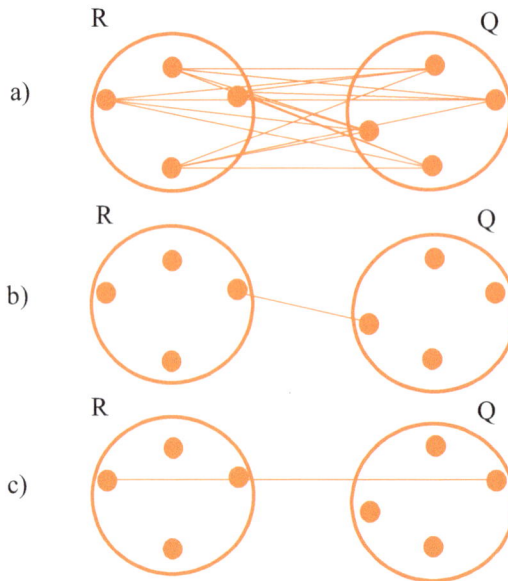

Fig. (2). Representation of some partition algorithms: **(a)** Distance Average Method. **(b)** Simple Connection Method. **(c)** Complete Connection Method. Source: Kaufman and Rousseeuw [31].

The hierarchical techniques are interesting because they allow the visualization of the groupings at each stage through the dendogram, offering a range of solutions from a single simulation. However, in this method, once a grouping is defined, it is never separated in the process. On the contrary, in non-hierarchical partition techniques, there is no tree formation in each cluster and this happens interactively. From seed points, objects are distributed in clusters concurrently and, at the end of the partition, some objects are relocated to other groups, until no object resembles another group than the one in which it is allocated [27]. The partition algorithms used in non-hierarchical methods are the sequential reference, the parallel reference and the optimization procedure (k-means), the latter being the most used, as it allows the reassignment of observations [27].

The last step would then be to describe the characteristics of each group. According to Bussab *et al.* [24], "the result of a cluster analysis must be a set of groups that can be consistently described through their characteristics, attributes and other properties".

The cluster analysis was applied in this study to identify different user profiles, presenting the particular characteristics of each group.

METHOD

The method applied in this study was carried out in three steps: (1) obtaining the data, (2) verifying the existence of an association between the variables occupancy and operation of openings in the same room and (3) identification of existing patterns in the sample. For the first stage, an existing database was used on the occupancy and operation routines in a sample of low-income housing. The data were filtered and treated, from which descriptive analyses were applied, such as contingency tables and association measures, in order to verify the level of association between the variables, from different mathematical measures (second step). Finally, in the third step, a database containing the association values found in the previous step was subjected to cluster analysis, which allowed the identification of different patterns of behaviour regarding the occupancy and operation of windows in the sample.

The following sections describe each step in more detail.

Obtaining the Data

The first step of this method deals with obtaining the data. The data used in this study were obtained from an existing database, created during the research project "Rational Use of Water and Energy Efficiency in Housing of Social Interest" [32], carried out in Southern Brazil. Within the scope of this project, four questionnaires were applied to each dwelling, referring to information such as socioeconomic characteristics of the residents (Questionnaire 1), geometry and materials and construction systems (Questionnaire 2), building operation (Questionnaire 3, on the use of equipment, lighting, occupancy of rooms and opening of doors and windows) and energy consumption (Questionnaire 4, with surveys of installed equipment, energy monitoring and energy bills.

The object of study of the aforementioned research project was the low-income housing located in Greater Florianópolis, southern Brazil, which met at least one of the following criteria:

- Have a family income equal to or less than three minimum wages (reference April / 2012: minimum wage = R$ 622.00);
- Owning a house financed by *"Minha Casa Minha Vida"* programme or another public housing programme for low-income families;
- To be inserted in an area destined to social interest, located within the limits of Greater Florianópolis (includes the municipalities of Biguaçu, São José and Palhoça), since family income does not overtake the amount of R$ 5,000.00 (based on the maximum limit accepted by the *Minha Casa Minha Vida* Programme in April / 2012).

In this study, data obtained from Questionnaires 1 (socioeconomic data) and 3 (operation) were used, with 107 questionnaires answered.

The main data collected in Questionnaire 1 were: number of residents, family structure (couple or adult, with or without children), family income and income per capita and the form of acquisition of the house (rented or owned, private or government resources). The questionnaire was filled in by the researcher according to the information provided by the resident regarding the socioeconomic status of the family. Such data were collected to characterise the user of the object of study of this research, that is, the characteristics of the residents of the low-income dwellings under study.

The operation data (Questionnaire 3) shows the way users operate the house and its systems. Therefore, they refer to the pattern of use of equipment and lamps, the pattern of occupancy of the rooms and the pattern of opening and closing doors and windows. This questionnaire was completed by the researcher according to the information provided by the resident responsible for the house about the family's routines within the dwelling. It is based on the hourly record of the use of equipment and lamps, the occupancy of rooms and the opening of doors and windows. The questionnaire is composed of a table that divides the day into 24 hours. For each hour, the reference value of the action was recorded.

Regarding the pattern of occupancy of rooms, the existing rooms were noted in each row (only the long-stay rooms were considered, such as living room, dining room and bedrooms). Next to each room, the maximum number of residents that occupy that room was identified, and then, how many people occupy that room at each hour of the day. Finally, in relation to the pattern of operation of doors and windows, it was noted, for each opening, the room to which it belonged and the hours when it remained open or closed.

The operation data used for the present study address information about occupancy and window operation routines. The data contained in the questionnaires were organized in an electronic spreadsheet and recorded on an hourly basis, representing each hour of the day, resulting in 24 intervals. Data on the occupancy of the rooms were noted as binary data, where 0 represented the unoccupied room and 1, the occupied room. As for the operation of the openings, 0 referred to the closed window, while 1 referred to the open window. Only information related to the living room, main bedroom and secondary bedroom settings for weekdays were selected.

Descriptive Data Analysis

In this step, it was verified if there was a relationship between the occupancy of a room and the operation of its openings. In other words, we tried to verify whether the opening or closing of the windows at a certain time is subject to the presence of a resident in the room and the corresponding time. To make this verification, methods of descriptive analysis were adopted to objectively assess the degree of association between these two variables.

After organising the data in an electronic spreadsheet, descriptive analysis began with the classification of the pairs of correspondence between the variables of a given room and time of the day, assuming the values "a" for the occurrence of the two attributes (occupied room and open window), "b" for occurrence of occupancy only (occupied room, but window closed), "c" for occurrence of operation only (unoccupied room, but window open) and "d" for no occurrence (unoccupied room and closed windows). These data were computed in a matrix that served as the basis for the descriptive analysis.

Initially, the descriptive analysis was performed using visual methods, such as frequency histograms and bar graphs representing the percentage of occurrence of each combination (a, b, c or d). Contingency tables were also used, in order to numerically observe the relationship of occurrence in each situation. This type of analysis makes it possible to clearly and visually identify which type of action or behaviour is most recurrent in the sample for each time, in each room, throughout the day, providing beforehand an initial perception of the existing relationships between occupancy and operation of windows.

In a second step, measures of association between the variables occupancy and operation of windows were adopted in each dwelling. Association measures are coefficients that were used to quantify the degree of correspondence between the variables. There are several coefficients or measures of association, each presenting a different way of measuring the degree of association between two variables. These measures use the criterion of presence or absence of a certain attribute, so that the data matrix formed in the previous step was used (classified as "a", "b", "c" or "d" for each interval of hour). The measures differ mainly in terms of the inclusion or exclusion of mutual absences (none of the characteristics present) in the numerator and/or denominator and the weight of mutual presence (characteristics of the two variables present) in relation to the other pairs. Due to the use of binary variables in this study, preference was given to the use of association measures for dichotomous variables. Eight measures of association were used in this study. Table **3** presents such measures, their formulas and criteria.

Table 3. Binary association measures and their formulas.

Measures	Formulas	Criteria
Russel and Rao (RR)	$\dfrac{a}{(a + b + c + d)}$	Mutual absences excluded from the numerator and included in the denominator
Simple Agreement (SM)	$\dfrac{(a + d)}{(a + b + c + d)}$	Mutual absences included in the numerator and in the denominator
Sokal and Sneath (SS1)	$\dfrac{2(a + d)}{[2(a + d) + b + c]}$	Mutual absences included in the numerator and in the denominator, double weight for concordances
Rogers and Tanimoto (RT)	$\dfrac{(a + d)}{[a + 2(b + c) + d]}$	Mutual absences included in the numerator and in the denominator, double weight for disagreements
Jaccard (JACCARD)	$\dfrac{a}{(a + b + c)}$	Mutual absences excluded from the numerator and denominator
Dice (DICE)	$\dfrac{2a}{(2a + b + c)}$	Double weight for mutual presence

(Table 3) cont.....

Measures	Formulas	Criteria
Sokal and Sneath 2 (SS2)	$\dfrac{a}{[a + 2(b + c)]}$	Mutual absences excluded from the numerator and in the denominator, double weight for disagreements
Soreson (SR)	$\dfrac{2a}{(2a + b + c)}$	Mutual absences excluded from the numerator and in the denominator, double weight for mutual presence

As a logical matter, observations with higher coefficients indicate a greater association between the variables, while lower coefficient values indicate a lower degree of association (except for the coefficients that do not consider agreements in the numerator – such as the Sokal Binary Distance – representing, therefore, the reverse: the higher the coefficient, the higher the degree of association).

The importance of using different measures to assess the association between variables relies in the different perceptions that one can have from the coefficients obtained. For example, measures that consider only the mutual presence in the nominator signal those in which the user is in the room and performs the action of opening the window. Those that do not include mutual absences are also interesting because, although the occurrence of no occupancy combined with closed opening represents an agreement, this measure, by disregarding situations in which the occupant is not in the room, represents only situations where there are possibility of occupant intervention (he/she was in the room and kept the window open, he/she was in the room and chose to keep the window closed, he/she was not in the room, but somehow another user took the action to open the window).

After applying the association measures for dichotomous variables, the data were summarized and submitted to cluster analysis, using special statistical similarity measures for data of this type.

Cluster Analysis Application

Cluster analysis step was carried out with two objectives: (1) identification of different patterns of behaviour in the sample, together with a description of their characteristics and (2) selection of an occupant profile for each group, which may be applied in computational simulation.

The technique adopted for the cluster analysis was based on the proposal by Hair *et al.* [27] and Schaefer and Ghisi [33], in which hierarchical and non-hierarchical clustering techniques were combined. Hierarchical techniques were used to define the number of clusters to be formed, while non-hierarchical techniques were used for the final formation of clusters.

For the application of the hierarchical cluster analysis technique, it was necessary to define two aspects: the similarity measure and the partition algorithm. As already shown in the introduction, the similarity measure represents a mathematical measure that calculates the degree of similarity between the elements, while the partition algorithm defines the rules for how groups will be formed from the distance between objects. In this study, the square Euclidean distance was defined as the measure of similarity, while the Ward Method was defined as the partition algorithm to be used.

The square Euclidean distance is defined by the sum of the squares of the differences between each variable, of all pairs of objects. The average Euclidean distance was adopted, as it has two interesting properties: with this measure, it is possible to consider cases even with the absence of some coordinates (missing values), in addition to allowing the accumulation of empirical evidence on the levels of similarity [24]. This measure is calculated for each pair of objects, thus obtaining a new data matrix, called the similarity matrix. The square Euclidean distance is calculated according to eq. (1).

$$d_{AB} = \sum_{i=1}^{p} (x_{i_A} - x_{i_B})^2 \tag{1}$$

where d_{AB} is the square Euclidean distance from object A to object B, is the value of A for each variable, and is the value of B for each variable.

For the partition step, Ward's method was selected, which defines as the best solution the combination that minimizes the residual increase in squares over all variables, in all clusters.

After having defined the distance measure and the partition algorithm, the hierarchical cluster technique was applied. In this type of partition, the objects were combined according to the distance value obtained with the selected similarity measure and the partition rules of Ward's algorithm. The objects that had the shortest distance from each other were the first to be grouped.

At each stage of the grouping process, two different objects (whether singular or groupings formed in any of the previous stages) were combined to form a new cluster. For each new grouping, the value of the distance between the grouped objects was calculated, indicating the similarity between the objects in the cluster. The shorter the distance, the greater the similarity.

The use of the hierarchical method of grouping allowed the construction of a dendrogram, which is a tree-shaped graph in which the levels of similarity obtained for each new cluster formation are observed.

After the construction of the dendrogram and with the aid of the distances between the clusters formed, it was possible to determine the preliminary number of distinct clusters present in the evaluated sample. The determination of the number of distinct clusters can be carried out from the evaluation of the evolution of the distances between each cluster at each new union, since a very large distance indicated the union of two not very similar clusters. Then, it was possible to find the "cutting line" (refers to the distance found at the point from which the distances between the new grouped objects became much greater than the previous distances).

When determining the ideal number of clusters, the non-hierarchical procedure was applied, using the k-means algorithm. In this stage, first the seed points were defined, which are reference points where the clustering process begins. They represent the initial centroid of the cluster, which changes as new objects are joined, and were defined using the sampling method (random selection). Then, the sample objects were distributed simultaneously in the clusters. As new objects were grouped, the centroid of the cluster was displaced, causing some objects previously assigned to a certain cluster to become part of another cluster. The distribution of objects was repeated as many times as necessary until no more objects were assigned to a new cluster due to the change in its centre, thus achieving convergence. Using this process, the final clusters were formed.

In order to determine a representative occupant profile of each cluster, a comparison was made between the values of the elements within the cluster and the centroid of the cluster, with the representative case being the closest one to the centroid of its cluster. For comparison with the centroid, for each variable, the module of the difference between the mean value and the value of the element was computed. After computing these differences, a sum of the values was made. The element with the smallest sum was adopted as a representative case. For the case in which there were only two elements within a cluster, the representative case was chosen as the one with the greatest distance from the other clusters centres, which results in a greater distance between occupant profiles.

Descriptive analysis resources were used to present the characterisation of each cluster, such as bar graphs and boxplot. The clusters were described based on their variables of occupancy and operation of openings, in order to explain the most frequent behaviour of that cluster. In a complementary way, the socioeconomic patterns of the clusters were also presented, allowing to verify if there are trends between the different social groups.

Occupant profiles were described in 24-hour routines, both for occupancy and for operation of the openings, in each room. Such routines can be inserted in computer simulation programmes, representing different patterns of occupant behaviour. It should be noted, however, that these routines should be used only as a base (and not exclusive) behaviour, and should be integrated in the simulation programme with other influencing factors regarding the issue of occupant behaviour (such as, for example, environmental conditions). Other studies should be consulted to determine the limits imposed by the other conditions.

RESULTS

In this section, the main results regarding the occupant behaviour are shown. First, the relationship between the variables occupancy and window operation were investigated. Then, different occupant profiles were identified in the sample and presented. The subsections below describe the main results obtained for each of these objectives:

Descriptive Analysis

The descriptive analysis aimed to investigate if there is (or not) of a relationship between the occupancy of a room and the operation of its openings.

Figs. (**3-5**) show the number of occurrences of occupancy and windows that were open in each hour of the day for the living room, bedroom 1 and bedroom 2, respectively. The red line represents the number of cases of which a room was occupied in each hour. The blue bars represent the number of cases that the window of a room was open, for each hour. It is possible to notice a certain incompatibility between the occupied hours and those when the window was open, especially for bedrooms (Figs. **4** and **5**). In the living room, the number of occurrences of occupancy and opening of the window (Fig. **3**) shows a similar behaviour, increasing during daytime hours and decreasing at night. Even so, it is possible to identify a higher occurrence of open windows than occupied room, in the period between 7 a.m. and 6 p.m.. In the following hours, there is a reduction in the number of open windows and an increase in occupancy, the latter also decreasing after 8 p.m.. In the period between midnight and dawn, there are times with the highest correspondence between occupancy and open windows, probably due to the fact that it corresponds to the period that people sleep and, therefore, occupy the other rooms.

As for the bedrooms (Figs. **4** and **5**), the inverse relationship between the variables occupancy and window opening is clear. While in the night period the occurrence of occupancy is high and the windows tend to remain closed, as the day begins, the occupancy rate starts to decrease and the windows start to be opened, keeping it that way throughout the day. At night (around 8 p.m.), the relationship starts to reverse again.

This simple analysis allows us to observe that the opening of windows does not seem to be conditioned to the occupancy of an environment, but rather, more related to a daily routine (such as getting out of bed and going back to sleep).

It is important to highlight that such conclusions have no statistical significance, being only a description of the sample raised. Even so, the relationships between the variables presented above can be clearly observed.

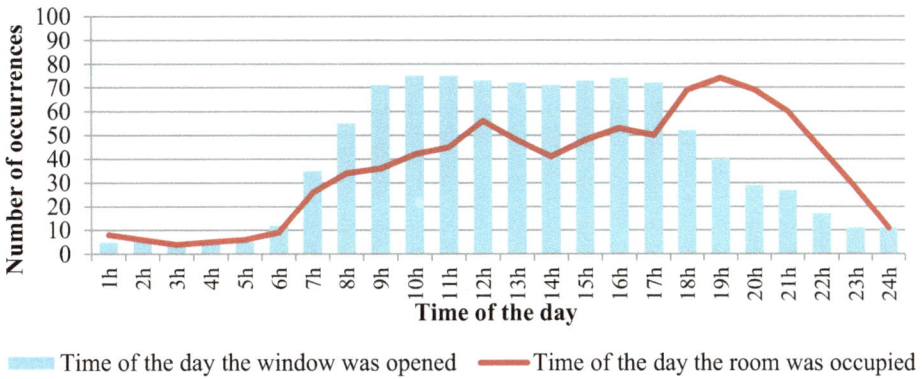

Fig. (3). Number of occurrences of occupancy and windows that were open in each hour of the day in the living room.

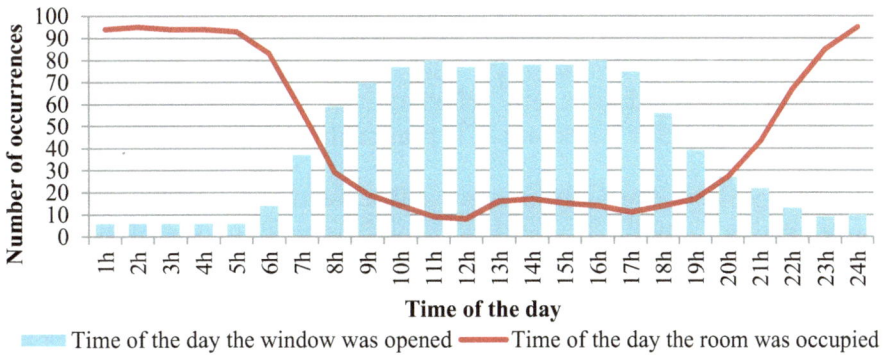

Fig. (4). Number of occurrences of occupancy and windows that were open in each hour of the day in bedroom 1.

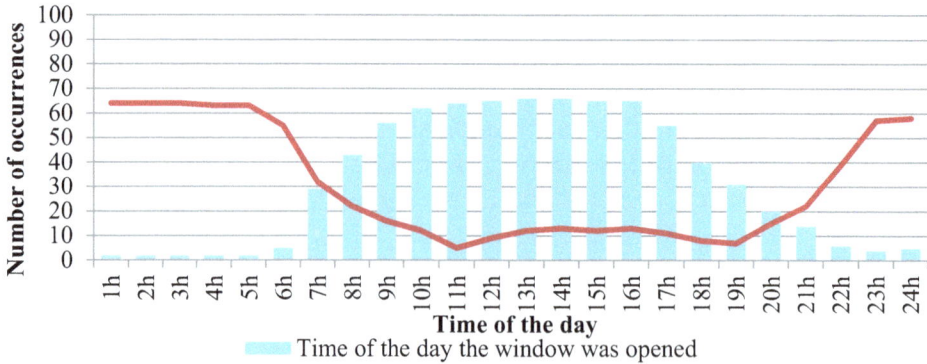

Fig. (5). Number of occurrences of occupancy and windows that were open in each hour of the day in bedroom 2.

Figs. (**6-8**) show the percentage of occurrence for the classification of the correspondence pairs between the variables of a given room and time of the day: the occurrence of the two attributes (occupied room and open window), the occurrence of occupancy only (occupied room, but window closed), occurrence of operation only (unoccupied room, but window open) and no occurrence (unoccupied room and closed windows). One can see that the most common correspondence of occupancy and window operation occurs at night in the living room, when the room is not occupied and windows are closed. For all rooms, it is possible to notice that the percentage of occurrences is higher for those that do not refer to a concordance between the variables. The percentage of occurrence of occupancy and open window is only relevant in the living room in the period from 8 a.m. to 8 p.m., while for bedrooms it is practically irrelevant all day long.

Tables **4** and **5** show the Contingency Table for the variables occupancy and window operation. Table **4** displays the values for midnight to 12 a.m. and Table **5** displays the values for 1 p.m. to midnight, for all rooms. The Contingency Table shows the percentage of occurrence of each concordance and disagreement in the sample. The total lines correspond to the sum of occurrences of the same attribute. For example, for the living room, from midnight to 1 p.m., in 88.4% of the sample there is no occupancy and window is close, and in only 6.3% of cases there is occupancy in the room and the window is open, corresponding to 94.7% of cases in which the window is closed. The darker the green tone, the smaller the percentage of occurrence. The values presented in the contingency table also show a tendency of a non-correspondence between the occupancy and window operation (higher values for no-yes pairs, for most of the day).

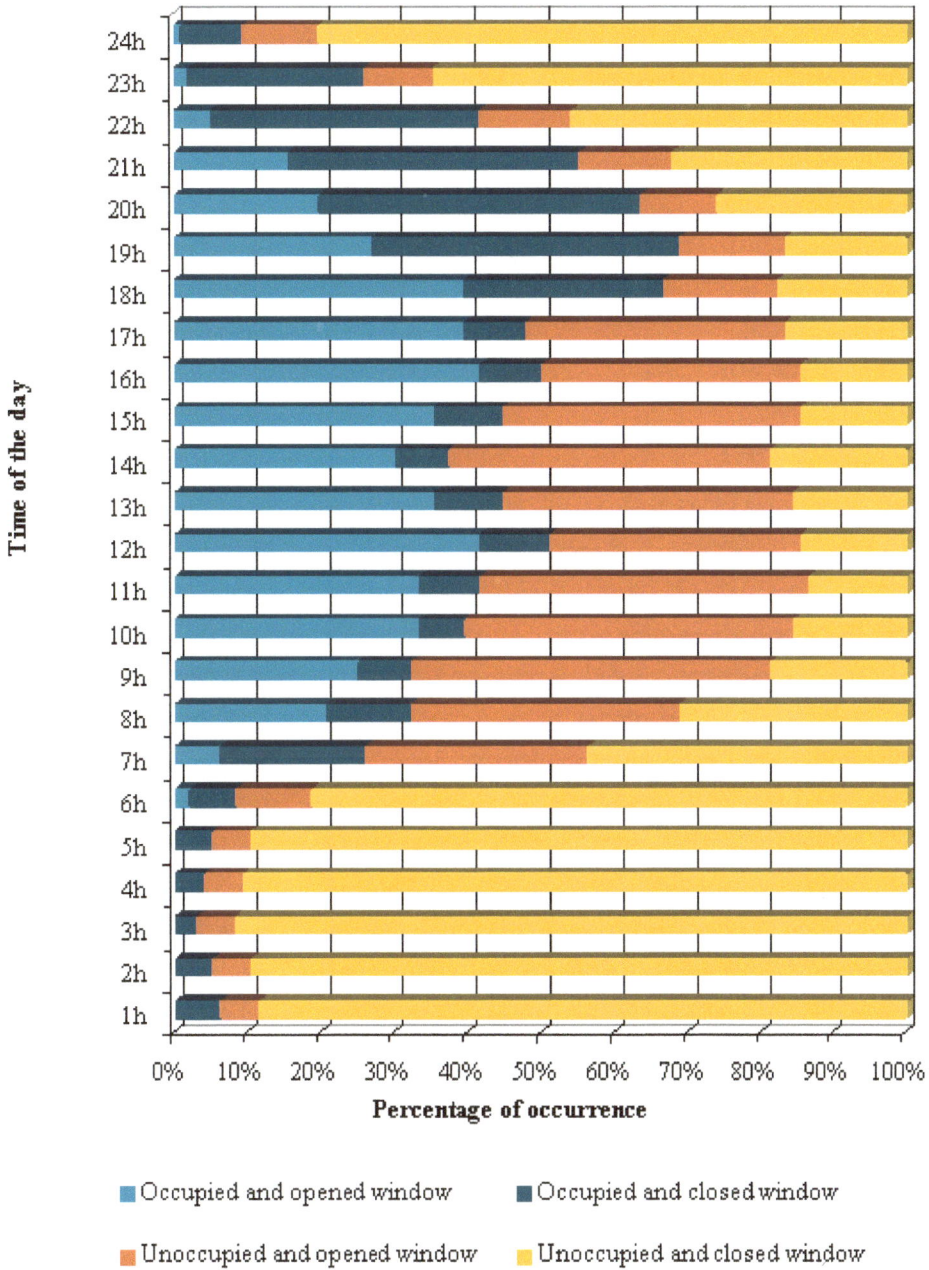

Fig. (6). Percentage of occurrences of correspondences between occupancy and open windows in each hour of the day in the living room.

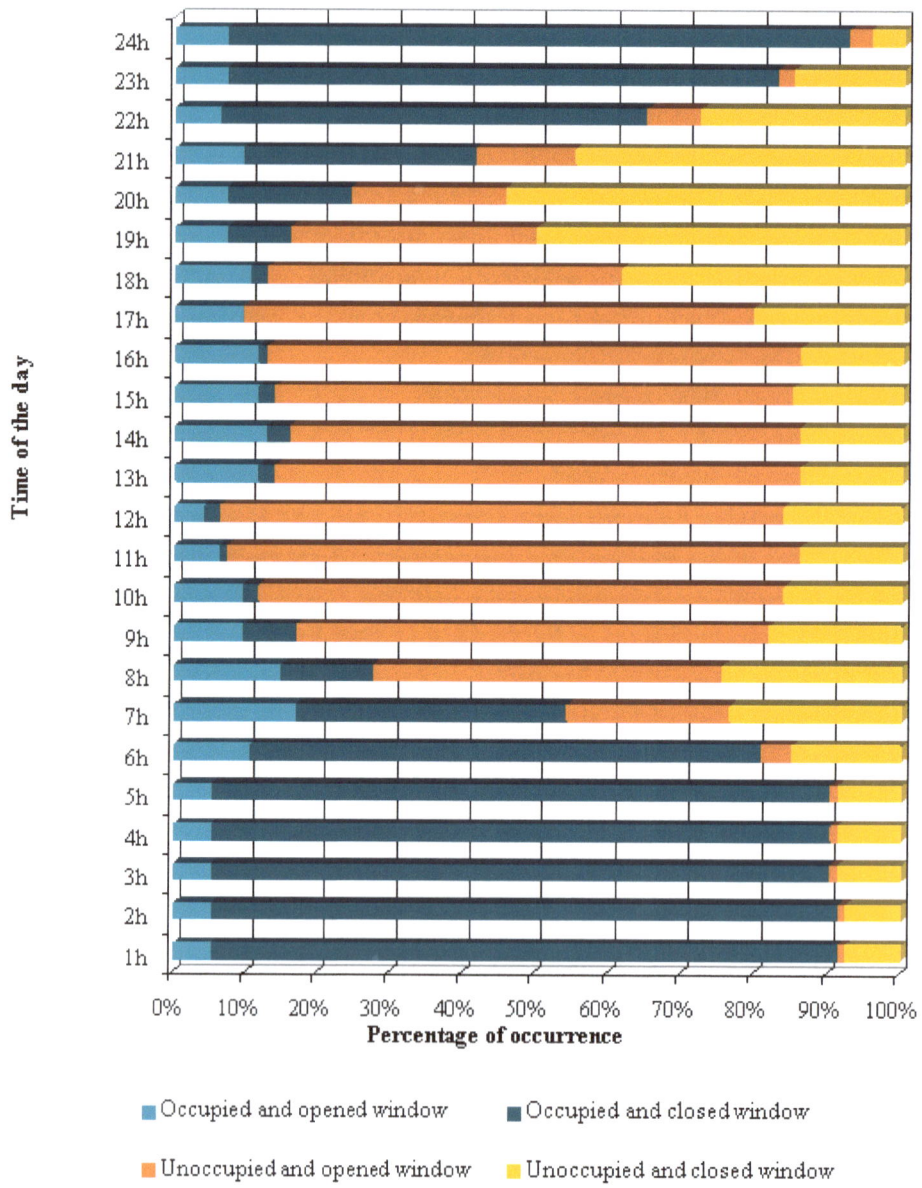

Fig. (7). Percentage of occurrences of correspondences between occupancy and open windows in each hour of the day in bedroom 1.

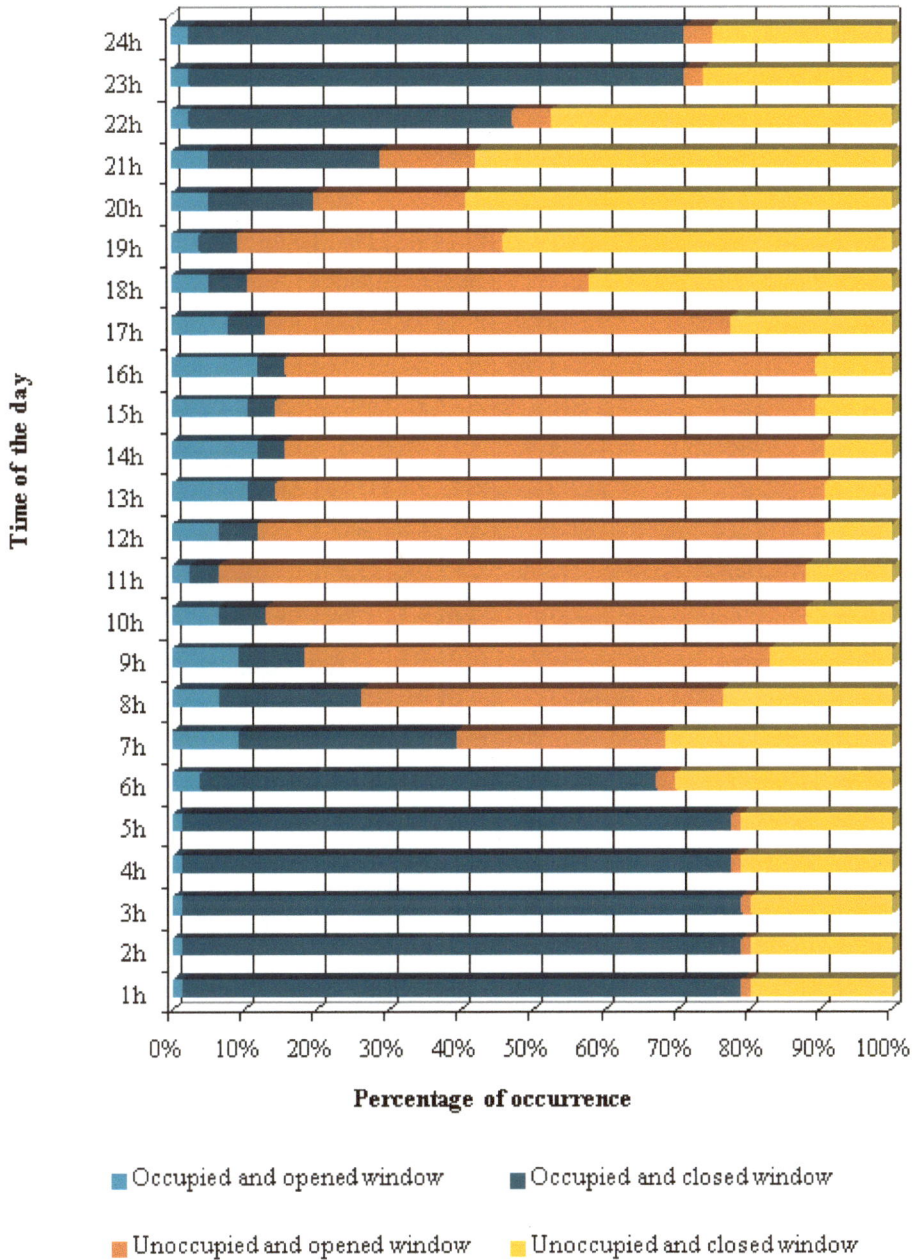

Fig. (8). Percentage of occurrences of correspondences between occupancy and open windows in each hour of the day in bedroom 2.

Table 4. Contingency Table for the variables occupancy and window operation for the midnight to 12 a.m. period.

Time			Living Room Occupancy (%)			Bedroom #1 Occupancy (%)			Bedroom #2 Occupancy (%)		
			No	Yes	**Total**	No	Yes	**Total**	No	Yes	**Total**
1a.m.	Window (%)	No	88.4	6.3	**94.7**	7.5	86.2	**93.6**	19.7	77.6	**97.4**
		Yes	5.3	0.0	**5.3**	1.1	5.3	**6.4**	1.3	1.3	**2.6**
		Total	**93.7**	**6.3**	**100.0**	**8.5**	**91.5**	**100.0**	**21.1**	**79.0**	**100.0**
2a.m.	Window (%)	No	89.5	5.3	**94.7**	7.5	86.0	**93.6**	19.7	77.6	**97.4**
		Yes	5.3	0.0	**5.3**	1.1	5.4	**6.5**	1.3	1.3	**2.6**
		Total	**94.7**	**5.3**	**100.0**	**8.6**	**91.4**	**100.0**	**21.1**	**79.0**	**100.0**
3a.m.	Window (%)	No	91.6	3.2	**94.7**	8.5	85.1	**93.6**	19.7	77.6	**97.4**
		Yes	5.3	0.0	**5.3**	1.1	5.3	**6.4**	1.3	1.3	**2.6**
		Total	**96.8**	**3.2**	**100.0**	**9.6**	**90.4**	**100.0**	**21.1**	**79.0**	**100.0**
4a.m.	Window (%)	No	90.5	4.2	**94.7**	8.5	85.1	**93.6**	21.1	76.3	**97.4**
		Yes	5.3	0.0	**5.3**	1.1	5.3	**6.4**	1.3	1.3	**2.6**
		Total	**95.8**	**4.2**	**100.0**	**9.6**	**90.4**	**100.0**	**22.4**	**77.6**	**100.0**
5a.m.	Window (%)	No	89.5	5.3	**94.7**	8.5	85.1	**93.6**	21.1	76.3	**97.4**
		Yes	5.3	0.0	**5.3**	1.1	5.3	**6.4**	1.3	1.3	**2.6**
		Total	**94.7**	**5.3**	**100.0**	**9.6**	**90.4**	**100.0**	**22.4**	**77.6**	**100.0**
6a.m.	Window (%)	No	81.1	6.3	**87.4**	14.9	70.2	**85.1**	30.3	63.2	**93.4**
		Yes	10.5	2.1	**12.6**	4.3	10.6	**14.9**	2.6	4.0	**6.6**
		Total	**91.6**	**8.4**	**100.0**	**19.2**	**80.9**	**100.0**	**32.9**	**67.1**	**100.0**
7a.m.	Window (%)	No	44.2	20.0	**64.2**	23.4	37.2	**60.6**	31.6	30.3	**61.8**
		Yes	29.5	6.3	**35.8**	22.3	17.0	**39.4**	29.0	9.2	**38.2**
		Total	**73.7**	**26.3**	**100.0**	**45.7**	**54.3**	**100.0**	**60.5**	**39.5**	**100.0**
8a.m.	Window (%)	No	31.6	11.6	**43.2**	24.5	12.8	**37.2**	23.7	19.7	**43.4**
		Yes	36.8	20.0	**56.8**	47.9	14.9	**62.8**	50.0	6.6	**56.6**
		Total	**68.4**	**31.6**	**100.0**	**72.3**	**27.7**	**100.0**	**73.7**	**26.3**	**100.0**
9a.m.	Window (%)	No	19.2	7.5	**26.6**	18.1	7.5	**25.5**	17.1	9.2	**26.3**
		Yes	48.9	24.5	**73.4**	64.9	9.6	**74.5**	64.5	9.2	**73.7**
		Total	**68.1**	**31.9**	**100.0**	**83.0**	**17.0**	**100.0**	**81.6**	**18.4**	**100.0**
10a.m.	Window (%)	No	15.8	6.3	**22.1**	16.0	2.1	**18.1**	11.8	6.6	**18.4**
		Yes	45.3	32.6	**77.9**	72.3	9.6	**81.9**	75.0	6.6	**81.6**
		Total	**61.1**	**39.0**	**100.0**	**88.3**	**11.7**	**100.0**	**86.8**	**13.2**	**100.0**

(Table 4) cont.....

11a.m.	Window (%)	No	13.7	8.4	**22.1**	13.8	1.1	**14.9**	11.8	4.0	**15.8**
		Yes	45.3	32.6	**77.9**	78.7	6.4	**85.1**	81.6	2.6	**84.2**
		Total	**59.0**	**41.1**	**100.0**	**92.6**	**7.5**	**100.0**	**93.4**	**6.6**	**100.0**
12a.m.	Window (%)	No	14.7	8.4	**23.2**	16.0	2.1	**18.1**	9.2	5.3	**14.5**
		Yes	34.7	42.1	**76.8**	77.7	4.3	**81.9**	79.0	6.6	**85.5**
		Total	**49.5**	**50.5**	**100.0**	**93.6**	**6.4**	**100.0**	**88.2**	**11.8**	**100.0**

Table 5. Contingency Table for the variables occupancy and window operation for the 1 p.m. – 12 p.m. period.

Time			Living Room Occupancy (%)			Bedroom #1 Occupancy (%)			Bedroom #2 Occupancy (%)		
			No	Yes	Total	No	Yes	Total	No	Yes	Total
1p.m.	Window (%)	No	15.8	8.4	**24.2**	13.8	2.1	**16.0**	9.2	4.0	**13.2**
		Yes	40.0	35.8	**75.8**	72.3	11.7	**84.0**	76.3	10.5	**86.8**
		Total	**55.8**	**44.2**	**100.0**	**86.2**	**13.8**	**100.0**	**85.5**	**14.5**	**100.0**
2p.m.	Window (%)	No	19.0	6.3	**25.3**	13.8	3.2	**17.0**	9.2	4.0	**13.2**
		Yes	44.2	30.5	**74.7**	70.2	12.8	**83.0**	75.0	11.8	**86.8**
		Total	**63.2**	**36.8**	**100.0**	**84.0**	**16.0**	**100.0**	**84.2**	**15.8**	**100.0**
3p.m.	Window (%)	No	13.8	8.5	**22.3**	14.9	2.1	**17.0**	10.5	4.0	**14.5**
		Yes	41.5	36.2	**77.7**	71.3	11.7	**83.0**	75.0	10.5	**85.5**
		Total	**55.3**	**44.7**	**100.0**	**86.2**	**13.8**	**100.0**	**85.5**	**14.5**	**100.0**
4p.m.	Window (%)	No	14.7	7.4	**22.1**	13.8	1.1	**14.9**	10.5	4.0	**14.5**
		Yes	35.8	42.1	**77.9**	73.4	11.7	**85.1**	73.7	11.8	**85.5**
		Total	**50.5**	**49.5**	**100.0**	**87.2**	**12.8**	**100.0**	**84.2**	**15.8**	**100.0**
5p.m.	Window (%)	No	16.8	7.4	**24.2**	20.2	0.0	**20.2**	22.4	5.3	**27.6**
		Yes	35.8	40.0	**75.8**	70.2	9.6	**79.8**	64.5	7.9	**72.4**
		Total	**52.6**	**47.4**	**100.0**	**90.4**	**9.6**	**100.0**	**86.8**	**13.2**	**100.0**
6p.m.	Window (%)	No	17.9	27.4	**45.3**	38.3	2.1	**40.4**	42.1	5.3	**47.4**
		Yes	15.8	39.0	**54.7**	48.9	10.6	**59.6**	47.4	5.3	**52.6**
		Total	**33.7**	**66.3**	**100.0**	**87.2**	**12.8**	**100.0**	**89.5**	**10.5**	**100.0**
7p.m.	Window (%)	No	16.8	42.1	**59.0**	50.0	8.5	**58.5**	54.0	5.3	**59.2**
		Yes	13.7	27.4	**41.1**	34.0	7.5	**41.5**	36.8	4.0	**40.8**
		Total	**30.5**	**69.5**	**100.0**	**84.0**	**16.0**	**100.0**	**90.8**	**9.2**	**100.0**
8p.m.	Window (%)	No	26.3	43.2	**69.5**	54.3	17.0	**71.3**	59.2	14.5	**73.7**
		Yes	9.5	21.1	**30.5**	21.3	7.5	**28.7**	21.1	5.3	**26.3**
		Total	**35.8**	**64.2**	**100.0**	**75.5**	**24.5**	**100.0**	**80.3**	**19.7**	**100.0**

(Table 5) cont.....

9p.m.	Window (%)	No	32.6	39.0	**71.6**	44.7	31.9	**76.6**	57.9	23.7	**81.6**
		Yes	11.6	16.8	**28.4**	13.8	9.6	**23.4**	13.2	5.3	**18.4**
		Total	**44.2**	**55.8**	**100.0**	**58.5**	**41.5**	**100.0**	**71.1**	**29.0**	**100.0**
10p.m.	Window (%)	No	46.3	35.8	**82.1**	27.7	58.5	**86.2**	46.1	46.1	**92.1**
		Yes	12.6	5.3	**17.9**	7.5	6.4	**13.8**	5.3	2.6	**7.9**
		Total	**59.0**	**41.1**	**100.0**	**35.1**	**64.9**	**100.0**	**51.3**	**48.7**	**100.0**
11p.m.	Window (%)	No	65.3	23.2	**88.4**	14.9	75.5	**90.4**	25.0	69.7	**94.7**
		Yes	9.5	2.1	**11.6**	2.1	7.5	**9.6**	2.6	2.6	**5.3**
		Total	**74.7**	**25.3**	**100.0**	**17.0**	**83.0**	**100.0**	**27.6**	**72.4**	**100.0**
12p.m.	Window (%)	No	81.1	7.4	**88.4**	4.3	85.1	**89.4**	23.7	69.7	**93.4**
		Yes	10.5	1.1	**11.6**	3.2	7.5	**10.6**	4.0	2.6	**6.6**
		Total	**91.6**	**8.4**	**100.0**	**7.5**	**92.6**	**100.0**	**27.6**	**72.4**	**100.0**

At last, Figs. (**9-11**) show a boxplot of all association measures applied to the sample. The association measures were applied in order to quantify the degree of similarity between the occupancy and window operation of a given room. Values closer to one indicate a greater degree of association between variables, while the closer to zero, the lower the degree of association.

As it is shown in Figs. (**9-11**), most of the association measures are described by a low median value, it is, close to zero. That means that, for this sample, the relationship between these two variables is low, for all rooms. An exception is for the measures Simple Coincidence (SM) and Sokal and Sneath 1 (SS1) in the living room (up to 0.60). The increase in the values of these measures in relation to the others is probably due to the consideration of mutual absences (when there is no occupancy or open window) in the numerator. In the living room, for example, less occupancy was observed in the sample at night, when the windows are also more frequently closed, thus increasing the occurrences of mutual absences and, consequently, the value of the association measures in question. The Roger and Tanimoto (RT) measure also considers mutual absences in the numerator; however, it assigns double weight to disagreements in the denominator, thus not resulting in an increase in the coefficient of association. In general, lower values of association between occupancy and opening of windows in the bedrooms were observed. This result is in line with the hypothesis that the opening of windows in this room is not conditioned by the presence of an occupant, probably occurring often in accordance with the routine habits of the residents.

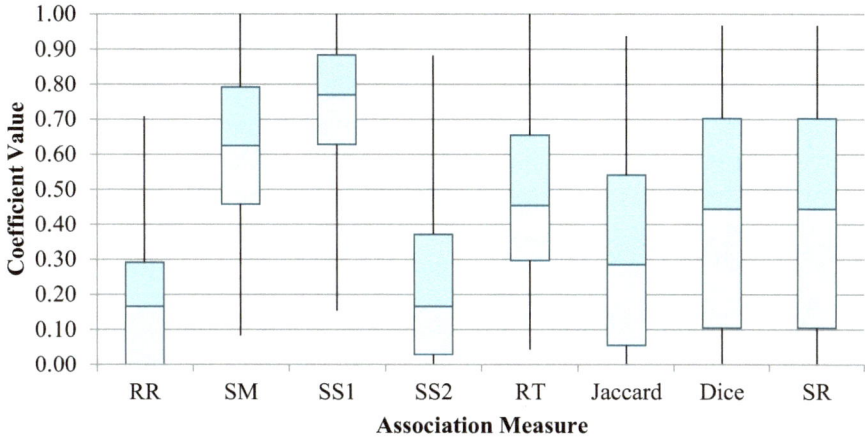

Fig. (9). Boxplot of association measures for variables occupancy and window operation in the living room.

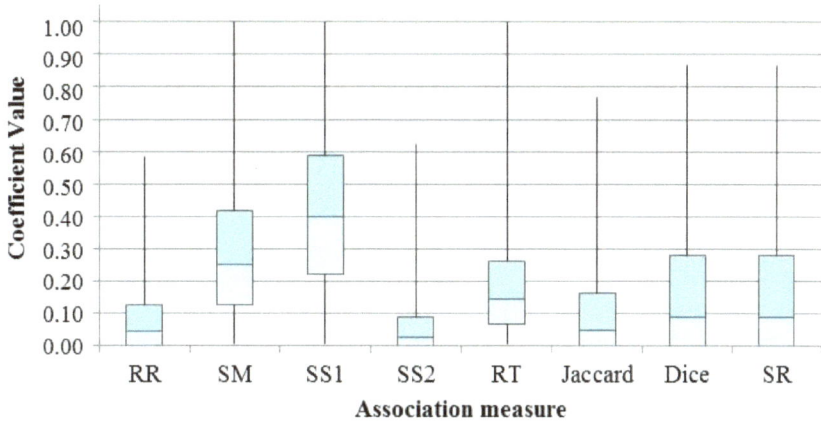

Fig. (10). Boxplot of association measures for variables occupancy and window operation in bedroom #1.

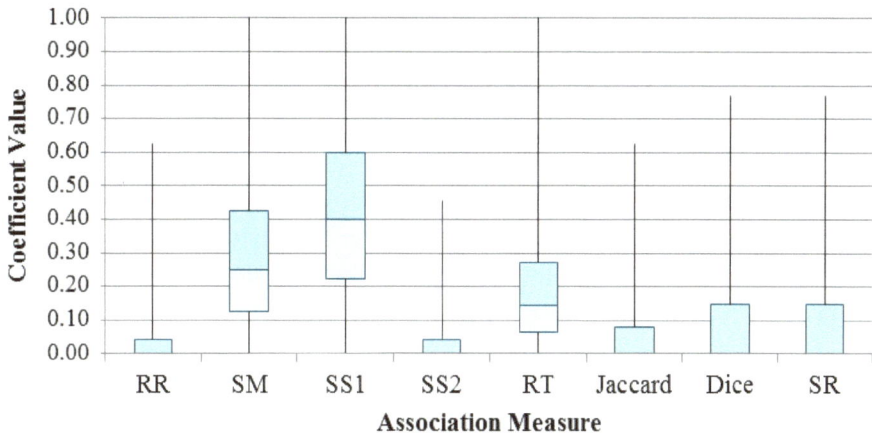

Fig. (11). Boxplot of association measures for variables occupancy and window operation in bedroom #2.

Cluster Analysis

The cluster analysis started with the hierarchical technique, from which a dendogram was obtained (Fig. **12**). The dendogram shows on the horizontal axis all the elements involved in the analysis (occupant profiles), while on the vertical axis, it shows the level of similarity obtained at each junction of a new element to an already formed cluster. An increase in the level of similarity relatively higher than the previous step indicates a point where the cut line (or stop rule line) should be made.

This line indicates the moment from which the joining of two existing clusters will produce a new cluster with high heterogeneity, an unwanted effect for the formation of good clusters. Thus, the ideal number of clusters to be formed is that referring to the number of clusters existing below the cut line. In the case of this study (Fig. **12**), the cut line must pass around the similarity level close to five, resulting in the formation of four clusters.

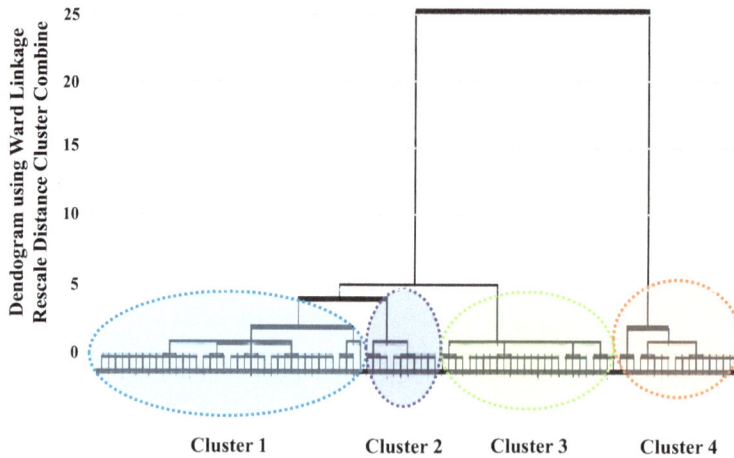

Fig. (12). Dendogram obtained through the application of hierarchical cluster analysis.

The non-hierarchical cluster analysis was performed for the number of clusters found in the hierarchical analysis, which, in this study, corresponded to four clusters. For each cluster, an occupant representative profile was designated, which corresponds of those who have shown the smallest distance to the centroid of its cluster.

Tables **6** and **7** present the description of daily routines for the variables occupancy (Table **6**) and window operation (Table 7). For the representative profile of cluster 1, the room is occupied from early morning until night, while the windows are closed for the entire period. Bedrooms 1 and 2 are occupied until 6 a.m. and after 9 p.m.. However, the windows are only open in bedroom 2, between 6 a.m. and 8 a.m.. Profile 2 shows occupancy only for the living room (5 p.m. – 10 p.m.) and bedroom 1 (until 8 a.m. and after 10 p.m.), with bedroom 2 remaining unoccupied for the entire period. The opening of the windows is constant, occurring in all rooms from 8 a.m. to 10 p.m., with the exception of bedroom 2, which is closed at 9 p.m.. Profile 3 has no occupancy of the living room, only in the bedrooms, occurring until 5 a.m. and after 9 p.m., in both bedrooms. The windows remain open from 5 a.m. to 8 a.m., in all rooms. Finally, profile 4 shows occupancy in the living room only in the late afternoon and early evening (from 5 p.m. to 9 p.m.), while bedroom 1 is occupied for most of the morning (until 9 a.m.) and from 8 p.m.. The windows of the two rooms remain open between 7 a.m. and 8 p.m.. As for bedroom 2, there is no occupancy at any time and the window remains closed for the entire period.

Table 6. Time of the day the room was occupied.

Room	Profile 1	Profile 2	Profile 3	Profile 4
Living room	6 a.m. - 10 p.m.	5 p.m. – 10 p.m.	unoccupied the whole period	5 p.m. - 9 p.m.
Bedroom #1	0 a.m. - 6 a.m. 9 p.m. - 12 p.m.	0 a.m. - 8 a.m. 10 p.m. - 12 p.m.	0 a.m. - 5 a.m. 9 p.m. - 12 p.m.	0 a.m. - 9 a.m. 8 p.m. - 12 p.m.
Bedroom #2	0 a.m. - 6 a.m. 9 p.m. - 12 p.m.	unoccupied the whole period	0 a.m. – 5 a.m. 9 p.m. – 12 p.m.	unoccupied the whole period

Table 7. Time of the day the window remained opened.

Room	Profile 1	Profile 2	Profile 3	Profile 4
Living room	Closed the whole period	8 a.m. - 10 p.m.	5 a.m. - 8 p.m.	7 a.m. – 8 p.m.
Bedroom #1	Closed the whole period	8 a.m. - 10 p.m.	5 a.m. – 8 p.m.	7 a.m. – 8 p.m.
Bedroom #2	6 a.m. – 8 p.m.	8 a.m. – 9 p.m.	5 a.m. – 8 p.m.	Closed the whole period

DISCUSSION

As shown in the Introduction section, many studies associate the opening of windows to the presence of an occupant in the room. However, in practice, it is known that in single-family residential buildings from Florianopolis (subtropical climate with abundant wind conditions) there is a tendency for residents to keep the windows of all rooms open for many hours during a day, regardless of the presence of the occupant in the room (as long as there is no interference from external conditions, like, for example, the weather). Thus, this study aimed to investigate the dependency relationship between the rooms' occupancy schedule and the operation of openings in residential buildings.

Initially, data were obtained regarding the occupancy and operation routines of windows in the living room and bedrooms in an existing database. This database was built from field surveys, thus constituting real data about occupant behaviour. Such consideration becomes important when it is desired that these data serve as input data in studies of thermal and energy performance of buildings, as stated by Silva and Ghisi [18] and Haldi and Robison [19]. Although the method for obtaining the data presents limitations (they were obtained through semi-structured questionnaires, and not by electronic monitoring), it is still possible to obtain relevant information about the occupant behaviour, as they indicate general trends in behaviour and how daily routines are perceived by the residents themselves.

In the next step, the data obtained were subjected to a sequence of exploratory analyses, such as the application of descriptive analysis (by means of frequency graphs and histograms), measures of association between the variables occupancy and operation of windows and also with a cluster analysis application. The descriptive analyses, together with the application of the association measures, were intended to verify the degree of association between the occupancy of a given room and the opening of the window in that same room, at each hour of the day. Contrary to what is found in the literature, in which the operation of a window is usually conditioned by the presence of a resident in the room, with this analysis it was found that this association is not relevant most of the time, especially in the bedrooms, where occupancy occurs predominantly at night, while the opening of windows, during the day. Even in the living room, whose tendency is to be occupied and with windows open during the day, there was a disassociation between the occupancy and the period when the windows were open. This consideration is important because, although there are other influencing factors in the decision to open or not to open a window, this decision should not be associated exclusively with the presence of the resident, as this posture may cause errors to the results of studies on the thermal performance of this type of building.

As for the cluster analysis, it was applied in order to identify different patterns of behaviour in the sample. In general, four different clusters were found, with a behaviour pattern associated with each of them. In order to allow the consideration of these patterns in studies of thermal and energy performance, such as those that adopt computer simulation, a representative occupant profile was determined for each cluster, corresponding to the profile closest to the clusters' centroid. Each representative profile was described from its daily occupancy and window operation routine, for each of the rooms (living room and bedrooms). It was observed that the four profiles differ from each other in terms of their routines, whether in terms of room occupancy or opening and closing windows.

Some profiles showed rooms that were not occupied at any time of the day, or whose windows were kept closed for the entire period. Such behaviours may represent a situation in which the room is not really occupied (or its windows operated), or there is no such room in the building. For example, a living room that is not occupied can represent a room in a house where the kitchen is occupied most of the time, the living room being used only on special occasions. Likewise, a secondary bedroom (noted in this study as bedroom 2) may represent a room that is never occupied or a building that does not have a second bedroom. In addition, some hourly data may seem unconventional (such as opening windows at 5 a.m.), but it is important to understand that the object of research is a specific social group, whose habits and routines differ from other social classes.

It should be noted that the representative profiles presented here describe a common pattern of actions taken by an occupant in relation to opening the windows, within a specific sample. The pattern of behaviour presented is static and it is known that there are other factors that influence the decision to open or close the window, such as external and internal temperature, rain, noise, privacy, security, among others. Therefore, it is important that the representative profile of the occupant presented here would not be applied exclusively. It must also be conditioned to the other influencing factors. For example, if at a certain time when, according to the profile, the window is open, but the room does not meet certain comfort conditions, an action can be taken to solve the problem. That is, the window remains open at a certain time if the representative profile informs that it would be open at that time and if it meets the other conditions criteria (other factors). Such conditions were not investigated in this study, but can be found in other publications.

It is important to emphasize that this study does not intend to refute or contradict other studies that point out the presence of the user in a room as an important factor for the operation of windows, as certainly must have been observed in these studies, specifically for the sample analysed by them. What was observed here is that this is not a mandatory condition, that is, there are cases in which this action (opening and closing windows) can happen without the occupant remaining in the room for longer than he/she needs to carry out the action, as it was identified for the sample of this study. Certainly, these findings reflect the combination of specific social and climatic factors of the sample under study, inserted in a region of mild climate, with reasonable temperature fluctuation throughout the year and which does not present extreme temperatures, neither for the hot season nor for the cold season, and whose neighbourhood relationship works in favour for social interaction. Therefore, the data presented herein are not indicated for use in studies whose characterization of the object of study does not match those of the sample analysed herein, but it is shown that when establishing occupation as a mandatory condition for operating windows, inconsistences can be found.

CONCLUDING REMARKS

When investigating studies focusing on the occupant's behaviour pattern, there was a strong tendency to consider opening windows as conditioned to the presence of an occupant in the room. In this study, it was found that such a condition does not necessarily represent a real situation, at least for the sample analysed herein, where during most of the day the opening hours of windows did not correspond to times when the room was occupied, especially with regard to the bedrooms. However, this is a specific observation for the investigated sample and may not reflect the behaviour of the sample of users from other studies.

Previous studies on the determination of behaviour patterns have also highlighted the variability of human behaviour, which was also verified in this study. Thus, assuming that all occupants of a set of buildings have a single pattern of behaviour can lead to thermal and energy simulations presenting results that are inconsistent with reality. The determination of the representative profiles presented in this study can help to obtain more reliable simulation results. It must also be considered that, in order to have simulations with results consistent with reality, it is necessary to be sure that the other boundary conditions have been modelled according to the real conditions of the building as well.

It is also important to note that this study has limitations, such as dealing with a very specific sample of occupants, which may not reflect the behaviour pattern of occupants from other climatic regions or construction typologies, in addition to their social identity. Another limitation refers to obtaining a generalized routine, informed by the resident, which must be combined with other relevant factors in computer simulations in order to produce reliable data. Future studies should deepen the interference relationship between daily routines and other influencing factors pointed out in the literature.

CONSENT FOR PUBLICATION

Not Applicable.

CONFLICT OF INTEREST

The author confirms that this chapter contents have no conflict of interest.

ACKNOWLEDGEMENTS

The authors acknowledge with thanks the financial support of National Council for Scientific and Technological Development (CNPq) of Brazil.

REFERENCES

[1]   "UNEP, Buildings: Investing in energy and resource efficiency, In: UNEP (Ed.), Towards a Green Economy: Pathways to Sustainable Development and Poverty Eradication. United Nations Environment Programme, 2011, pp. 330–373",

[2]   IEA, *Key World Energy Statistics.*, 2012. http://www.iea.org/publications/freepublications /publication/kwes.pdf

[3]   EC, *Energy Efficiency Status Report.*, 2016. http://iet.jrc.ec.europa.eu /energyefficiency/sites/ energyefficiency/files/energy-efficiency-status-report-2016.pdf

[4]   DOE, *Buildings Energy Data Book.*, 2011.http://buildingsdatabook.eren.doe.gov/ChapterIntro1.aspx

[5]   BRASIL, "Ministério de Minas e Energia", *Balanço Energético Nacional (BEN): Relatório Final [National Energy Balance: Final Report],* Brasília, (in Portuguese), 2019.

[6]   E.J. Palacios-Garcia, A. Chen, I. Santiago, F.J. Bellido-Outeirino, J.M. Flores-Arias, and A. Moreno-Munoz, "Stochastic model for lighting's electricity consumption in the residential sector: Impact of energy saving actions", *Energy Build.,* vol. 89, pp. 245-259, 2015. [http://dx.doi.org/10.1016/j.enbuild.2014.12.028]

[7]   IEA-EBC, http://www.iea-ebc.org/index.php?id=141

[8]   S. D'oca, and T. Hong, "A data-mining approach to discover patterns of window opening and closing behaviour in offices", *Build. Environ.,* vol. 82, pp. 726-739, 2014. [http://dx.doi.org/10.1016/j.buildenv.2014.10.021]

[9]   J. Page, D. Robinson, N. Morel, and J.L. Scartezzini, "A generalised stochastic model for the simulation of occupant presence", *Energy Build.,* vol. 40, no. 2, pp. 83-98, 2008. [http://dx.doi.org/10.1016/j.enbuild.2007.01.018]

[10]   D. Aerts, J. Minnen, I. Glorieux, I. Wouters, and F. Descamps, "A method for the identification and modelling of realistic domestic occupation sequences for building energy demand simulations and peer comparison", *Build. Environ.,* vol. 75, pp. 67-78, 2004. [http://dx.doi.org/10.1016/j.buildenv.2014.01.021]

[11]   V.L. Erickson, M.A. Carreira-Perpifian, and A.E. Cerpa, "Occupation modeling and prediction for building energy management", *ACM Trans. Sens. Netw.,* vol. 10, no. 3, pp. 42-69, 2014. [TOSN]. [http://dx.doi.org/10.1145/2594771]

[12]   M.J. Sorgato, A.P. Melo, and R. Lamberts, "The effect of window opening ventilation control on residential building energy consumption", *Energy Build.,* vol. 133, pp. 1-13, 2016. [http://dx.doi.org/10.1016/j.enbuild.2016.09.059]

[13]   L. Wang, and S. Greenberg, "Window operation and impacts on building energy consumption", *Energy Build.,* vol. 92, pp. 313-321, 2015. [http://dx.doi.org/10.1016/j.enbuild.2015.01.060]

[14]   S.T. Moghadam, F. Soncini, V. Fabi, and S. Corgnati, "Simulating window behaviour of passive and active users", *Energy Procedia,* vol. 78, pp. 621-626, 2015. [http://dx.doi.org/10.1016/j.egypro.2015.11.040]

[15]   M. Bonte, F. Thellier, and B. Lartigue, "Impact of occupant's actions on energy building performance and thermal sensation", *Energy Build.,* vol. 76, pp. 219-227, 2014. [http://dx.doi.org/10.1016/j.enbuild.2014.02.068]

[16]   K. Papakostas, and B.A. Sotiropoulos, "Occupational and energy behaviour patterns in Greek residences", *Energy Build.,* vol. 26, no. 2, pp. 207-2013, 2008. [http://dx.doi.org/10.1016/S0378-7788(97)00002-9]

[17]   S. Chen, W. Yang, H. Yoshino, M.D. Levine, K. Newhouse, and A. Hinge, "Definition of occupant behaviour in residential buildings and its application to behaviour analysis in case studies", *Energy Build.,* vol. 104, pp. 1-13, 2015.

[http://dx.doi.org/10.1016/j.enbuild.2015.06.075]

[18]   A.S. Silva, and E. Ghisi, "Uncertainty analysis of user behaviour and physical parameters in residential building performance simulation", *Energy Build.,* vol. 76, pp. 381-391, 2014.
[http://dx.doi.org/10.1016/j.enbuild.2014.03.001]

[19]   F. Haldi, D. Cali, R.K. Andersen, M. Wesseling, and D. Mueller, "Modelling diversity in building occupant behaviour: a novel statistical approach", *J. Build. Perform. Simul.,* vol. 10, no. 5, pp. 527-544, 2009.

[20]   R. Andersen, V. Fabi, J. Toftum, S.P. Corgnati, and B.W. Olesen, "Window opening behaviour modelled from measurements in Danish dwellings", *Build. Environ.,* vol. 69, pp. 101-113, 2013.
[http://dx.doi.org/10.1016/j.buildenv.2013.07.005]

[21]   R.K. Andersen, V. Fabi, and S.P. Corgnati, "Predicted and actual indoor environmental quality: Verification of occupants' behaviour models in residential buildings", *Energy Build.,* vol. 127, pp. 105-115, 2016.
[http://dx.doi.org/10.1016/j.enbuild.2016.05.074]

[22]   R. Ramos, *Qual a Importância da Análise Exploratória de Dados?,* 2015. https://oestatistico.com.br/analise-exploratoria-de-dados/

[23]   P.I.R.C. Dias, M.W. Hemais, and F.C. Borelli, ">Utilização de Variáveis Binárias como Método de Seleção para Pesquisas Qualitativas: um Estudo sobre Carros", *IV Encontro de Marketing da ANPAD, Florianópolis,* Brazil,(in Portuguese), 2007.

[24]   W.O. Bussab, E.S. Miazaki, and D.F. Andrade, "Introdução à análise de agrupamentos [Introduction to cluster analysis]", *IX Simpósio Nacional de Probabilidade e Estatística,* São Paulo: Brazil,(in Portuguese), 1990.

[25]   H. Finch, and H. Huynh, " Comparison of similarity measures in cluster analysis with binary data". In: Annual Meeting - American Educational Research Association, New Orleans, EUA, 2000.,

[26]   S. Azambuja, Estudo e implementação da análise de agrupamento em ambientes virtuais de aprendizagem*,* 2005.

[27]   J.F. Hair, R.E. Anderson, R.L. Tatham, and W.C. Black, *Análise multivariada de dados.* 6th ed. Bookman: Porto Alegre, 2009. Multivariate data analysis (in Portuguese)

[28]   A.K. Jain, M.N. Murty, and P.J. Flynn, "Data clustering: a review", *ACM Comput. Surv.,* vol. 31, pp. 264-323, 1999.
[http://dx.doi.org/10.1145/331499.331504]

[29]   S.A. Mingoti, *Análise de dados através de métodos de estatística multivariada: uma abordagem aplicada.* Ed. da UFMG: Belo Horizonte, 2005. Data analysis using multivariate statistical methods: an applied approach (in Portuguese).

[30]   *Mathworks,* 2014.

[31]   L. Kaufman, and P.J. Rousseeuw, *Finding groups in data: an introduction to cluster analysis.* John Wiley: New Jersey, 2005.

[32]   E. Ghisi, A. S. Vieira, A. Schaefer, A. K. Marinoski, A. S. Silva, B. F. Balvedi, and L. S. S. Almeida, "Uso racional de água e eficiência energética em habitações de interesse social – Hábitos e indicadores de consumo de água e energia", *Relatório Técnico de Pesquisa,* vol. 1, Florianópolis, (in Portuguese), 2015.

[33]   A. Schaefer, and E. Ghisi, "Method for obtaining reference buildings", *Energy Build.,* vol. 128, pp. 660-672, 2016.
[http://dx.doi.org/10.1016/j.enbuild.2016.07.001]

CHAPTER 7

Investigating the Uncertainties of Occupant Behaviour in Building Performance Simulation: A Case Study in Dwellings in Brazil

Arthur Santos Silva[1,*]

[1] *Laboratory of Analysis and Development of Buildings, Federal University of Mato Grosso do Sul, Campo Grande, Mato Grosso do Sul, Brazil*

Abstract: The literature states that the occupancy and related operational characteristics in buildings are key variables that cause the gap between the estimate and actual thermal and energy performance. To address such issue, the objective of this study is to investigate the uncertainties of occupant behaviour in building performance simulation through a probabilistic approach. This case study considers a model of a low-income dwelling in southern Brazil using five different construction for the envelope and with natural or hybrid ventilation. Field survey provided a dataset of uncertainties of the occupant behaviour, which was related to the occupancy of the rooms, operation of openings and use of electric appliances. The EnergyPlus programme was used to conduct the simulations and the R Studio was used for data processing, analysis, and treatment. A global sensitivity analysis was performed, along with an uncertainty analysis. The results showed that the number of occupants, the schedules of occupancy of the bedrooms, the setpoint temperatures for operating the openings, the cooling setpoint of the HVAC and the limits for operative temperatures of the rooms were the most influent variables for the thermal and energy performance, especially in the heating period. The uncertainty was up to 65.6% for estimating the degree-hours for heating (in the natural ventilation mode) and up to 59.3% for estimating the total electricity consumption with HVAC (in the hybrid ventilation mode), indicating that these operational uncertainties had a great impact in the simulation results.

Keywords: Building simulation, EnergyPlus, Energy consumption, Global sensitivity, Operational uncertainty, Performance evaluation, Sensitivity analysis, Sobol' indices, Thermal performance, User behaviour, Uncertainty analysis.

---

* **Corresponding author Arthur S. Silva:** Laboratory of Analysis and Development of Buildings, Federal University of Mato Grosso do Sul, Campo Grande, Mato Grosso do Sul, Campo Grande, Mato Grosso do Sul, Brazil; Tel: +556733457477; E-mail: arthur.silva@ufms.br

Enedir Ghisi, Ricardo Forgiarini Rupp and Pedro Fernandes Pereira (Eds.)

INTRODUCTION

The global energy demand, one of the main contemporary challenges, tends to grow with the population increase and economy expansion as it can be seen in the 12.5% increase from 2010-2018, according to IEA [1]. The buildings' sector represented, in 2018, up to 29.3% of the total energy consumption and 49.3% of the total electricity consumption worldwide (by grouping residences, commercial and public services together). The energy efficiency investments at a global scale were, according to the Energy Efficiency Report from IEA [2] at 2018 base, 58% orientated to the buildings' sector (30% of it was related to the envelope, followed by the HVAC systems).

In this context, energy efficiency is considered the "first fuel" and could address some challenges such as the climate change, energy security and economic growth, without damaging the environment or negatively affecting social aspects. Energy efficiency in buildings can be defined as the reduction in energy consumption by maintaining the comfort and productivity of the occupants at adequate levels, to enable, in a macro scenario, a reduction in the growth rate of energy demand. However, as the building context becomes more complex with the emergence of new materials and technologies, the performance of buildings also becomes hard to predict or estimate. Indeed, innovations and strategies for energy efficiency are becoming difficult to model and evaluate, in the mathematical point of view, which make the strategies (*i.e.* performance measures) not trivial to implement.

The simulation programmes are key tools to analyse the performance of the building in different phases, since the conceptual, design, operational or maintenance. They also brought many advantages in analysing the energy performance and the behaviour of buildings. There are some general applications for the use of building simulation in early design [3,4], or operation and maintenance [5]. Nevertheless, despite using advanced algorithms, physical models, empirical databases and being powered with research results all over the world, this type of tool continues to be a mathematical model. Thereby, every mathematical model depends on their algorithms and logical structures in an attempt to emulate or simulate a real physical phenomenon. Another issue is that every statistical inference made on mathematical models is susceptible to Type I and Type II errors for tests of hypotheses, in which Type I can be defined as the incorrect rejection of a true null hypothesis, while Type II error is failing to reject a false null hypothesis [6].

It is common to understand the building performance analysis as a mathematical model that depends on at least three major groups of information: the building model, the external factors and the internal factors. Usually, the building model is set by defining the envelope and the building systems; the external factor represents the climate within all related variables; the internal factors are related to the occupant, operation of the systems and indoor comfort conditions and requirements. The latter is particularly interesting for this study as, since 1999, Degelman [7] stated that, despite the huge development in the thermal (and energy) processes of the mathematical models in the simulation programmes, from the 1970s until nowadays, studies on operational patterns do not achieve the same development of knowledge. The author also stated that this is, somehow, ironic as the operational characteristics of a building could exert a greater impact in the building performance than the physical properties of the envelope. However, in the last decades, one may find many publications regarding building performance and the effects of operational characteristics, especially the occupant behaviour, in all sorts of building performance aspects, as stated by Balvedi *et al.* [8].

The literature shows that similar buildings, with similar architectural aspects (shape, floor area and materials) and with reasonable similar occupancy and electric appliances possessions, can have very distinct energy consumption due to the operational characteristics of the users and the buildings' systems. This aspect is called the "performance gap" in many papers [9 - 11], and have become, itself, a whole field of research. Jia *et al.* [12] described two main purposes for studying the occupant behaviour of a building: (1) for understanding the performance gap between the simulated and actual energy use of buildings; or (2) for performing a more robust optimization of the buildings, especially by setting control and operation parameters more accurately.

A major research project has been developed, in this area, by the International Energy Agency in respect to the EBC/IEA Annex 66 [13]. The main purpose was to develop a framework for dealing with the occupant behaviour in building simulation by considering data collection, modelling and evaluation, as well as integrating these processes into the programmes for supporting designers, stakeholders, policies and other areas. This annexe also developed an approach to reduce the performance gap related to occupant behaviour.

Jia *et al.* [12] categorized five different areas for occupant modelling in a building. The (1) agent-based modelling is related to the occupant's perception and their interaction with each other and the outdoor environment, which depends on modelling their actions by algorithms powered with measured data. The statistical analyses (2) try to establish a numerical relation between the occupant behaviour and other related variables; occupant profiles can be created by grouping some influent variable in the sample. The data mining approach (3) is another statistical tool that enables the correlation of the occupant behaviour with other related aspects, such as energy consumption or thermal performance; in this sense, the algorithms should learn how to behave. The stochastic approach (4) is related to modelling of the occupant status with time. And there is another generic category (5) which depends on different types of estimation for different purposes. Still in this subject, Hong *et al.* [11] stated that building performance can be quantified by two approaches: the first is to compare the energy performance of similar buildings with different occupant patterns, and the second is to simulate the building performance with a different sort of occupant pattern profiles. While the first approach enabled a direct comparison of the effects of occupant behaviour, the second depends on the act of modelling the occupant's behaviour based on research data and mathematical models.

By giving this succinct introduction, this study intent is to investigate the operational characteristic of the occupants regarding the performance of residential buildings, which is already recognised as the major aspect that influences the uncertainty in the building performance [14]. This investigation followed a statistical procedure to develop operational probabilistic profiles, which is not the same as modelling or emulating the occupancy, as stated by some authors [12]. In this sense, the followed approach seeks the second type described by Hong *et al.* [11], which depends on empirical data to understand the uncertainties by giving, a priori, the type of sample and building characteristics, which were not directly correlated with other variables but treated separately as a separated phenomenon.

The objective of this chapter is to describe the development of operational schedules for houses, from a case study in southern Brazil, and to perform building simulation experiments to understand the effects of user behaviour in the building performance by using advanced statistical techniques.

LITERATURE REVIEW

There are some review studies regarding the subject of occupant behaviour related to its effects on the performance of buildings, specially concerned with the development and validation of occupancy profiles. Hong *et al.* [11] performed an interesting review by answering some questions related to the occupant's behaviour, such as how to measure the occupant behaviour or how occupancy can be modelled for simulation purposes. By analysing how the occupancy models are dealt with by the simulation programmes, the authors stated that most of the programmes adopt deterministic input values to represent the occupant behaviour. Thus, if a researcher wants to analyse and investigate the occupancy by advanced methods, the model should be perturbed to generate the desired mathematical experiment, by using the native deterministic input strategy. Balvedi *et al.* [8] reviewed some methods for collecting user behaviour information and how they are used in building simulation tools. The authors addressed the monitoring methods for occupant behaviour, the development of behavioural models and the application of these models in building performance simulation.

Hong *et al.* [10] stated that addressing the building performance gap is one important challenge of the building simulation community. All the building models specification (climate, envelope, services and systems, operation, and indoor environmental requirements) are susceptible to uncertainties, which can affect the energy performance results. However, this fact causes some scepticism or lack of credibility regarding the building simulation initiatives by the various engineering sectors. Thus, the gaps need to be properly identified and analysed, especially those related to the occupancy. De Wilde [9] listed up to seven issues that need to be addressed in further attempts to study the performance gap. To cite some of them, the authors described the need to consider a broader sample of buildings to understand the uncertainties in the occupancy, while understanding that each building would have a unique pattern. Another aspect is that the simulation outcomes must be treated as uncertain, especially due to those variables that cannot be known at the design stage (actual occupancy, weather, control of HVAC); in this sense, there is a need to consider the results with a probabilistic approach, which implicates in the use of advanced data treatment and experiments, and considering confidence intervals for both the inputs and outputs.

Pfafferott and Herkel [15] showed the statistical occupant behaviour modelling in a case study for the Fraunhofer Institute for Solar Energy Systems (Germany). The measurements were performed for one month in 2003 considering sixteen offices inside the building. The response function was the indoor air temperatures and their variations in time, which were analysed by the mean and standard deviation. Azar and Menassa [16] analysed 30 typical office buildings in five main climate zones of the United States, by varying the occupant behaviour on the eQuest simulation programme. The construction characteristics of buildings were extensively detailed through a survey.

The authors defined nine input variables to investigate the energy consumption, such as the hours of equipment and lighting usage, the temperature setpoints during the occupied and unoccupied hours, among others. A one-at-a-time sensitivity analysis was performed to understand the impact of such variables on the output. For some buildings, the heating setpoint during the occupied periods was the most influent variable, while for others, it was the building schedule. They concluded that the occupancy parameters have a significant influence in the energy consumption, and the information on what parameter matter the most, the stakeholder or building designers could understand this patter to help to make better decisions.

Lindberg *et al.* [17] determined load profiles (for heat and electric appliances) based on hourly measurements of 114 non-residential buildings. These profiles can be used as a reference for buildings before implementing energy efficiency measures, and are dependent on many variables, by using regression analysis. Gucyeter [18] compared the energy consumption of three offices by using different occupancy profiles (deterministic, semi-deterministic and probabilistic). The author stated that the inclusion of detailed occupancy patterns led to higher accuracy of the model and reduced the performance gap, as the deterministic profiles led to higher internal gains in comparison with the actual occupant behaviour. Thus, deterministic approaches overestimate the occupants' presence, especially in office buildings. Feng *et al.* [19] dealt with a stochastic occupancy modelling for building simulation, by using an empirical model of colling behaviour depending on the indoor temperature (as a probability of switching on/off). The authors performed repeated simulations using the occupancy stochastic model and developed a sort of probability density estimations. They recommend some timestep combinations along with repeated runs to address the confidence interval derived from the stochastic model. D'Oca and Hong [20] used a data mining approach to create occupancy patterns from a specific database and was used for a case study of 16 offices in Germany. The profiles were dependent on the probability of occupancy by the season of the year, the day of the week and the time of day. Through cluster analysis, the study discovered some specific patterns that were different depending on the time the occupants came to office, the actual working time, the lunch period, the afternoon working time and the time people leave the office.

Marszal-Pomianowska *et al.* [21] developed a high-resolution household demand profile using a probabilistic bottom-up approach. The authors used statistics of power consumption and daily usage patterns of individual appliances of an 89 and 16 household samples from Denmark, with different hourly resolutions. Jeong *et al.* [22] studied the window control patterns in 20 houses during different seasons. The windows operation were dependent on the activities of the room and time of day, and the main purposes were for desiring fresh air or for specific activities. The differences in the schedules for the heating or cooling seasons were remarkable. The authors also discovered that the indoor temperatures and temperature drop were more adequate as explanatory variables for windows operation than the outdoor temperatures (based on correlation results). Firląg *et al.* [23] studied five control algorithms for dynamic windows regarding the shading operation. The algorithms were based on heating/cooling, simple rules, perfect citizen, heat flow or predictive weather, and were implemented in EnergyPlus through the EMS (Energy Management System) object for the building simulation. These control algorithms represent different patterns and motivations for operating the shades and resulted in different energy consumption results, such as reducing by 21.6% the energy consumption when an automatically controlled shade is compared with no shading.

From an international point of view, the ASHRAE Standard 90.1 [24] provides some normative occupancy profiles for different building types to be used for simulation purposes. These profiles have the advantage of being easy to interpret, to use and to include in the building simulation programmes, as they usually are represented by a 24-hour profile. The possibility of replication is also important, as different researchers or analysts can compare the performance results to identify common patterns and efficiency measures, as the occupancy would be the same. Brazil, for instance, also has some standardised documents that deal with building simulation for energy efficiency labelling purposes. There is the RTQ-C regulation [25], that deals with the commercial, services and public buildings, and the RTQ-R regulation [26] that deals with residential buildings. Only the residential energy efficiency regulation (RTQ-R) fixes the occupancy profiles for simulation purposes, mainly the occupancy, the opening of windows and doors controlled by temperature, and the use of lighting and appliances. However, the new proposals of development of these regulations are bringing some standardised profiles for commercial and residential buildings, to help the energy labelling and enable the comparison and replicability aspect.

The need for a database for occupancy and other operational characteristics seem to be the first step to understand the magnitude of the uncertainties in building performance simulation. In this sense, the first step of this study was to develop "operational schedules" to use in the simulation programmes as the first set of input variables needed to evaluate the energy performance of buildings.

DEVELOPING OPERATIONAL SCHEDULES

This section deals with the description of the data acquisition, data treatment and representation of the schedules of user behaviour of dwellings. For terminology purposes, the author decides to use the term "operational schedules" instead of using "user behaviour pattern" or "user behaviour routines". The data sample comes from a relatively small case study in southern Brazil and was a part of a larger research project funded by FINEP ("Financiadora de Estudos e Projetos" in Portuguese-BR), a public agency which provides funding for researches over the country through public notices.

Some previous work reported partial results obtained through this research. Silva *et al.* [27] and Silva *et al.* [28] have performed the data treatment of the electricity end-uses of the sample of houses; Schaefer and Ghisi [29] analysed the architectural and construction data to develop some reference buildings for building simulation purposes. However, much of the crude information was published only in internal reports for the funding agencies and/or in the Portuguese language. Thus, this chapter first intent is to bring the information needed to understand the process of data acquisition, data treatment and the development of the schedules into one single document.

The research was performed in the region of Florianópolis, capital of Santa Catarina (southern Brazil) through the year of 2012-2014. The chosen buildings were classified as low-income houses because they attended some of the criteria related to the total family wage, they have the house funded by social housing programmes or they were located in critical areas/neighbourhoods destinated to social houses. Up to 120 houses were surveyed for this research; however, the size of the sample was different (*i.e.* lower) for each subject of the research, which will be shown further in this chapter.

Fig. (1) shows a flowchart describing the whole subjects of the original research, which considered the study of the socioeconomic characteristics of the families, the study of the building characteristics for building-stock purposes, the study of the materials and construction components, the measurements and estimation of the energy and water end-uses. Finally, there is the study of the operational schedules, which, for this chapter's matter, is the user behaviour and related patterns in the sample of the houses.

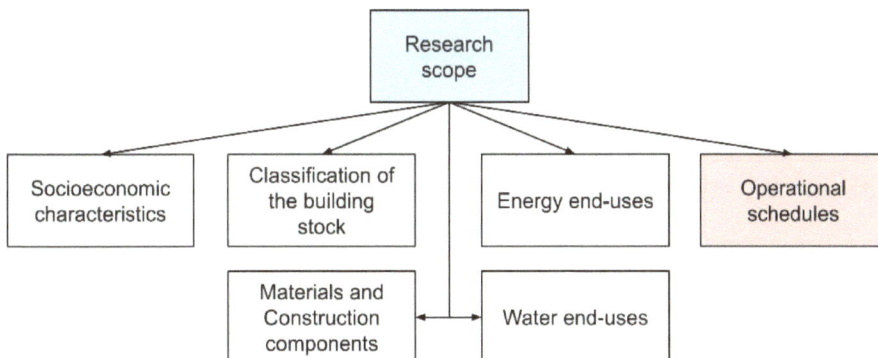

Fig. (1). The scope of the original research, highlighting the subject covered by this chapter, which is the "operational routines".

Data Acquisition

For the data acquisition, semi-structured questionnaires were used along with measurement equipment for recording the energy consumption of electric appliances. The aim was to understand some characteristics of a sample of houses regarding the operational constraints of occupancy, natural ventilation and use of electric appliances.

Up to four types of questionnaires were used: (1) a first that dealt with the socio-economic characteristics of the occupants; (2) the second dealt with the architectural features of shape, form, construction materials and fenestration; (3) the third dealt with the determination of routines of occupancy, operation of doors and windows for natural ventilation and use of electric appliances; (4) the fourth dealt with the register of all the appliances and lamps characteristic per house.

All questionnaires were filled out along with information given by the house owner.

Measurements of the Average Power of Equipment and Lighting

All the electric appliances were registered from each dwelling, including information of the room that the appliance is located, name, manufacturer and model, rated power, and the energy measurement information. All the lamps and luminaires were also registered, including information of the room which they were located, type and rated power. Some additional information regarding the data acquisition was already reported by Silva *et al.* [28].

The dwelling was visited at least twice: once to register the information, interview the occupants with questionnaires, and install the monitoring equipment; and later once to remove the equipment and register the measured energy data. The main measurement equipment used was the T8 plug load meter (from Northmeter), and one equipment was installed per each electric appliance in each dwelling. The equipment registered the energy consumption and time of use, which enables the calculation of the average power of each appliance.

The procedure of the data acquisition can be summarized as follows:

- All electric appliances were registered, and the average power was measured for each appliance in each dwelling.
- The electric appliances were, but not limited to: refrigerator, washing machine, microwave, television, computer, iron, fans, cafeteria, hair drier, cellphone charger, among others.
- All electric appliances and lamps were organized by the dwelling identification,

the room which they were located, and the average power calculated through the measurements.

- A questionnaire was applied for the dwelling owner regarding the routine of usage of each equipment and lighting, especially concerning the time of day, duration of events and frequency of use during one typical month.

The sample size for the determination of average power was 53 dwellings with complete data. The measurements and interviews were conducted in 2012. For this study purpose, as the building simulation deals with the long-permanence rooms of the house (*i.e.* bedrooms and living room), the electric shower information and consumption were excluded from the data treatment and the simulation experiments.

Questionnaires on the Operational Schedules

The operational schedules are related to the way the occupants use the dwelling and its systems; thus, they concern the patterns of use of the appliances, the lighting, the occupancy of the rooms and the way the windows and doors are operated for natural ventilation.

The questionnaire about the operational schedules was filled out together with the house owner, and the answers were related to the duration of events of using each appliance of the house, the periods of the day which windows and doors are opened for natural ventilation, the routine of use of the rooms based on the behaviour of the family members, and the patterns of using the lighting of the rooms. The researcher divided the answers by routines for the weekdays and weekends and based on the subjective answer of the owner.

The procedure of filling out the questionnaires can be summarized as follows:

- As each type of equipment was already registered considering the previous step of the research, the list of appliances and lighting was used to estimate the total duration of events through the answers of the owner.
- The same was performed for the routines of opening the windows and internal doors of each room for natural ventilation purposes.
- The occupancy of each room was also estimated, as each family have a specific pattern of using the rooms, time of absence, period of working and total hours of sleep.

The sample size of the houses for this subject was 17 dwellings during the summer period and 34 during the winter period. It is worth to state that all houses of the sample were naturally ventilated or used some electric fans; no dwelling had air-conditioning to provide artificial cooling or heating for achieving thermal comfort.

Data Treatment

After the data acquisition, the next step was to develop a dataset of operational schedules in the format required by the building simulation programmes. All the data gathered from the use of electric appliances, door and window operation for natural ventilation and room occupancy were treated with different statistical methods.

This section aims to achieve the so-called "diversity profiles", which represent a historical hourly base data related to the topics mentioned above throughout one typical year. Fig. (2) shows a scheme of the schedules developed for this study, based on the data acquisition.

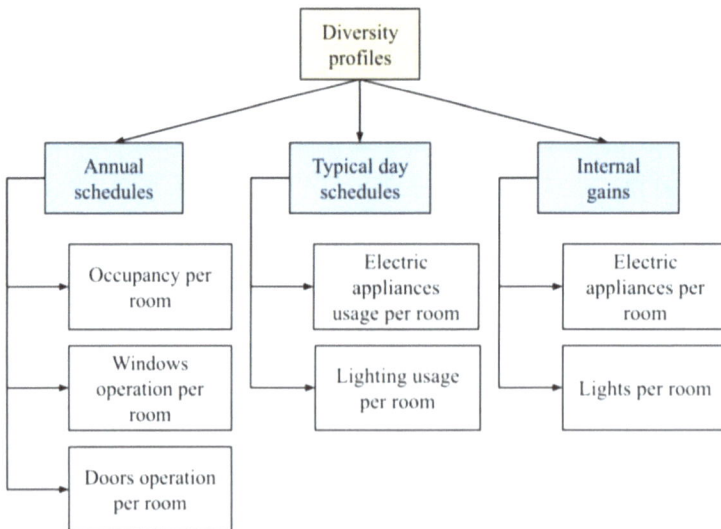

Fig. (2). Schedules and datasets developed from the data acquisition.

Developing Schedules for Occupancy and Openings Operation

For the development of diversity profiles related to the occupancy, operation of windows and doors, the Wilcoxon statistical test was used. It is a non-parametric test for calculating confidence intervals based on the median of the sample rather than the mean. Non-parametric tests are useful for models that deal with behavioural data [30] or when the shape of the statistical distribution does not follow a known parametric model, such as normal or uniform distribution.

The Wilcoxon test uses the signal (direction) and the magnitude of the difference between the data and the median. The confidence interval is obtained from the Walsh averages of the sample, as being calculated according to eq. (**1**). The Walsh averages are ordered and the median is calculated: when the size of the data is odd, the median is the intermediate value; when the size of the data is even, the algorithm of Johnson and Mizoguchi [31] is used. The confidence level achieved is calculated from the normal approximation of Wilcoxon according to eq. (**2**).

$$W_k = \frac{X_i + X_j}{2} \tag{1}$$

$$Z_W = \frac{\left| W_k - \frac{n(n+1)}{4} \right| - 0,5}{\sqrt{\frac{n(n+1)(2n+1)}{24}}} \tag{2}$$

Where: W_K are the Walsh averages; X_i and X_j are the sample data; n is the size of the random sample; Z_W is the number of deviation from the normal approximation.

The procedure can be summarized as follows:

- All the data obtained by the survey were organized for treatment: occupancy of each room of the dwelling, for winter or summer periods, for weekdays or weekends.
- The Wilcoxon test was applied to each hour of the day grouped by the variables "room", "summer or winter" and "weekdays or weekends". In this sense, the median and confidence interval calculated was performed for each hour from 1 a.m. to 24 p.m. for a typical day.
- For generating the non-parametric confidence intervals, the 80% confidence level was used, as it could represent an adequate amplitude range of uncertainties.

It should be emphasized that, for data treatment purposes, other parametric distributions were tested to verify the adherence of the data based on some degree of significance. No parametric distribution could overcome, satisfactorily, the variability of the sample.

The data treatment outcomes were a set of diversity profiles with 80% confidence range, which indicated the possibility of working with a probabilistic approach.

Indeed, some annual schedules were created for the occupancy and operation of doors and windows based on the following procedure:

- The profile values were considered as the probability of occurrence (*i.e.* occupation or operation); thus, some random days were generated based on this assumption.
- The random days were generated by using the Latin Hypercube Sampling algorithm with the function "LHS" from the R package "pse" developed by [32].
- The annual schedule respected the summer/winter and weekdays/weekends characteristics and was generated for each room (bedrooms, living room and kitchen) and each confidence interval segment (lower limit, median and upper limit).

Developing Schedules for Equipment and Lighting Usage

Regarding this subject of appliances, the data obtained from the questionnaires were organized for each hour of the day for both summer and winter conditions. The duration of use of each appliance was the known variable.

The following process was executed:

- The average power of the electric appliance was calculated by having the electricity consumption and duration of use; this was calculated for each appliance for each dwelling according to eq. (**3**).
- The average power of the type of electric appliance was calculated for all dwellings, by having the average power in each dwelling and the number of appliances in the sample of dwellings.
- Each set of values ranging from 1 a.m. to 24 p.m. represents a schedule of use of equipment for each room (bedrooms, living room and kitchen), as can be seen in eq. (**4**).
- The fraction of power usage represents daily schedules for appliances and lighting. For the appliances, only a median schedule was obtained due to the difficulty of treating large amounts of data. For the lighting schedules, the same procedure for occupancy and operation of openings was adopted to create an 80% confidence interval for the median profile.

$$AvgPwD_i = \sum_{j=0}^{n} \frac{AvgPw_{ij}}{n} \qquad\qquad (3)$$

$$FPw_k = \frac{\sum\limits_{i=0}^{m} FPw_{ik} \times AvgPwD_i \times n_i}{n \times \sum\limits_{i=0}^{m} AvgPw_i} \tag{4}$$

Where: $AvgPw_{ij}$ is the average power of each appliance i in each dwelling j [W]; is the average power of each appliance i considering all sample of dwellings [W]; n is the sample size of dwellings [non-dimensional]; FPW_{ik} is the fraction of power of each appliance i in each hour of the day k, for each room of the dwelling [non-dimensional]; m is the total number of appliances; n_i is the number of dwelling that has the device i ; FPW_K is the fraction of power of each room for the hour of the day K. For the values of the average power with electric appliances and lighting, the normal distribution was used to calculate the confidence intervals considering 90% confidence according to the Student t distribution.

Representation of the Schedules

This section is intended to show the graphs of the schedules, *i.e.* the diversity profiles, based on the above procedures of data acquisition and data treatment.

Schedules of Occupancy

The schedules of occupancy were created for each room, by summer or winter conditions, and by Saturday, Sunday and Weekdays. The treatment was performed according to the Wilcoxon non-parametric test considering all the houses sample and 80% confidence interval for the Median. Fig. (**3**) shows the schedules of occupancy for the summer period, while Fig. (**4**) shows the same for the winter period.

For numerical purposes, the values of the schedules can represent the fraction of occupancy in each room for each hour of the day. For a year, the schedules can also represent the probability of occupancy of each room in each hour of the day.

The schedules reported here were the daily profiles; the annual schedules cannot be reported in a written document due to its size (each one has 8760 rows of random data). It can be noticed that the schedules had a lower amplitude for the bedrooms than for the other rooms (living room and kitchen). This represents the higher uncertainty for dealing with this type of data in building simulation experiments. The bedroom profiles showed that, for weekdays, the median behaviour is the occupancy from 24 p.m. to 5 a.m. with higher confidence;

however, due to the sample diversity, the bedroom can be occupied up to 10 a.m. for summer (see Fig. **3**), with a low probability of occurrence. In the same sense, it is least probable that the bedroom would be occupied from 12 a.m. to 9 p.m. for weekdays over summer. Similar behaviour was observed for the winter period, but with relative different values in the profiles.

For the living room and kitchen, the profiles are very uncertain, *i.e.*, show high amplitude ranges (see Fig. **3**). The confidence intervals showed that there is some probability of low occupancy in these rooms over one typical day, and it is very hard to be occupied by all people at the same time (which is referred to the value of 1.0). For weekdays at the living room the median value is 1.0 at 9 p.m., but lower values for the rest of the day.

Schedules of the Operation of Windows and Doors

The schedules of the operation of windows and doors were treated separately for each room and summer and winter periods. It was considered the Wilcoxon non-parametric test with 80% confidence interval for the median. Fig. (**5**) shows the operation of windows and doors for the summer period, while Fig. (**6**) shows the same for the winter period.

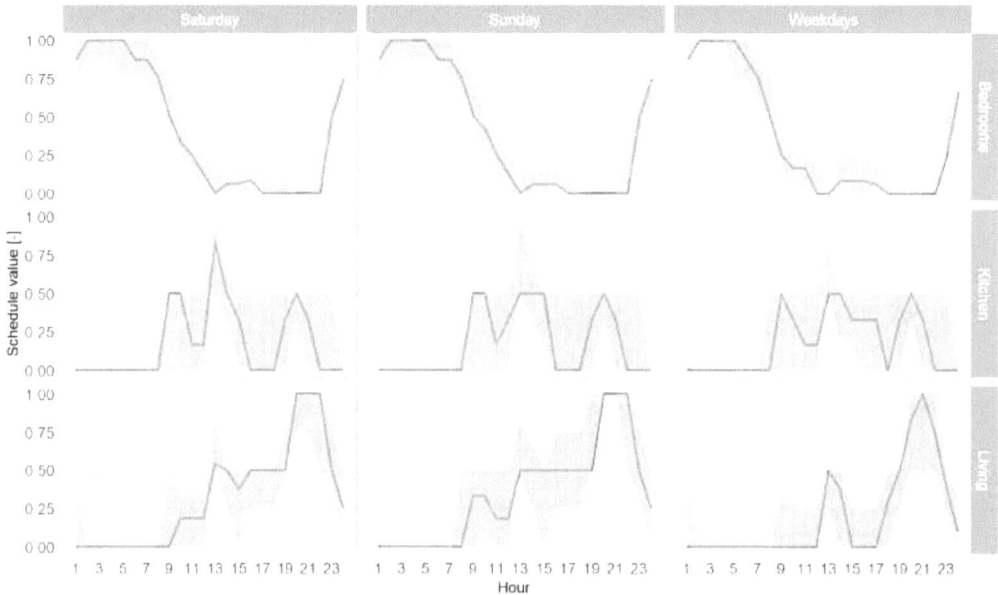

Fig. (3). Schedules of occupancy of the rooms for each day type of Saturday, Sunday and Weekdays for the Median, along with the 80% confidence interval (the grey shade) for summer.

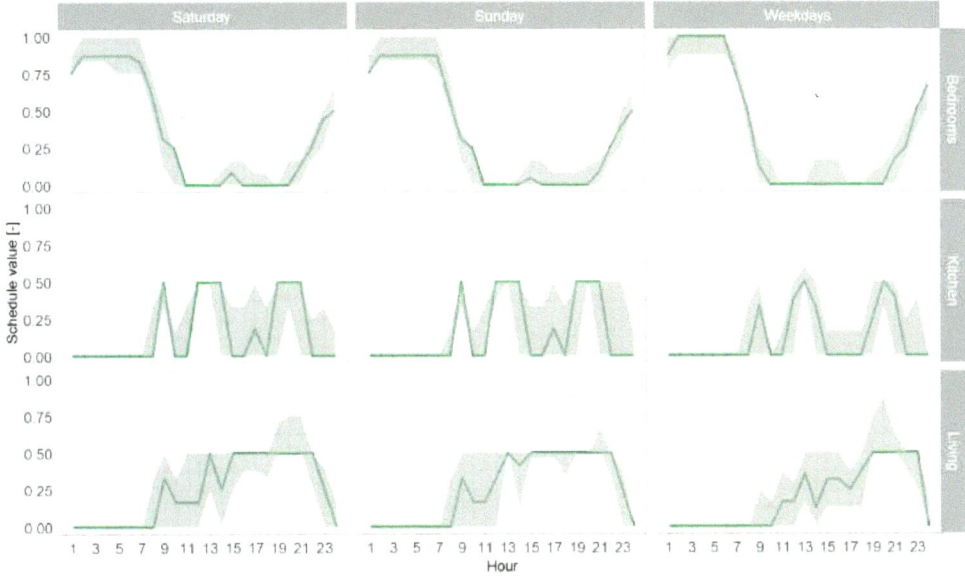

Fig. (4). Schedules of occupancy of the rooms for each day type of Saturday, Sunday and Weekdays for the Median, along with the 80% confidence interval (the grey shade) for winter.

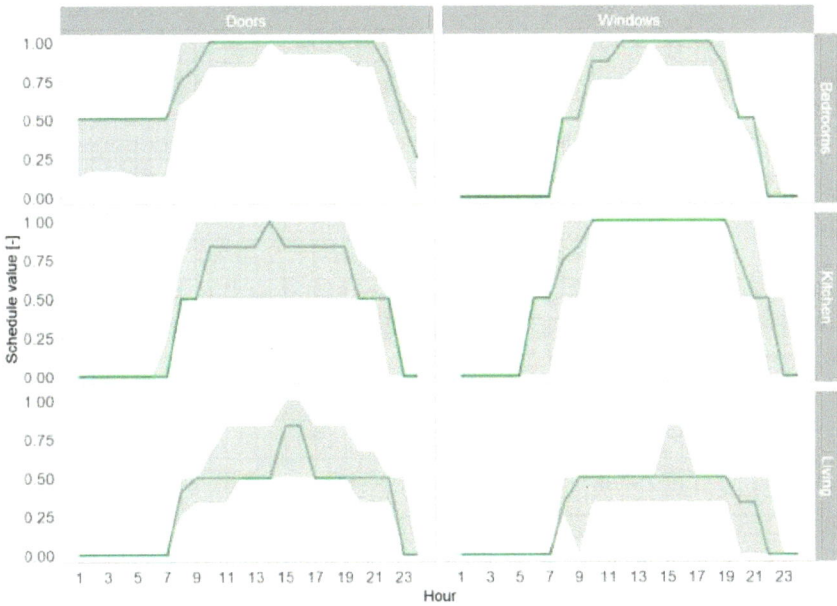

Fig. (5). Schedules of operation of doors and windows of the rooms for the Median, along with the 80% confidence interval (the grey shade) for summer.

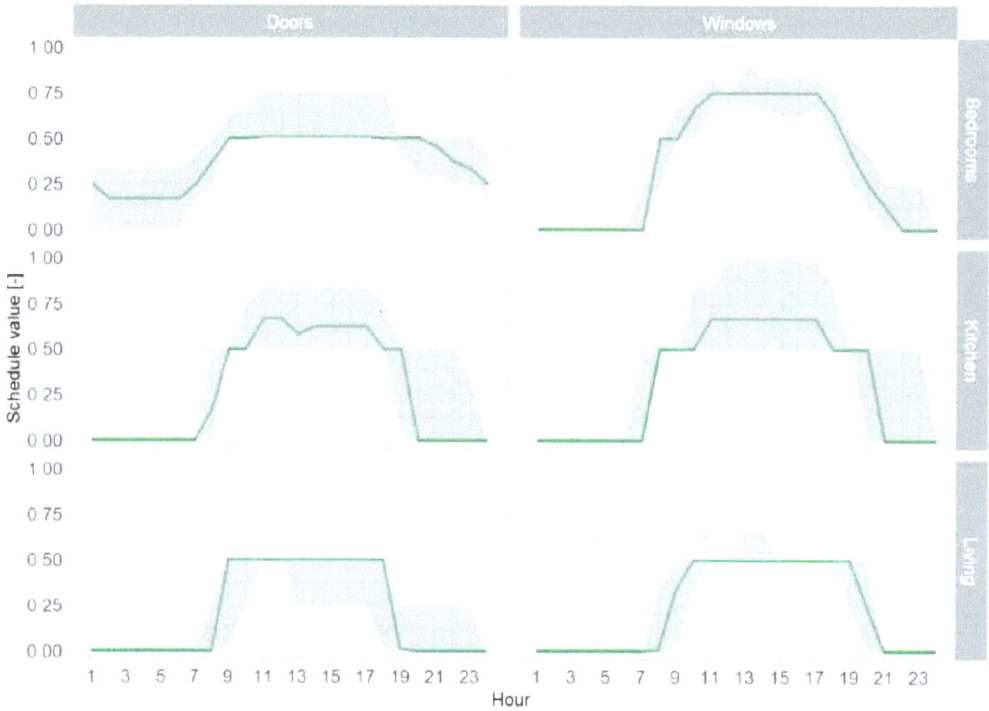

Fig. (6). Schedules of operation of doors and windows of the rooms for the Median, along with the 80% confidence interval (the grey shade) for winter.

Annual schedules were also generated based on the daily profiles but cannot be reported here due to the space required.

In general, the profiles state that the windows and doors are opened during the day, not during the night. And the probability of opening is lower in the winter condition than in the summer period.

Equipment and Lighting Schedules

The schedules of electric appliances usage, according to the data treatment, was performed for each room and is considered a median profile for all the sample of dwellings. The schedules are shown in Fig. (7), and they represent a fraction of the power used in each hour of the day in that room. It should be emphasized that these schedules are not supposed to be used separately from the average power obtained by this research.

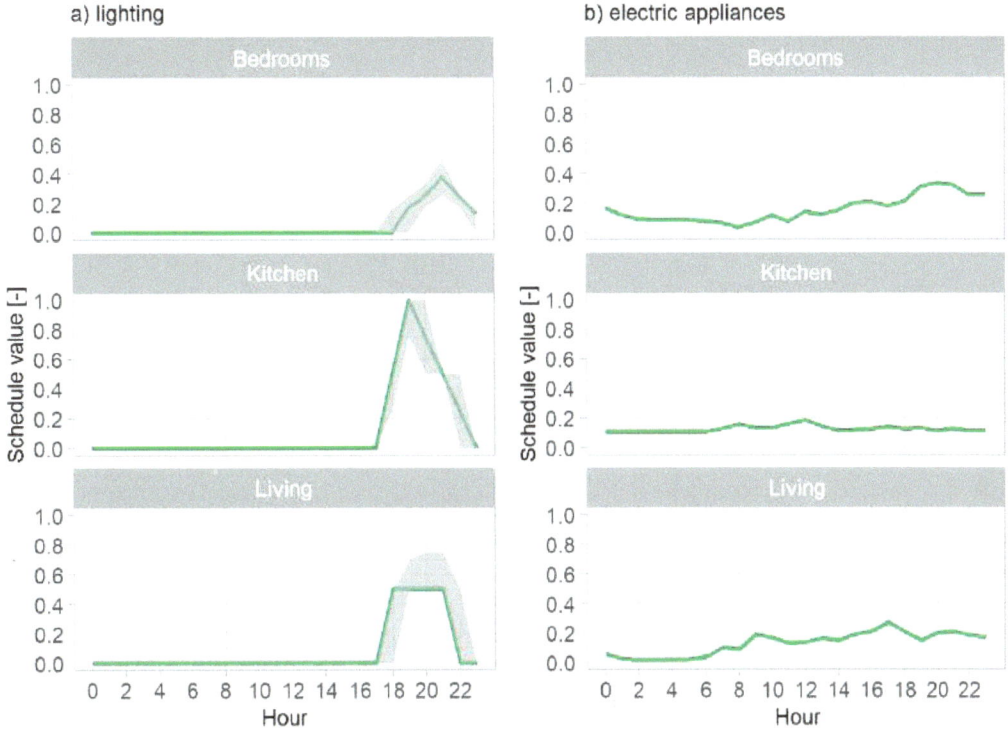

Fig. (7). Schedules of the usage of the (a) lighting and (b) electric appliance for each room, the grey shade show the 80% confidence interval (only for the lighting).

The schedules of lighting usage were obtained for each room and by the statistical procedure of the non-parametric Wilcoxon test considering all the sample of dwellings. Thus, the median curve is shown along with the limits represented by the grey shade in the graph. Fig. (**7a**) shows the schedules of lighting usage for each room, in which the values represent a fraction of power used in each hour of the day.

From the electric appliances schedule (Fig. **7b**), it can be noticed that in all rooms there will be some electricity consumption occurring every hour. It is common in this type of house to have television or another electronic device in standby mode, in the bedrooms so as in the living room.

It should be remembered that the values obtained for the fraction of usage of electric appliances are low (always lower than 0.4) because they represent a fraction of the total average power installed in each room. In this sense, it is hard to have all appliances turned on at the same time.

The same is observed for lighting usage (see Fig. **7a**). The values of the profile represent a fraction of the total average power installed with lights in that room. The low values indicate that, for that specific hour of the day, the lights were not turned on the entire hour – but only for a fraction of it. Another observation is that, for this sample of dwellings and the 80% confidence interval, there were no events of use of lighting during the day – only in the night period. However, this observation does not indicate that the daylighting is effectively used by the dwellings or that there would be adequate illuminance levels for the activities during the day.

Internal Loads with Equipment and Lighting

The installed power with appliances and lighting were calculated in W/m^2, to generalize the power per floor area of each room. This indicator is called "power density" and was calculated by using the t-test with a 90% confidence interval for the mean. In this sense, these values of power density only make sense when used together with the schedules showed above.

Table **1** shows the values of the power density for the electric appliances and lighting for each room, based on the lower, mean and upper limits with 90% confidence.

Table 1. 90% confidence interval for the appliances and lighting power density in each room.

Level	Appliances Power Density [W/m^2]			Lighting Power Density [W/m^2]		
	Kitchen	Bedrooms	Living room	Kitchen	Bedrooms	Living room
Lower	59.30	10.21	12.51	2.16	3.35	1.62
Mean	71.89	18.28	19.31	2.59	3.82	2.01
Upper	84.47	26.36	26.10	3.03	4.29	2.40

BUILDING PERFORMANCE SIMULATION

This section begins with a description of the building performance simulation procedures. While the previous section showed the operational schedules, this section shows the settings for the simulation experiments by considering those schedules as input variables, along with other operational inputs as well.

Building Simulation Programme

The EnergyPlus™ v. 9.3 [33] was used as the simulation engine to perform the calculations. This programme was developed from other simulation programmes (*e.g.* BLAST and DOE-2) by the U.S. Department of Energy's (DOE) and is maintained by the National Renewable Energy Laboratory (NREL). The programme is a mathematical model with a complex formulation of the physical, thermal, mechanical, optical phenomena to simulate the performance of a building submitted to external/internal conditions.

The Sketchup tool and the OpenStudio programme [34] were used to develop the geometry of the building model with some prior configurations. The IDF Editor form EnergyPlus was used to set all configurations needed to simulate the required models.

For all simulation runs, some specific classes of objects were used from EnergyPlus, such as:

- Air Flow Network: this class of objects can simulate the airflow through thermal zones based on the input data from the openings and pressure coefficients for the external surfaces. The classes of "SimulationControl", "MultiZone:Zone", "MultiZone:Surface" and "MultiZone:Component:DetailedOpening" were used.
- Ground Domain: this class of objects was used to simulate the heat transfer through the ground with the floor of each thermal zone of the building. The "Undisturbed Finite Difference" method was used to estimate the initial ground temperatures. Some properties of the ground materials should be informed to perform the calculation.
- HVAC PTHP: this object was used only in the "Hybrid Ventilation" mode. This represents a template for air-conditioning based on some information of the fans and coils.
- EMS (Energy Management System): this class of objects was used to provide specific controls regarding the hybrid ventilation, such as the setpoints for open the windows and the conditions for operating the HVAC system based on the occupancy and the operative temperatures.

Table **2** shows the general settings used for the building performance simulation using the EnergyPlus™ programme.

Table 2. General simulation settings used for building performance simulations.

- EnergyPlus version = 9.3.
- Solar beam distribution = FullExterior
- Timestep = 6.
- Location: Latitude = -27.6°, Longitude = -48.55°, TimeZone = -3, Elevation = 2.0m.
- Surface convection algorithm = TARP (inside) and DOE-2 (outside).
- Heat balance algorithm = Condution Transfer Function (CTF).
- Ground reflectance = 0.2 (for all months).
- Wind pressure coefficients = Swami and Chandra [35] equations for low rise buildings.

Building Model

The chosen building model was a reference building for the Brazilian social housing programme called "Minha Casa Minha Vida" of the Federal Government. This Government social programme established limits for the building size and cost to provide public funding with low-interest rates. The building design was given by the Brazilian Bank "Caixa Econômica Federal", responsible for funding the construction of social housing through the country. The building has 41,87m² of useful area, with two bedrooms, living room, kitchen, and a bathroom.

The actual building design has an attic with a gable roof condition. Fig. (8) shows the design floor plan, along with the North axis and the roof projection. It should be emphasized that the construction materials and layers are defined in the simulation programme as numerical information, not as a visual aspect of the model. That is why the roof was simplified as a flat roof in the drawing, but considered with all its thermal properties of a gable roof with an air gap in the model. This model is considered as a reference model. In the further sections, the simulation experiment will be explained along with the changing of some parameters for each scenario.

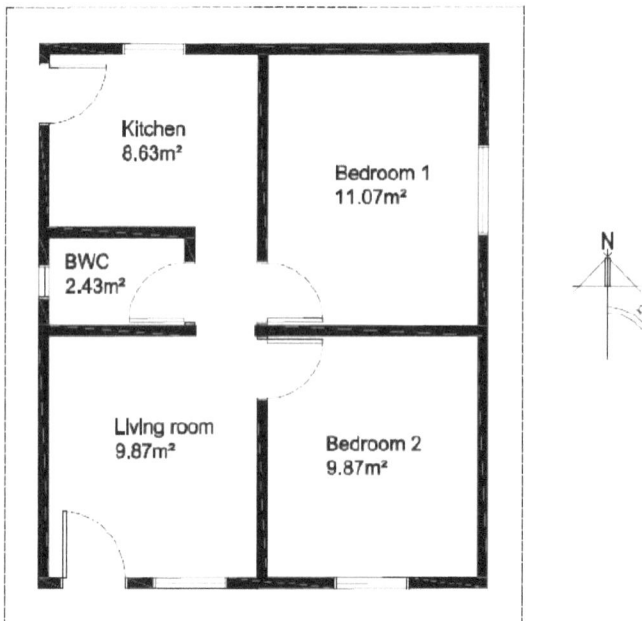

Fig. (8). Building Model **(a)** building floor plan and **(b)** Open Studio 3D model.

CONSTRUCTION AND MATERIALS

The reference model was varied into five different construction systems, *i.e.*, different construction components for its walls, roof and floor. The variation of the construction components aims to observe, in the further simulation experiment, what would be the effects of the operational uncertainties by varying the construction and materials.

Some materials were selected for developing a set of different construction components. Table **3** shows the materials used, in which construction they were considered, and information regarding the density (ρ), thermal conductivity (λ) and specific heat (c). Table **4** shows the construction components created for this study, for walls, roofs and floor with the calculated properties of thermal transmittance (U), thermal capacity (Ct), solar absorptance (α) and total thickness of the element. Table **5** shows the construction system used in each scenario and which construction component is considered in each case.

Table 3. Materials used for creating the construction, along with thermal properties.

Material	Construction	ρ (kg/m³)	λ (W/mK)	c (kJ/kgK)
Asbestos tile	Roof_1, Roof_2	2000.0	0.95	0.84
Cast concrete	Wall_4a	2400.0	1.75	1.00
Cast concrete	Wall_4b, Roof_3, Roof_4, Floor_1	2200.0	1.75	1.00
Cement board	Wall_2	1500.0	0.65	0.84
Cement mortar	Wall_3	2100.0	1.15	1.00
Ceramic (for brickwork)	Wall_3	1300.0	0.90	0.92
Ceramic (for brickwork)	Wall_5a, Wall_5b	2000.0	1.05	0.92
Ceramic (for floor)	Floor_1	1200.0	0.90	0.92
Ceramic (for roof tile)	Roof_3, Roof_4	1400.0	0.90	0.92
OSB	Wall_1, Roof_1	550.0	0.12	2.30
Plasterboard	Wall_2, Roof_2	900.0	0.35	0.84
Rock wool insulation	Wall_2, Wall_5a, Roof_2, Roof_4	80.0	0.05	0.75

Where: ρ is the density, λ is the thermal conductivity and c is the specific heat.

Table 4. Construction components and layers for walls, roofs and floor, along with information of thermal properties.

Construction	Layers	U [W/m²K]	Ct [kJ/m²K]	α [-]	Thickness [cm]
Wall_1	OSB + air gap + OSB	1.72	37.9	0.5	5.0

(Table 4) cont.....

Wall_2	Cement board + rock wool insulation + plasterboard	0.90	30.4	0.5	6.7
Wall_3	Cement mortar + hollow brickwork + cement mortar	2.59	94.2	0.5	11.5
Wall_4a	On site cast concrete	3.91	360.0	0.5	15.0
Wall_4b	On site cast concrete	4.64	176.0	-	8.0
Wall_5a	Massive brickwork single + rock wool insulation + Massive brickwork double	0.81	352.0	0.5	23.0
Wall_5b	Massive brickwork double	3.30	257.6	-	14.0
Roof_1	Asbestos tile + air gap + OSB ceiling	1.96	22.7	0.6	-
Roof_2	Asbestos tile + air gap + rock wool insulation + plasterboard ceiling	0.74	21.0	0.6	-
Roof_3	Ceramic tile + air gap + concrete slab	2.10	188.9	0.6	-
Roof_4	Ceramic tile + air gap low-e + rock wool insulation + concrete slab	0.57	191.3	0.6	-
Floor	Ceramic + concrete	5.08	187.0	-	9.0

Where: U is the thermal transmittance, Ct is the thermal capacity, α is the solar absorptance.

Table 5. Construction system used in the simulation experiment.

Construction	System Setting	Type	External Wall	Internal Wall	Roof	Floor	Windows	Doors
1	Wood framework	Light	Wall_1	Wall_1	Roof_1	Floor_1	Simple transparent 3mm glass	Wooden door
2	Light steel framing	Light and insulated	Wall_2	Wall_2	Roof_2			
3	Masonry of hollow brickwork	Medium	Wall_3	Wall_3	Roof_3			
4	Cast concrete	Heavy	Wall_4a	Wall_4b				
5	Masonry of double massive brickwork with insulation	Heavy and insulated	Wall_5a	Wall_5b	Roof_4			

The values for each thermal property and the calculation methods were estimated based on the Brazilian standard NBR 15220-2 [36].

Thus, five different construction systems were considered, as being (1) wood framework – lightweight, (2) Light steel framing (light and insulated), (3) masonry of hollow brickwork – medium weight, (4) cast concrete – heavyweight, and (5) masonry of double massive brickwork with insulation – heavy and insulated. The same composition of the floor, window glass and doors was considered for all construction systems.

Climate Consideration

As the data acquisition was performed in the Florianópolis region, this study considered only this specific climate for the simulation experiments. Florianópolis has a humid subtropical climate and is located at the littoral region of Santa Catarina, southern Brazil. According to the Köppen-Geiger [37] climate classification, the weather has the Cfa type (subtropical humid).

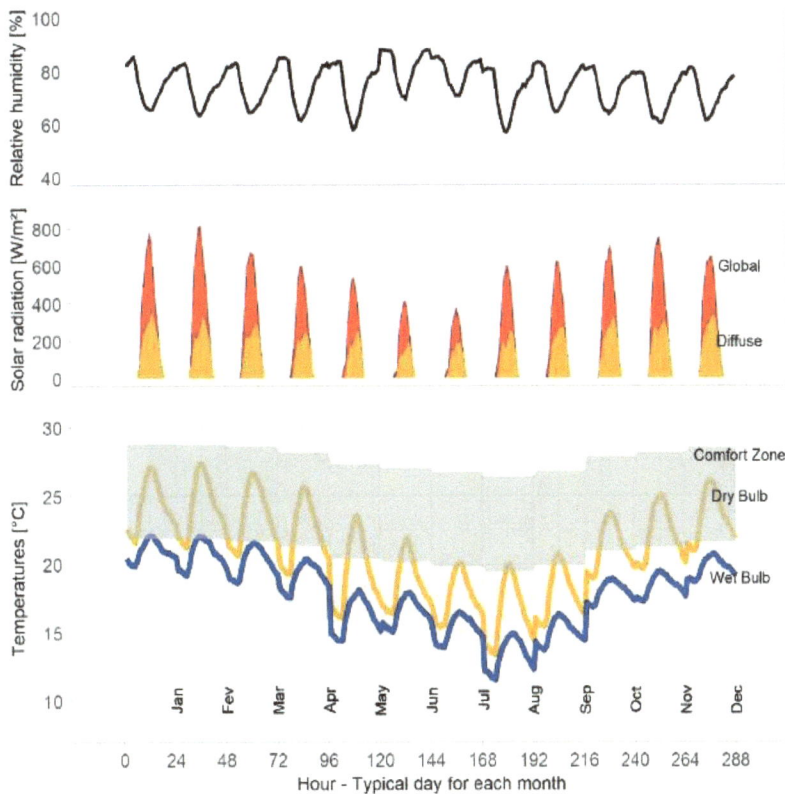

Fig. (9). Weather data for Florianópolis-SC: dry bulb, wet bulb and comfort zone temperatures, global and diffuse solar radiation and relative humidity for an average day for each month – according to the INMET weather file.

Based on the INMET (Brazil's National Institute of Meteorology) weather file of Florianópolis [38], some descriptive statistics were calculated for understanding this climate. Fig. (9) shows the relative humidity, global and diffuse solar radiation, dry bulb and wet bulb temperatures, and the thermal comfort interval for the operative temperatures. The climate has well-defined seasons (*i.e.* summer and winter); the humidity is high for the entire year, always between 60%-80% for the average day in each month; the solar radiation varies among the seasons, achieving the highest value in February and the lowest in July. The thermal comfort interval for each month was calculated by using the ASHRAE Standard 55 adaptive method [39] for 80% acceptability.

Fig. (**10**) shows information about wind direction and speed. The North and South directions showed the greatest frequency of occurrence of wind, while the highest speed was noticed in the Northeast and South directions. This information is relevant because the building is naturally ventilated and the "Air Flow Network" object from EnergyPlus™ considers all information of the wind to provide the calculation of air changes in the thermal zones. In the same sense, the building was oriented in the North-South axis, which showed the most frequency of wind for this weather file.

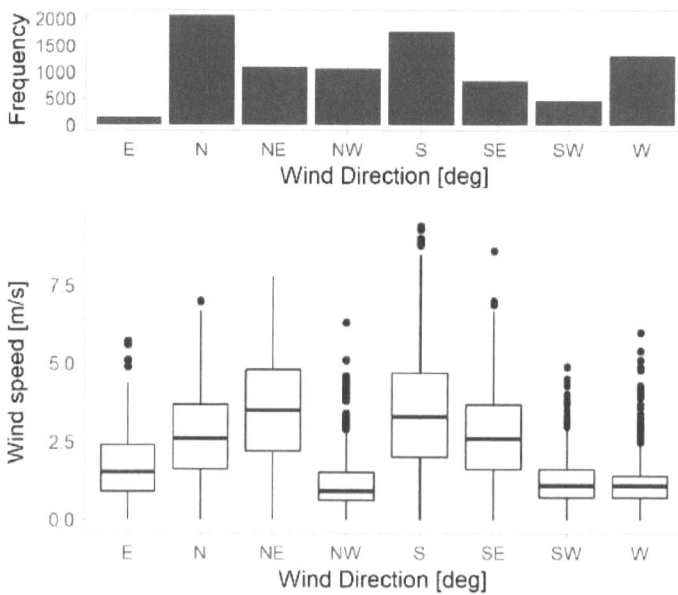

Fig. (10). Weather data for Florianópolis-SC: wind direction and wind speed for each main orientation – according to the INMET weather file.

Simulation Experiments

The simulation experiment comprehended a set of simulation with different purposes concerning the identification of influent variables on the performance criteria and the estimation of the amplitudes of uncertainty due to the occupant behaviour. Two modes of environmental conditioning were adopted: natural ventilation and hybrid ventilation modes. The objects of "Air Flow Network" and "Energy Management System" were used from EnergyPlus™ computer programme to properly configure these modes.

Fig. (11) shows the flowchart of the simulation experiments. For each mode (natural ventilation or hybrid ventilation) two experiments were performed: global sensitivity analysis and uncertainty analysis. Each experiment demanded a separated sample of inputs to be simulated in the EnergyPlus programme and different data treatment strategies.

Different approaches were needed due to the different purposes of the analyses. The global sensitivity analysis with the variance-based methods is useful to calculate total sensitivity indices with high accuracy, by considering the first-order effects and interaction between the input variables [40]. The uncertainty analysis performed with random sampling techniques is needed to estimate the confidence interval for the outputs due to the variation of the input variables, which can be defined as a forward propagation (model prediction), according to Tian *et al.* [41].

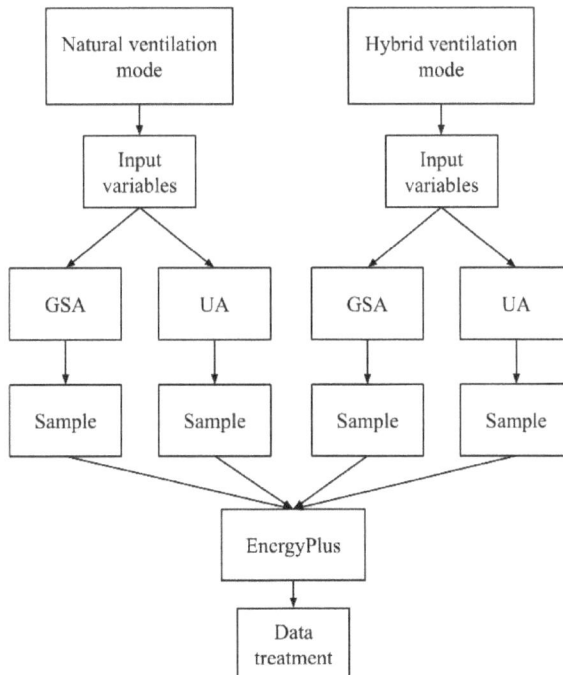

Fig. (11). Flowchart of the simulation experiment. Where: GSA is "Global Sensitivity analysis" and UA is "Uncertainty Analysis".

Natural Ventilation Mode

For the natural ventilation mode, the openings were set to be operated based on the outdoor air temperatures, the indoor air temperatures and on the availability schedule for each thermal zone (*i.e.* room).

Fig. (**12**) shows the algorithm of the Air Flow Network for the natural ventilation mode. Three conditions must be fulfilled in each timestep of the simulation to enable natural ventilation. The first is that the dry-bulb temperature should be higher than a set point temperature (T_{db} > SetpointNV); the second is the dry-bulb temperature should be lower than the indoor air temperature of the thermal zone (T_{db} < $T_{air,zone}$); the third is that the availability schedule should be different of 0 (zero). If any of these conditions are not fulfilled simultaneously for each timestep, the natural ventilation is set off.

This mechanism is based only in the verification of air temperatures, not operative temperatures, and represents the "native" configuration from EnergyPlus. In this case, the availability schedules are the schedules created for operating the windows and doors for each room. Some values were taken from the new proposal of the Brazilian regulation for residential buildings [42], such as: the air mass flow coefficient when the opening is closed was 0.001 kg/s.m; the air mass flow exponent when the opening is closed was 0.66; the discharge coefficient for 1.0 opening factor was 0.60. The SetpointNV input is considered as an independent variable for the experiment.

Table **6**, shows the input variables considered for the natural ventilation mode, with the levels defined for the two experiments (global sensitivity analysis and uncertainty analysis). 30 input variables were defined, according to Table **6**, all of them are related to operational characteristics of the building. Some details of the input variables are shown below:

Table 6. Input variables for natural ventilation mode for the two experiments: global sensitivity analysis (GSA) and uncertainty analysis (UA).

Parameter	ID	Unit	Reference	Global SA	LHS UA
Number of occupants (house)	NumOcc	occupants	4	D{2; 4; 6}	D{(2; 4; 6)(0.2; 0.6; 0.2)}
Radiant fraction of occupants (house)	RadFOcc	-	0.6	D{0.27; 0.435; 0.60}	T{0.27; 0.435; 0.60}
Activity level of occupants (bedroom)	ActOccBed	W/person	72	D{72; 76.5 ; 81}	T{72; 76.5 ; 81}

(Table 6) cont.....

Parameter	ID	Unit	Reference	Global SA	LHS UA
Schedules of occupancy (bedroom)	SchOccBed	hours/year	3229	D{2684; 3229; 4009}	D{(2684; 3229; 4009)(0.2; 0.6; 0.2)}
Activity level of occupants (living room)	ActOccLiv	W/person	108	D{81; 103.5; 126}	T{81; 103.5; 126}
Schedules of occupancy (living room)	SchOccLiv	hours/year	1785	D{889; 1785; 2996}	D{(889; 1785; 2996)(0.2; 0.6; 0.2)}
Activity level of occupants (kitchen)	ActOccKit	W/person	126	D{108; 117; 126}	T{108; 117; 126}
Schedules of occupancy (kitchen)	SchOccKit	hours/year	1623	D{498; 1623; 3072}	D{(498; 1623; 3072)(0.2; 0.6; 0.2)}
Average power of equipment (bedroom)	PwEquipBed	W/m^2	18.28	D{10.21; 18.28; 26.36}	T{10.21; 18.28; 26.36}
Average power of equipment (living room)	PwEquipLiv	W/m^2	19.31	D{12.51; 19.31; 26.10}	T{12.51; 19.31; 26.10}
Average power of equipment (kitchen)	PwEquipKit	W/m^2	71.89	D{59.30; 71.89; 84.47}	T{59.30; 71.89; 84.47}
Radiant fraction of equipment (house)	RadFEquip	-	0.5	D{0.3; 0.55; 0.8}	T{0.3; 0.55; 0.8}
Average power of lighting (bedroom)	PwLightBed	W/m^2	3.82	D{3.35; 3.82; 4.29}	T{3.35; 3.82; 4.29}
Schedules of lighting (bedroom)	SchLightBed	hours/day	1.167	D{0.584; 1.167; 1.686}	D{(0.584; 1.167; 1.686)(0.2; 0.6; 0.2)}
Average power of lighting (living room)	PwLightLiv	W/m^2	2.01	D{1.62; 2.01; 2.40}	T{1.62; 2.01; 2.40}
Schedules of lighting (living room)	SchLightLiv	hours/day	2	D{1.5; 2.0; 3.2}	D{(1.5; 2.0; 3.2)(0.2; 0.6; 0.2)}
Average power of lighting (kitchen)	PwLightKit	W/m^2	2.59	D{2.16; 2.59; 3.03}	T{2.16; 2.59; 3.03}
Schedules of lighting (kitchen)	SchLightKit	hours/day	3	D{2.0; 3.0; 3.5}	D{(2.0; 3.0; 3.5)(0.2; 0.6; 0.2)}
Radiant fraction of luminaires (house)	RadFLum	-	0.72	D{0.37; 0.545; 0.72}	T{0.37; 0.545; 0.72}
Set point temp. for ventilation control (except over winter)	SetpointNV	°C	20	D{18; 20; 22; 24}	D{(18; 20; 22; 24)(0.1, 0.4, 0.4, 0.1)}

(Table 6) cont.....

Parameter	ID	Unit	Reference	Global SA	LHS UA
Set point temp. for ventilation control over winter	SetpointNVwinter	°C	22	D{18; 20; 22; 24}	D{(18; 20; 22; 24)(0.1, 0.4, 0.4, 0.1)}
Window opening factor	OpenFactorW	-	0.5	D{0.3; 0.5; 0.7; 0.9}	T{0.3; 0.6; 0.9}
Temp. difference lower limit for maximum opening factor	TempDiffOpenLL	°C	0	D{0; 2; 4}	D{(0; 2; 4)(0.2; 0.6; 0.2)}
Temp. difference upper limit for minimum opening factor	TempDiffOpenUL	°C	40	D{10; 20; 30; 40}	D{(10; 25; 40)(0.2; 0.6; 0.2)}
Availability sch. operation of windows (bedroom)	AvailWBed	hours/year	3683	D{2954; 3683; 4206}	D{(2954; 3683; 4206)(0.2; 0.6; 0.2)}
Availability sch. operation of windows (living room)	AvailWLiv	hours/year	2239	D{1627; 2239; 2739}	D{(1627; 2239; 2739)(0.2; 0.6; 0.2)}
Availability sch. operation of windows (kitchen)	AvailWKit	hours/year	4058	D{3330; 4058; 5200}	D{(3330; 4058; 5200)(0.2; 0.6; 0.2)}
Availability sch. operation of doors (bedroom)	AvailDBed	hours/year	5099	D{3799; 5099; 5978}	D{(3799; 5099; 5978)(0.2; 0.6; 0.2)}
Availability sch. operation of doors (living room)	AvailDLiv	hours/year	2434	D{1630; 2434; 3384}	D{(1630; 2434; 3384)(0.2; 0.6; 0.2)}
Availability sch. operation of doors (kitchen)	AvailDKit	hours/year	3255	D{2232; 3255; 4510}	D{(2232; 3255; 4510)(0.2; 0.6; 0.2)}

- The NumOcc variable represents the number of occupants of the house, which was divided by the rooms. When the input is equal to 2, each room receives 1 occupant and when the value is 6, each bedroom receives 3 occupants. This consideration is based on the uncertainty in the number of occupants in the dwelling.

- The RadFOcc represents the radiant fraction of the occupants, which was set between 0.27 – 0.60 according to the Handbook of Fundamentals [43]. This variable is related to the thermal heat transfer by the radiation phenomena and interferes in the heat balance of the room.

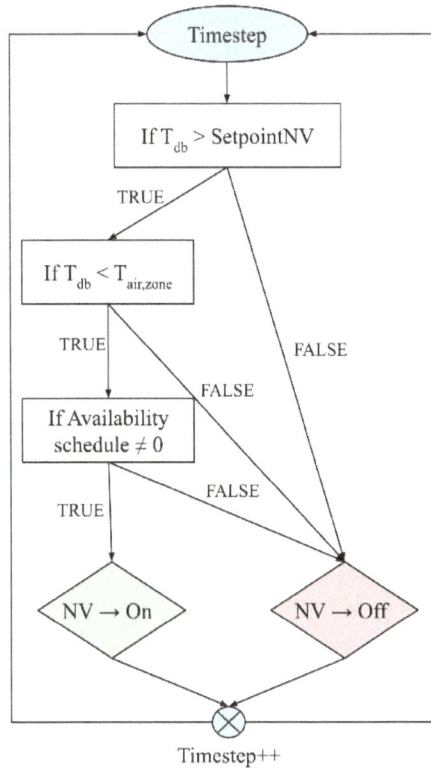

Fig. (12). Natural ventilation function settings for Air Flow Network object. Where: SetpointNV is the air temperature setpoint for natural ventilation; NV is natural ventilation.

- The activity level (*i.e.* metabolic rate) was set for each room based on the activity. An interval from $72 - 81$ W/person was set for the bedroom, $81 - 126$ W/person for the living room and $108 - 126$ W/person for the kitchen. These values are based on the tables from [43] and are related to the uncertainty in the internal heat gains with the occupants.

- The radiant fraction of equipment and luminaires were set as a reference value of 0.50 and 0.72, respectively, based on the recommendation from the EnergyPlus Input/Output Reference manual. Different intervals were set for each experiment to represent uncertainties regarding the types of equipment and luminaires disposal.

- The windows opening factor was varied from $0.3 - 1.0$ to represent the fraction of the window opened, which depends on the type of fenestration and behavioural patterns. In this sense, two setpoint variables were defined to modulate this opening factor for summer or winter.

- The setpoint for operating the openings and enable natural ventilation was varied from 18°C – 24°, separately for summer and winter periods. This is the SetpointNV variable from Fig. (**12**) and enables natural ventilation when the outdoor air temperature is higher than the temperature setpoint value.

- All the further variables were defined previously, and are represented by the lower, mean/median and upper values: SchOccBed, SchOccLiv, SchOccKit, SchLightBed, SchLightLiv, SchLightKit, AvailWBed, AvailWLiv, AvailWKit, AvailDBed, AvailDLiv, AvailDKit, PwEquipBed, PwEquipLiv, PwEquipKit, PwLightBed, PwLightLiv, PwLighKit.

Hybrid Ventilation Mode

The hybrid ventilation algorithm was adopted from the Brazilian new proposal of an energy efficiency labelling for residential buildings [42]. Fig. (**13**) shows the algorithm; a first verification of the zone operative temperature is performed, and the natural ventilation is only enabled when the operative temperature is within the comfort range (lower than EMSOperativeUL and greater than EMSOperativeLL, which were independent variables for this mode). In this case, the verification of the outdoor air temperature with the setpointNV and with the zone air temperature is also performed, the same way as for the previous mode. If the zone operative temperature is outside the comfort range (TRUE arrow), the verification of the occupancy schedule is performed: if there is occupancy (TRUE) the HVAC is set on, if there is no occupancy (FALSE) the HVAC is set off. This verification is performed for each timestep and each room separately.

The same properties for the windows, doors and pressure coefficients were maintained in this hybrid ventilation mode such as in the natural ventilation mode.

Table **7** shows the input variables considered for the hybrid ventilation mode. Most of the variables are the same as the natural ventilation mode, except those related to the EMS settings and the air-conditioning objects.

Table 7. Input variables for hybrid ventilation mode for the two experiments: sensitivity analysis (GSA) and uncertainty analysis (UA).

Parameter	ID	Unit	Reference	Global SA	LHS UA
Number of occupants (house)	NumOcc	occupants	4	D{2; 4; 6}	D{(2; 4; 6)(0.2; 0.6; 0.2)}
Radiant fraction of occupants (house)	RadFOcc	-	0.6	D{0.27; 0.435; 0.60}	T{0.27; 0.435; 0.60}
Activity level of occupants (bedroom)	ActOccBed	W/person	72	D{72; 76.5 ; 81}	T{72; 76.5 ; 81}

(Table 7) cont.....

Parameter	ID	Unit	Reference	Global SA	LHS UA
Schedules of occupancy (bedroom)	SchOccBed	hours/year	3229	D{2684; 3229; 4009}	D{(2684; 3229; 4009)(0.2; 0.6; 0.2)}
Activity level of occupants (living room)	ActOccLiv	W/person	108	D{81; 103.5; 126}	T{81; 103.5; 126}
Schedules of occupancy (living room)	SchOccLiv	hours/year	1785	D{889; 1785; 2996}	D{(889; 1785; 2996)(0.2; 0.6; 0.2)}
Activity level of occupants (kitchen)	ActOccKit	W/person	126	D{108; 117; 126}	T{108; 117; 126}
Schedules of occupancy (kitchen)	SchOccKit	hours/year	1623	D{498; 1623; 3072}	D{(498; 1623; 3072)(0.2; 0.6; 0.2)}
Average power of equipment (bedroom)	PwEquipBed	W/m^2	18.28	D{10.21; 18.28; 26.36}	T{10.21; 18.28; 26.36}
Average power of equipment (living room)	PwEquipLiv	W/m^2	19.31	D{12.51; 19.31; 26.10}	T{12.51; 19.31; 26.10}
Average power of equipment (kitchen)	PwEquipKit	W/m^2	71.89	D{59.30; 71.89; 84.47}	T{59.30; 71.89; 84.47}
Radiant fraction of equipment (house)	RadFEquip	-	0.5	D{0.3; 0.55; 0.8}	T{0.3; 0.55; 0.8}
Average power of lighting (bedroom)	PwLightBed	W/m^2	3.82	D{3.35; 3.82; 4.29}	T{3.35; 3.82; 4.29}
Schedules of lighting (bedroom)	SchLightBed	hours/day	1.167	D{0.584; 1.167; 1.686}	D{(0.584; 1.167; 1.686)(0.2; 0.6; 0.2)}
Average power of lighting (living room)	PwLightLiv	W/m^2	2.01	D{1.62; 2.01; 2.40}	T{1.62; 2.01; 2.40}
Schedules of lighting (living room)	SchLightLiv	hours/day	2	D{1.5; 2.0; 3.2}	D{(1.5; 2.0; 3.2)(0.2; 0.6; 0.2)}
Average power of lighting (kitchen)	PwLightKit	W/m^2	2.59	D{2.16; 2.59; 3.03}	T{2.16; 2.59; 3.03}
Schedules of lighting (kitchen)	SchLightKit	hours/day	3	D{2.0; 3.0; 3.5}	D{(2.0; 3.0; 3.5)(0.2; 0.6; 0.2)}
Radiant fraction of luminaires (house)	RadFLum	-	0.72	D{0.37; 0.545; 0.72}	T{0.37; 0.545; 0.72}
Set point temp. for ventilation control	SetpointNV	°C	22	D{18; 20; 22; 24}	D{(18; 20; 22; 24)(0.1, 0.4, 0.4, 0.1)}
Window opening factor	OpenFactorW	-	0.5	D{0.3; 0.5; 0.7; 0.9}	T{0.3; 0.6; 0.9}

(Table 7) cont.....

Parameter	ID	Unit	Reference	Global SA	LHS UA
Temperature difference lower limit for maximum opening factor	TempDiffOpenLL	°C	0	D{0; 2; 4}	D{(0; 2; 4)(0.2; 0.6; 0.2)}
Temperature difference upper limit for minimum opening factor	TempDiffOpenUL	°C	40	D{10; 20; 30; 40}	D{(10; 25; 40)(0.2; 0.6; 0.2)}
Upper limit for operative temperature EMS	EMSOperativeUL	°C	26	D{24; 26; 28}	D{(24; 26; 28)(0.2; 0.6; 0.2)}
Lower limit for operative temperature EMS	EMSOperativeLL	°C	16	D{16; 18; 20}	D{(16; 18; 20)(0.2; 0.6; 0.2)}
Heating Thermostat HVAC	HeatingHVAC	°C	18	D{18; 19; 20}	D{(18; 19; 20)(0.2; 0.6; 0.2)}
Cooling Thermostat HVAC	CoolingHVAC	°C	23	D{23; 24; 25}	D{(23; 24; 25)(0.2; 0.6; 0.2)}

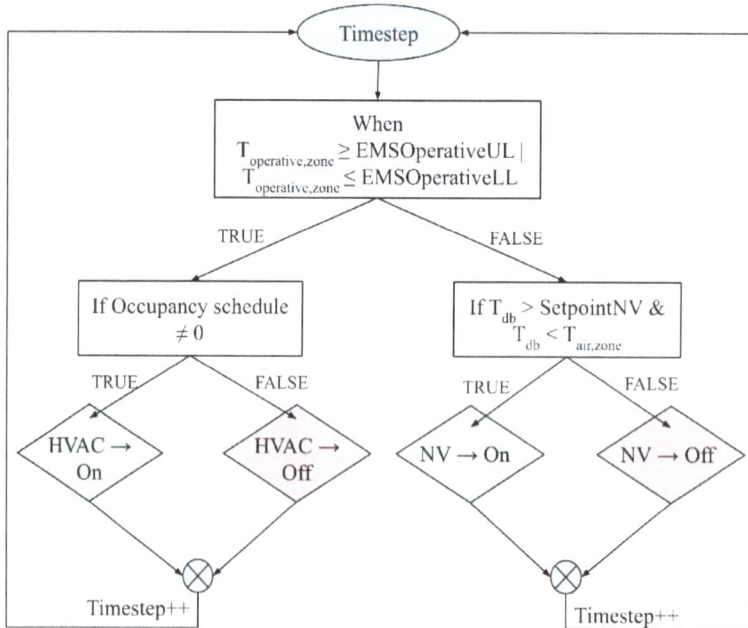

Fig. (13). Hybrid ventilation function settings for Air Flow Network, HVAC PTHP and EMS objects. Where: EMSOperativeLL is the operative temperature lower limit for the room; EMSOperativeUL is the operative temperature upper limit for the room; SetpointNV is the air temperature setpoint for natural ventilation; NV is natural ventilation.

Obs.: the code between brackets {} indicates the probability distribution set for each experiment. *E.g.* : D{2; 3; 4; 5; 6} indicates a discrete distribution and the respective levels; D{0.27 − 0.60} 10 lv indicates that the interval from 0.27 to

0.60 was divided into 10 equal discrete values; T{0.27; 0.435; 0.60} indicates the continuous triangular distribution.

- The EMSOperativeLL and EMSOperativeUL are input variables that can modulate the hybrid ventilation and they were varied in 16 – 22°C and 24 – 30°C, respectively, with a discrete distribution. The greater amplitude combination would be 16 to 30°C (*i.e.* the comfort zone), and the narrower combination would be 22 to 24°C of operative temperatures. When the comfort zone has greater amplitude, the HVAC tends to be less used.
- The heating and cooling thermostat for the HVAC was also varied, to represent different preferences for the occupants regarding the thermal adaptation and sensation. For the uncertainty analysis, the heating setpoint was varied from 18 – 20°C, and the cooling setpoint was varied from 23 – 25°C.

Output Variables

The output variables were different for each conditioning mode. For the natural ventilation mode, the EnergyPlus programme calculated the outdoor air dry-bulb temperatures, the room operative temperatures and the occupancy status for each timestep. This enables the calculation of four different performance criteria: the percentage of discomfort hours for heating, the percentage of discomfort hours for cooling, the degree-hours for heating and the degree-hours for cooling. Eqs. (**5 – 8**) shows the performance criteria calculated for each room, while eq. (**9**) shows the calculation of the weighted average applied to each criterion for the rooms weighted by the floor area.

The lower and upper limits for thermal comfort were determined using the adaptive method from the ASHRAE Standard 55 [39]. The prevailing outdoor air temperature was considered as the monthly averages of air temperatures for the Florianópolis weather file, and the lower and upper limits were calculated using the 80% acceptability equation. Fig. (**9**) has already shown the thermal comfort interval, along with the weather variables.

PDHH and PDHC are thermal comfort indices that represent the percentage of discomfort (weighted average) for the house. The value is a fraction between 0 and 1, and the sum of the two measures represents the total discomfort hours of the house concerning the occupied hours. The DHH and DHC are thermal comfort indices which represent both the discomfort period and the severity of discomfort. These indices represent long-term thermal comfort evaluation and were reported by ISO 7730 [44].

$$PDHH_{room} = \frac{\sum_{i=1}^{n} \begin{cases} if\ T_{op,i} < T_{ll,month}\ ;1 \\ if else\ ;0 \end{cases}}{n} \tag{5}$$

$$PDHH_{room} = \frac{\sum_{i=1}^{n} \begin{cases} if\ T_{op,i} > T_{ul,month}\ ;1 \\ if else\ ;0 \end{cases}}{n} \tag{6}$$

$$DHH_{room} = \sum_{i=1}^{n} \begin{cases} if\ T_{op,i} < T_{ll,month}\ ;\ T_{ll,month} - T_{op,i} \\ if else\ ;0 \end{cases} \tag{7}$$

$$DHC_{room} = \sum_{i=1}^{n} \begin{cases} if\ T_{op,i} > T_{ul,month}\ ;\ T_{op,i} - T_{ul,month} \\ if else\ ;0 \end{cases} \tag{8}$$

$$IndicesNV = \frac{IndicesNV_{room} \times A_{room}}{A_{room}} \tag{9}$$

Where: i is each occupied hour for each room; $T_{op,i}$ is the operative temperature of each room for each occupied hour; $T_{ll,month}$ is the monthly lower limit of operative temperature for comfort [°C]; is the monthly upper limit of operative temperature for comfort [°C]; $T_{ul,month}$ is the percentage of discomfort hours for heating of each room [-]; $PDHC_{room}$ is the percentage of discomfort hours for cooling of each room [-]; DHH_{room} is the degree-hours for heating of each room [°Ch]; DHC_{room} is the degree-hours for cooling of each room [°Ch]; A_{room} is the room floor area [m²]; $IndicesNV_{room}$ represent $IndicesNV_K$ each performance criteria for the rooms; represent each performance criteria for the house (*PDHH, PDHC, DHH, DHC*).

For the hybrid ventilation mode, the same performance criteria were calculated along with the additional criteria related to the HVAC. Eqs (**10 – 11**) show the thermal loads calculated for heating and cooling of the house and eq. (**12**) shows the total electricity consumption of the house due to the HVAC system.

$$TLH = \sum TLH_{room} \tag{10}$$

$$TLC = \sum TLC_{room} \tag{11}$$

$$TELC = \sum TELC_{room} \tag{12}$$

Where: TLH_{room} is the annual thermal load with heating for each room [kWh/year]; TLC_{room} is the annual thermal load with cooling for each room [kWh/year]; is the total electricity consumption of the HVAC for each room [kWh/year]; is the annual thermal load with heating for the house [kWh/year]; $TELC_{room}$ is the annual thermal load with cooling for the house [kWh/year]; TLH is the total electricity consumption of the HVAC for the house [kWh/year].

Statistical Methods and Data Treatment

Two statistical techniques were used for the experiment: global sensitivity analysis (GSA) and uncertainty analysis (UA). These techniques enable, in general, to understand the relative influence of each input variable in some specific performance criterion and to quantify the uncertainties due to the variation in the inputs. However, two different sensitivity analysis methods were coupled to explore, more deeply, the relations between inputs and output, which is also one of the capabilities of these methods that enable a better understanding of the model [45]. The global sensitivity analysis chosen was the Sobol'-Jansen method, proposed by Saltelli *et al.* [46] using the Jansen [47] estimator to compute the total sensitivity indices. The Sobol' approach is a variance-based technique, which is a model-free approach and can deal with non-linear effects and interaction between inputs. This approach was implemented with the R package "sensitivity" and the "soboljansen" function, developed by [48].

This approach is gaining space in the literature, especially due to some recent publications [49 - 51].

For the uncertainty analysis, the Latin Hypercube sampling technique was used by considering the input variables as being continuous (triangular distribution) or discrete. The sampling procedure was implemented with the R package "pse" and the function "LHS" developed by Chalom *et al.* [32] with the mathematical model from [52].

The data processing and data treatment was performed by using R programming and generating EnergyPlus input files for simulations. For the natural ventilation mode, 3200 simulation runs were performed for each construction system for the global sensitivity analysis, totalling 16,000 runs. For the uncertainty analysis, 3000 runs were performed for each construction system, totalling 15,000 runs. For the hybrid ventilation mode, 2900 runs were performed for the global sensitivity analysis for each construction, totalling 14,500 runs. And for the uncertainty analysis, 3000 runs were performed for each construction system, generating 15,000 runs.

For each analysis, the following procedure was sought:

- By having the reference model built on the Sketchup OpenStudio plugin, the "IDF" file is set for the two conditioning modes.
- Each "IDF" file is parametrized, *i.e.*, some inputs are treated as variables with specific key identification. These variables are different for each analysis (GSA and UA).
- For each analysis, an R script is run for creating the required sample by using the input variables distribution and sample size needed. These samples of inputs are formatted to stay at the exact format for using in the "IDF" file.
- Another R script is run to create as many "IDF" files as the sample size for each analysis, by having both the reference parametrized "IDF" and the sample with the key identification for each input variable.
- The group simulation tool is used to run many simultaneous simulations in the EnergyPlus programme. Many output variables in "csv" format are generated for each simulation run.
- An R script is run to treat all the "csv" files to calculate the required performance criteria for each run. In the end, a unique "csv" file is created with the summarized data for all the results of each analysis.
- The dependent variables calculated from the simulation are inserted in the R scripts for calculating or treating the results for each analysis: by creating charts and graphs, calculating sensitivity indices or descriptive statistics for the outputs.

RESULTS AND DISCUSSION

The results section was separated by the two conditioning modes (natural or hybrid ventilation) and by each performed experiment (global sensitivity and uncertainty analyses). It should be emphasized that many figures and tables were included to effectively transmit the integrity of the analyses and to help the reader to understand the capabilities and peculiarities of each one, although it can become quite extensive to interpret.

Natural Ventilation Mode

In this first conditioning mode (natural ventilation) one should remember that the performance criteria are only related to comfort conditions and no HVAC consumption are considered, and 30 input variables were considered. The performance criteria evaluated were: PDHH – the percentage of discomfort hours for heating; PDHC – the percentage of discomfort hours for cooling; DHH – degree-hours for heating; and DHC – degree-hours for cooling.

Global Sensitivity Analysis

The global sensitivity analysis was reported by showing the total sensitivity indices (St) from the Sobol'-Jansen method for each performance criteria and each construction system. One can see that, as expected, only a few input variables are responsible for the most variation in the performance criteria, especially for the heating case.

Fig. (**14**) shows the total sensitivity indices (St) for the variables in the PDHH criterion. It was clear that the number of occupants of the house (NumOcc) was the most influent variable. However, for the construction 5, the second most influent variable was the setpoint temperature for allowing natural ventilation in the winter (SetpointNVwinter), while the second in the rank for the other construction system was the schedule of occupancy in the bedrooms (SchOccBed). The same behaviour was observed for the DHH criterion (Fig. **16**).

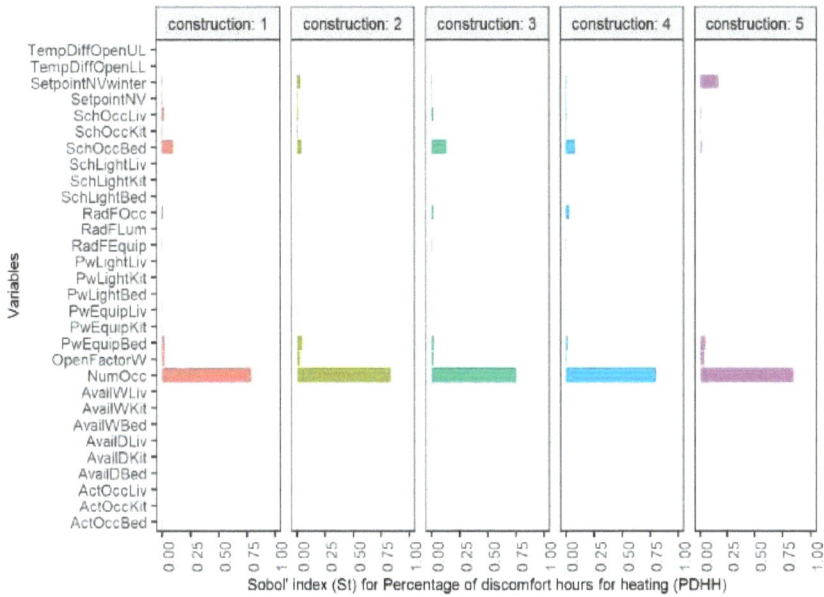

Fig. (14). Sobol' Total sensitivity indices (St) of the operational input variables for the Percentage of discomfort hours for heating (PDHH) for the global sensitivity analysis in the natural ventilation mode.

Fig. (**15**) shows the total sensitivity indices (St) for the PDHC, while Fig. (**17**) shows the same for the DHC criterion. For the PDHC criteria, the most influent variable was different for each construction system. For construction 5, the "OpenFactorW" was the most influent, followed by the "NumOcc"; for construction 2, the "OpenFactor" was the most influent followed by the SchOccBed; and for construction 1, 3 and 4 the most influent variable was the

"SchOccBed" followed by the OpenFactorW. This indicated a relatively second-order effect of the most influential variables with the construction system. The degree-hours for cooling (DHC) showed the same pattern of the PDHC criterion.

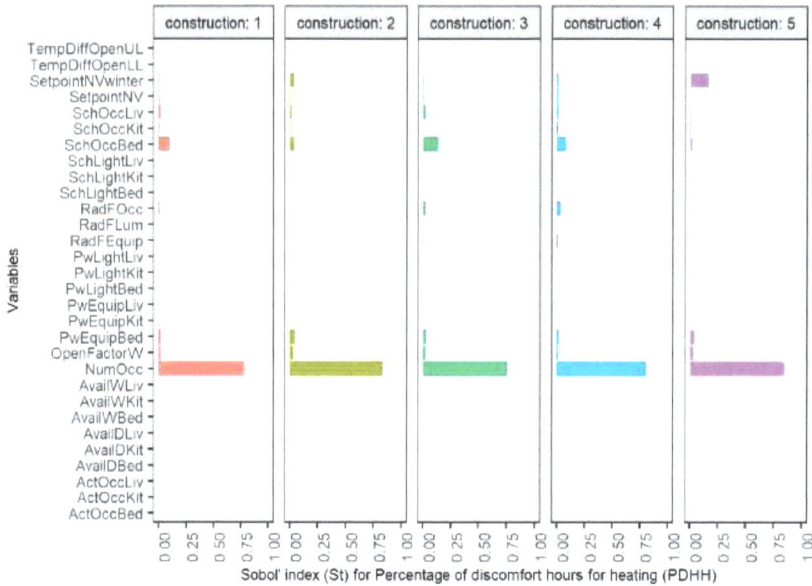

Fig. (15). Sobol' Total sensitivity indices (St) of the operational input variables for the Percentage of discomfort hours for cooling (PDHC) for the global sensitivity analysis in the natural ventilation mode.

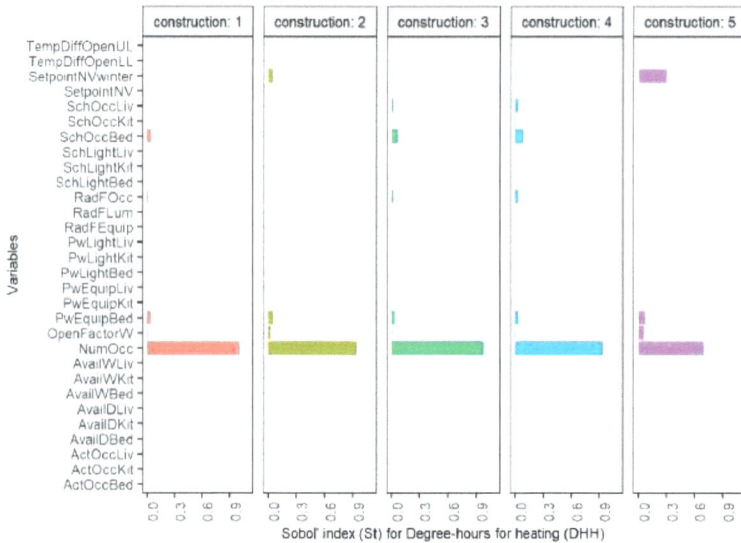

Fig. (16). Sobol' Total sensitivity indices (St) of the operational input variables for the Degree-hours for heating (DHH) for the global sensitivity analysis in the natural ventilation mode.

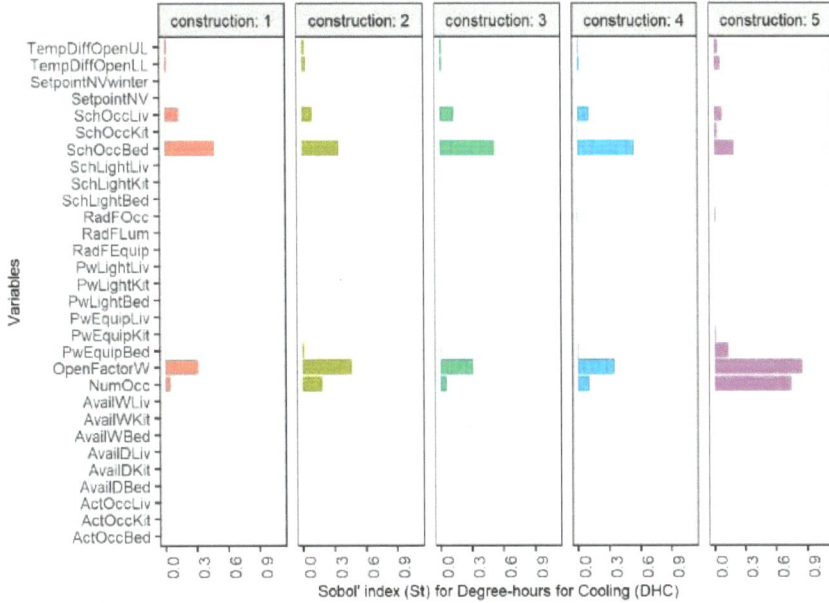

Fig. (17). Sobol' Total sensitivity indices (St) of the operational input variables for the Degree-hours for cooling (DHC) for the global sensitivity analysis in the natural ventilation mode.

Uncertainty Analysis

The uncertainty analysis was reported by showing the boxplot for the performance criteria, grouped by the construction system, the room of the house and the number of occupants (NumOcc) variable. This variable was chosen because it represented the most influent variable for the heating related criteria, which represented the most important criteria for this climate, and due to the sample space chosen, which was discrete.

Fig. (**18**) shows the boxplot of the uncertainty analysis for the PDHH criterion. This graph enables many considerations. First one can see that, by observing the median values of the grouping, the construction 5 showed better results (lower PDHH) and lower uncertainty (lower amplitudes of the values). Construction 2 also presented lower median values in some cases (especially for 6 occupants), but showed higher uncertainty of the data. Second, the amplitude of the data (*i.e.* the uncertainty) was higher for constructions 1-4 than for the construction 5, which indicate that the construction with higher thermal capacity and lower thermal transmittance was more robust to the operational uncertainties, even in the natural ventilation condition. Third, the living room presented the lower median values for PDHH for all construction, except in construction 5, which presented similar values compared to the other rooms. Fourth, the influence of the number of occupants in the uncertainty is noticeable, as the boxplot showed a significant difference for the median values among the 2, 4 and 6 levels of occupants. The range of uncertainty is quite high as it could be from 0.3 to 0.4 (*i.e.* 30% to 40%) of discomfort range due to the operational uncertainties.

Fig. (**20**) shows the boxplot for the DHH. Almost the same pattern was observed compared to the PDHH criterion, except that the differences from the construction among themselves were higher. The DHC values for construction 2 and 5 now are very distinct from the other construction. Construction 5 performed better, followed by the 2, 4, 1 and 3, based on the median value for the whole house.

Fig. (**19**) shows the boxplot for the PDHC, and Fig. (**21**) shows it for the DHC criterion. For the PDHC criterion, the boxplot showed many outlier points (*i.e.* observations very distant from the median), which represents a high uncertainty and discontinuity in the data. It seems that construction 4 showed the lower ranges of uncertainty, but construction 5 showed the lower median values in general. It should be noticed that the grouping variable (SchOccBed) was the most influent only for some construction, not for the construction 5. In this case, the SchOccBed did not show a well-defined difference for the living room (as expected) but was chosen due to its relevance for the whole house. The range of uncertainty is in the order of 0.05-0.10 (5-10%) of the percentage of discomfort for cooling, which is relatively high in comparison to the median values. For the DHC criterion, the same was observed, especially concerning the difference among rooms and the distance from the boxes in respect to the "SchOccBed". However, for this criterion, construction 4 and 5 showed the lowest values and uncertainties, which was different from the previous criterion PDHC. In this sense, one can observe that even if the percentage of discomfort hours could be higher in some cases, the severity of the discomfort (in respect to the degree-hour calculation) was not high for the construction 4 and 5. To summarize the findings:

- By looking at the different rooms, one can see that the living room always had lower values for the heating criteria (PDHH and DHH) than the other rooms, while the living room presented higher values for the cooling related criteria (PDHC and DHC), but it depends on the grouping variable.
- The amplitude range is high for all rooms and grouping variable for PDHH but is higher for construction 1, 3 and 4 for the DHH criterion. The same was observed for PDHC, which have high amplitude ranges for all cases but had lower ranges for construction 4 and 5 for DHC.
- The influence of "NumOcc" in the heating criteria is noticed by the grouping variable, as it occurred in the case of the "SchOccBed" for the cooling criteria.

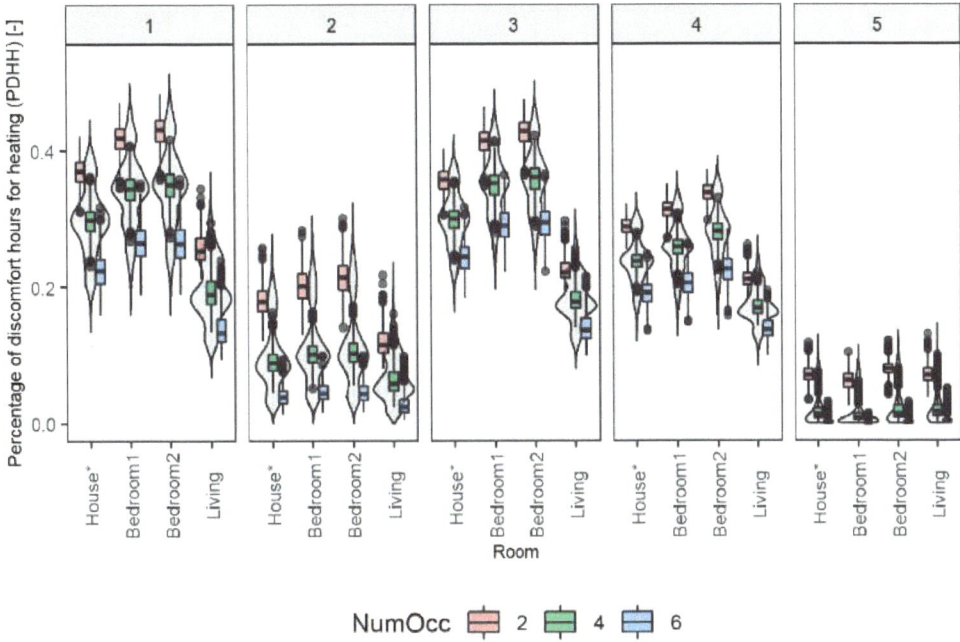

Fig. (18). Boxplot of the Percentage of discomfort hours for heating (PDHH) grouped by the construction and the number of occupants (NumOcc) for each room and the house.

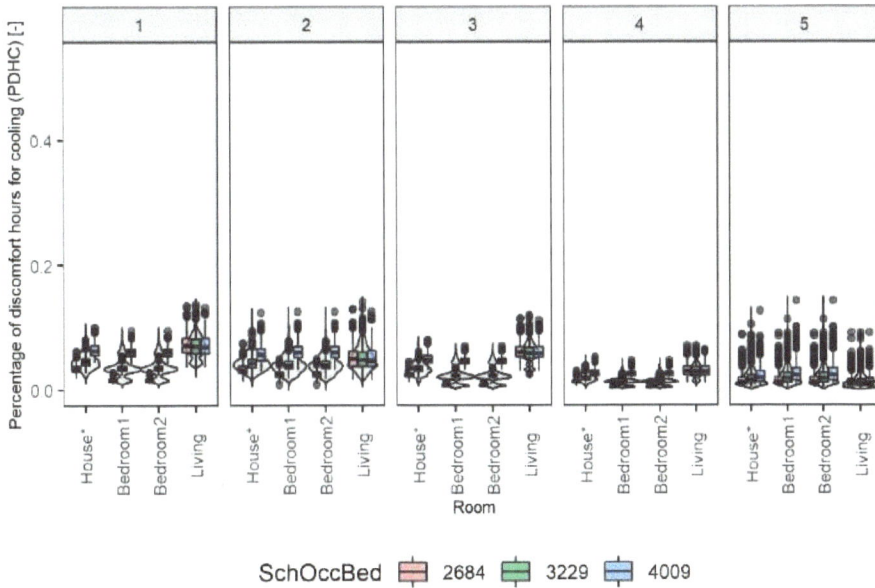

Fig. (19). Boxplot of the Percentage of discomfort hours for cooling (PDHC) grouped by the construction and the schedule of occupancy in the bedrooms (SchOccBed) for each room and the house.

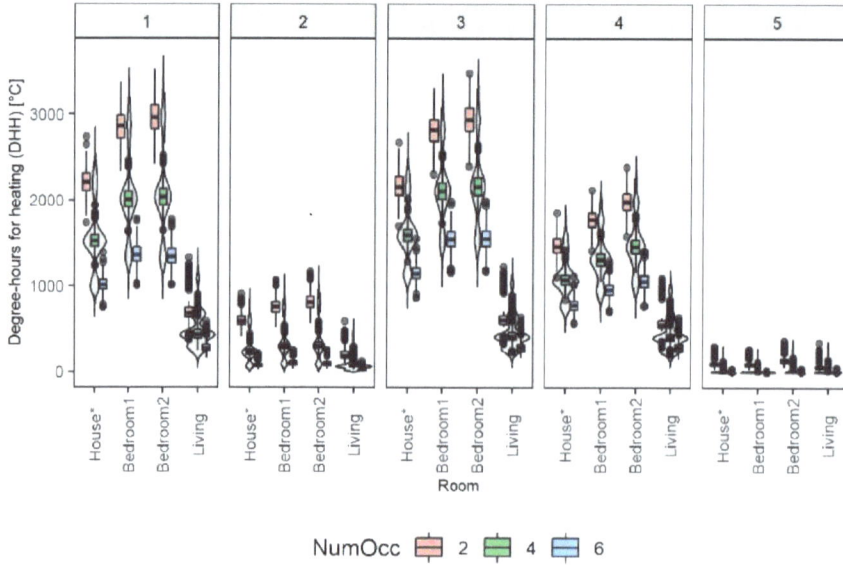

Fig. (20). Boxplot of the Degree-hours for heating (DHH) grouped by the construction and the number of occupants (NumOcc) for each room and the house.

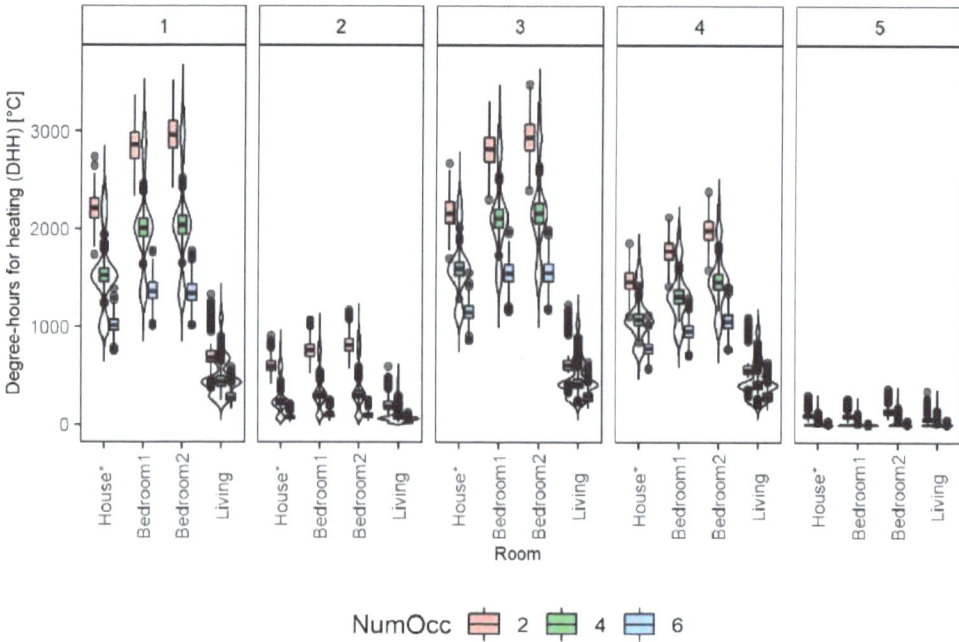

Fig. (21). Boxplot of the Degree-hours for heating (DHC) grouped by the construction and the schedule of occupancy in the bedrooms (SchOccBed) for each room and the house.

By calculating some statistical measures, Table **8** shows the main parameters for understanding the behaviour of the data, divided by each performance criterion and the construction system. Some considerations can be pointed out:

- For the PDHH, the higher amplitude range was obtained for construction 1 (0.2622), which indicates that there is an uncertainty of 26.22% for determining the percentage of discomfort hours for heating in the case of the wood framework system. The lowest range was obtained for construction 5, as being 11.79% of uncertainty in the comfort hours greatness. However, when looking at the coefficient of variation, which considers both the mean and the standard deviation, the higher value was obtained for the construction 5, as being 107.8%, followed by the construction 2 of 49.9%.
- For the PDHC, the higher amplitude range was obtained for the construction 5 (0.1229) and the lowest for the construction 4 (0.0466). The coefficient of variation was 27.0% for construction 1 and 68.9% for construction 5.
- In the case of DHH, the coefficient of variation was greater in construction 5 and lower in construction 3, achieving the highest range in the construction 1 (1982.1 °Ch).
- For the DHC, the coefficient of variation was greater in the construction 5 (89.0%) and lower in the construction 1 (50.0%).

Table 8. Descriptive statistics for the output variables of the house grouped by the construction for the uncertainty analysis of the natural ventilation mode.

Construction	Measure	PDHH [-]	PDHC [-]	DHH [°Ch]	DHC [°Ch]
1	min	0.1601	0.0186	753.9	29.4
	max	0.4223	0.0996	2736.0	779.5
	range	**0.2622**	0.0810	**1982.1**	**750.1**
	median	0.2978	0.0455	1522.6	187.8
	mean	0.2969	0.0469	1559.6	207.9
	std.dev	0.0510	0.0127	393.5	103.9
	coef.var	17.2%	27.0%	25.2%	50.0%
2	min	0.0125	0.0127	25.2	16.2
	max	0.2579	0.1225	910.9	593.6
	range	0.2454	0.1098	885.7	577.4
	median	0.0884	0.0420	229.3	107.2
	mean	0.0981	0.0437	275.8	122.7
	std.dev	0.0489	0.0139	180.9	68.7
	coef.var	**49.9%**	31.7%	65.6%	56.0%

(Table 8) cont.....

	min	0.1842	0.0121	865.3	15.5
	max	0.4031	0.0786	2673.4	496.4
	range	0.2189	0.0665	1808.1	480.9
3	median	0.2994	0.0330	1591.1	117.5
	mean	0.2987	0.0344	1620.0	131.3
	std.dev	0.0412	0.0107	339.7	68.2
	coef.var	13.8%	31.0%	21.0%	51.9%
	min	0.1363	0.0050	569.4	4.8
	max	0.3227	0.0516	1859.3	168.8
	range	0.1864	**0.0466**	1289.9	**164.0**
4	median	0.2373	0.0180	1072.3	36.6
	mean	0.2376	0.0189	1097.3	42.0
	std.dev	0.0342	0.0065	234.8	23.6
	coef.var	14.4%	34.6%	21.4%	56.2%
	min	0.0000	0.0021	0.0	1.5
	max	0.1179	0.1250	297.1	205.5
	range	**0.1179**	**0.1229**	**297.1**	204.0
5	median	0.0128	0.0147	8.5	13.1
	mean	0.0245	0.0177	28.0	17.9
	std.dev	0.0264	0.0122	41.6	16.0
	coef.var	**107.8%**	**68.9%**	**148.4%**	**89.0%**

Hybrid Ventilation Mode

In this second air conditioning mode, the hybrid ventilation allowed both the natural ventilation and the air conditioning system according to specific variable conditions throughout the timestep of the simulation. In this sense, the performance criteria were related to the thermal comfort (PDHH, PDHC, DHH and DHC) and the HVAC system (TLH – thermal loads with heating, TLC – thermal loads with cooling and TELC – total electricity consumption of the HVAC). Up to 26 input variables were considered.

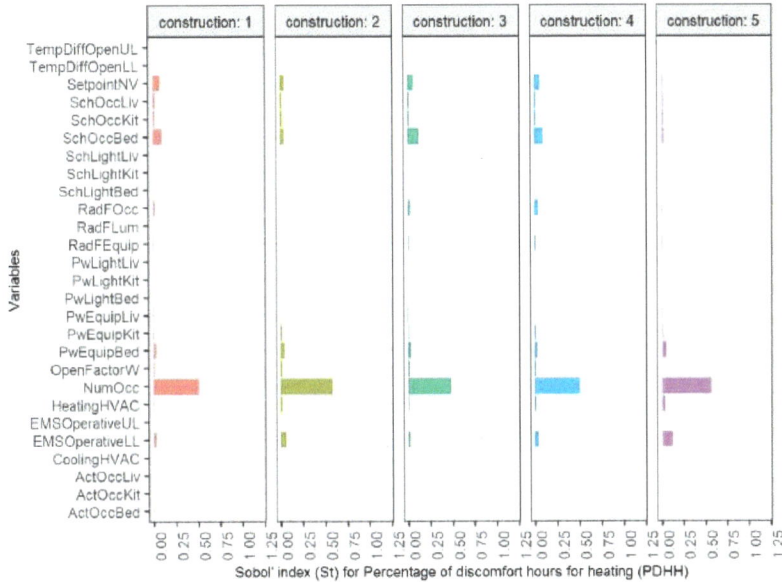

Fig. (22). Sobol' Total sensitivity indices (St) of the operational input variables for the Percentage of discomfort hours for heating (PDHH) for the global sensitivity analysis in the hybrid ventilation mode.

Global Sensitivity Analysis

The global sensitivity analysis, to be succinct, showed the following results:

- The most influent variable in the PDHH (Fig. **22**) was the number of occupants (NumOcc).
- For the DHH (Fig. **24**), the "NumOcc" was the most influent variable for constructions 1, 2 and 5, while the HeatingHVAC was the most influent for constructions 3 and 4. Other variables such as the "EMSOperativeLL" has some influence as well, especially in the construction 5.
- The most influent variable for the PDHC (Fig. **23**) was the "CoolingHVAC". The schedules of occupancy (SchOccBed, SchOccLiv) and the "EMSOperativeUL" also have some influence. The construction 5 had no thermal discomfort for cooling in this hybrid mode.
- In the DHC (Fig. **25**), the same variables showed to be influential, such as the "CoolingHVAC", "SchOccBed", "SchOccLiv" and the "EMSOperativeUL". Construction 5 had no thermal discomfort for cooling.
- For the TLH (Fig. **26**), the "HeatingHVAC", "EMSOperatuveLL" and "NumOcc" were the most influent variables for constructions 1, 3 and 4. In the case of constructions 2 and 5, the "EMSOperativeLL" was the most influent for the TLH criterion.
- The "EMSOperativeUL" was the most influent variable for the TLC (Fig. **27**),

followed by the "CoolingHVAC" and the "SchOccLiv". The TELC (Fig. **28**) performance criterion followed this same pattern.

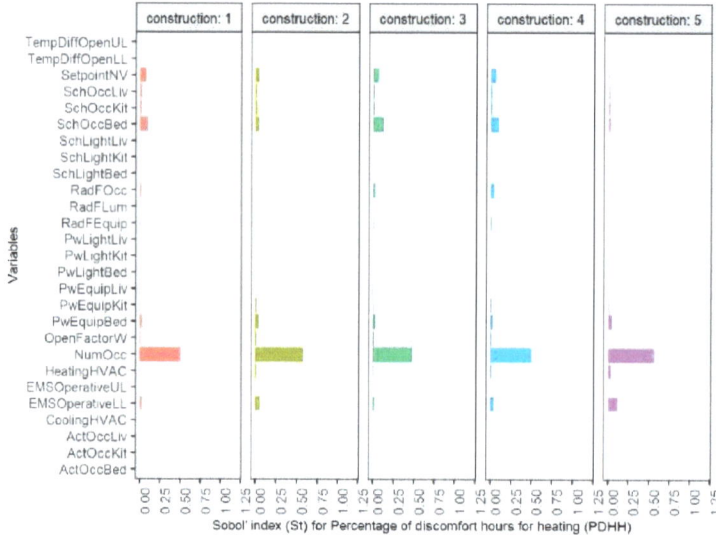

Fig. (23). Sobol' Total sensitivity indices (St) of the operational input variables for the Percentage of discomfort hours for cooling (PDHC) for the global sensitivity analysis in the hybrid ventilation mode.

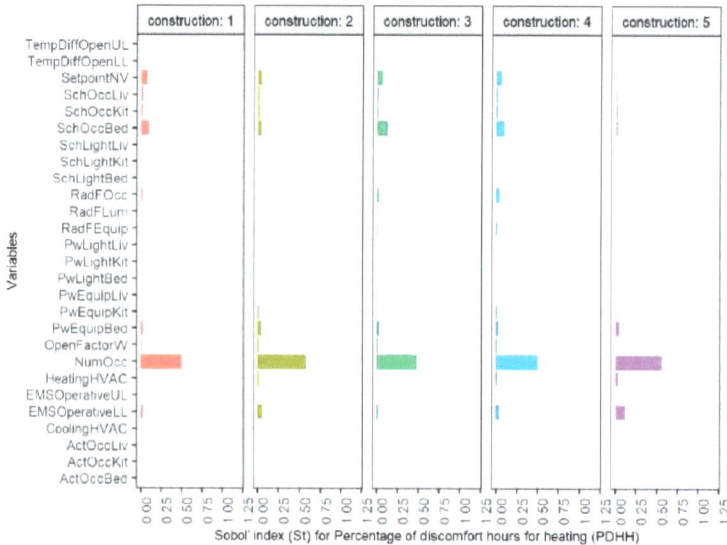

Fig. (24). Sobol' Total sensitivity indices (St) of the operational input variables for the Degree-hours for heating (DHH) for the global sensitivity analysis in the hybrid ventilation mode.

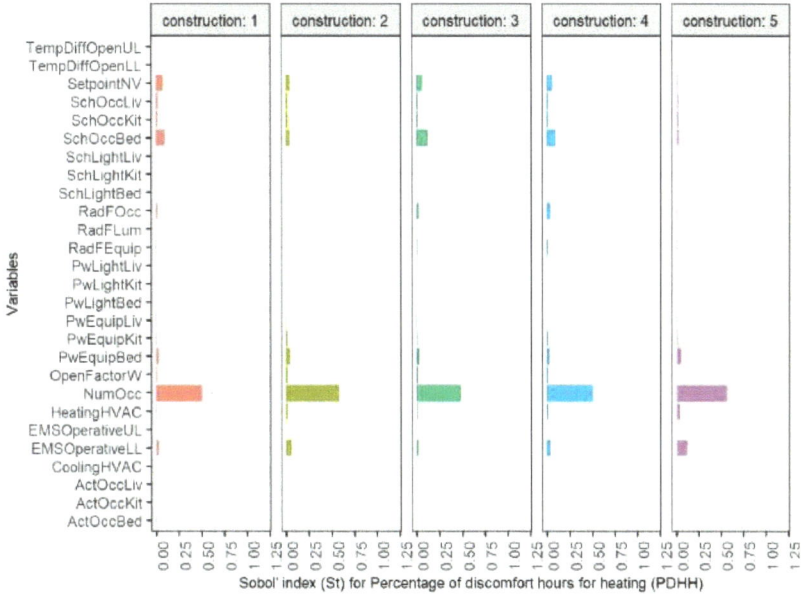

Fig. (25). Sobol' Total sensitivity indices (St) of the operational input variables for the Degree-hours for cooling (DHC) for the global sensitivity analysis in the hybrid ventilation mode.

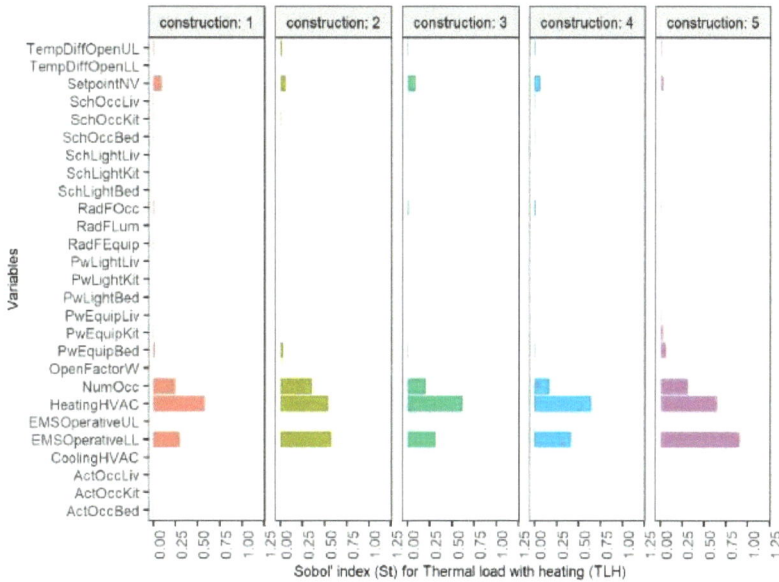

Fig. (26). Sobol' Total sensitivity indices (St) of the operational input variables for the Thermal load with heating (TLH) for the global sensitivity analysis in the hybrid ventilation mode.

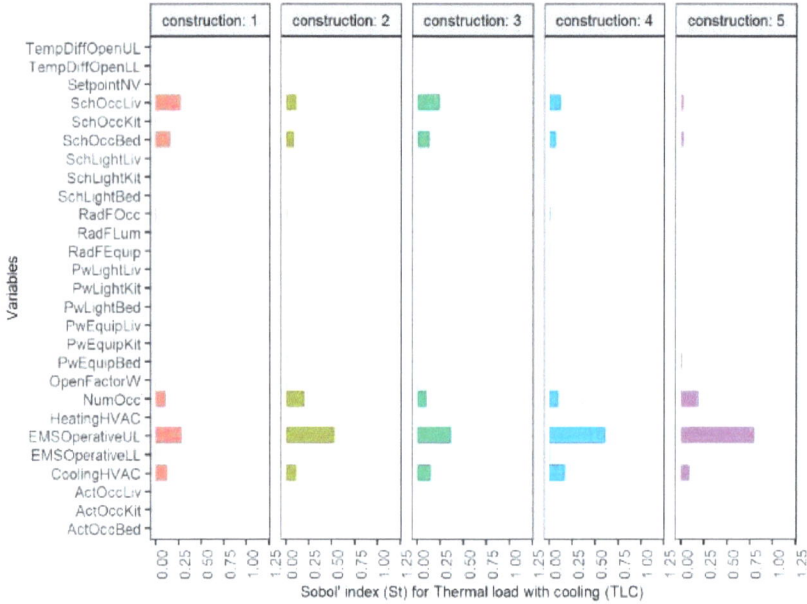

Fig. (27). Sobol' Total sensitivity indices (St) of the operational input variables for the Thermal load with cooling (TLC) for the global sensitivity analysis in the hybrid ventilation mode.

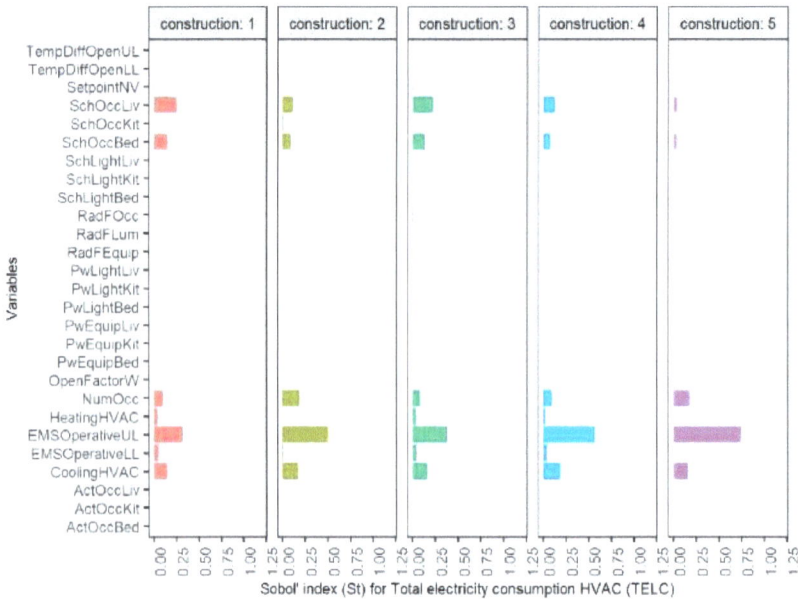

Fig. (28). Sobol' Total sensitivity indices (St) of the operational input variables for the Total electricity consumption of the HVAC (TELC) for the global sensitivity analysis in the hybrid ventilation mode.

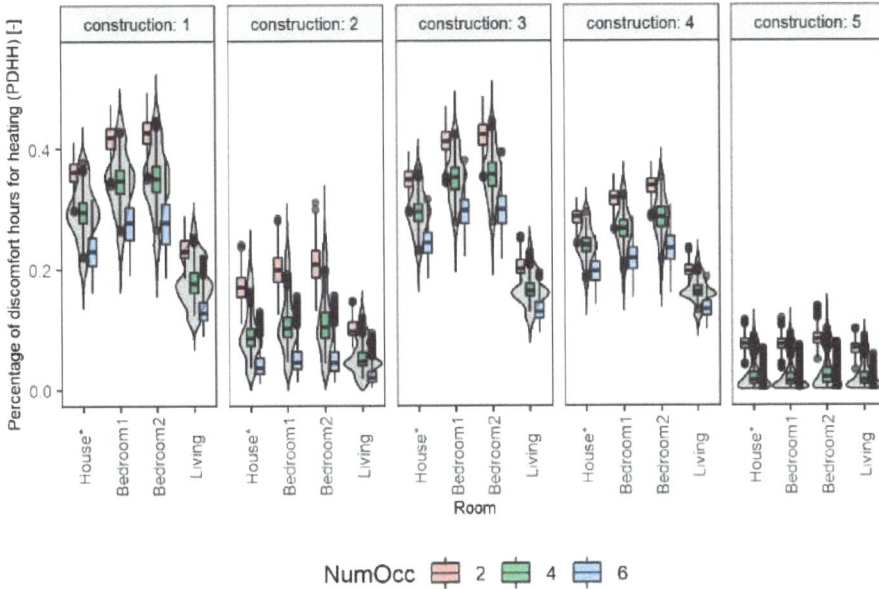

Fig. (29). Boxplot of the Percentage of discomfort hours for heating (PDHH) grouped by the construction and the number of occupants (NumOcc) for each room and for the house for the hybrid ventilation mode.

Uncertainty Analysis

For the hybrid ventilation mode, some general findings can be drawn:

- The PDHH (Fig. **29**) is much higher than the PDHC (Fig. **30**), which indicates that the hybrid ventilation effectively reduced the thermal discomfort for cooling. The same was observed for the DHH (Fig. **31**) which was very higher than the DHC (Fig. **32**). In this sense, the thermal discomfort for cooling criteria was not important for this hybrid ventilation mode.
- For the PDHH (Fig. **29**), the amplitude range was higher for constructions 1 and 3, and lower for construction 5. The same is valid for the DHH (Fig. **31**), where the degree-hours were almost none for construction 5 and had lower amplitude range. The living room always had lower levels of thermal discomfort for heating than the bedrooms.
- For the TLH (Fig. **33**), the values were much lower than the TLC Fig. (**34**). In the case of heating, the amplitudes were higher for constructions 1, 3 and 4, and lower for constructions 2 and 5.
- For the TLC (Fig. **34**), the amplitude ranges were higher for all cases, especially for construction 5. The data were grouped by the "EMSOperativeUL", which was the most influent in this case. When the value is 26°C for the operative temperature upper limit, the amplitudes and the medians are lower. The same pattern was observed for the TELC criterion (Fig. **35**).

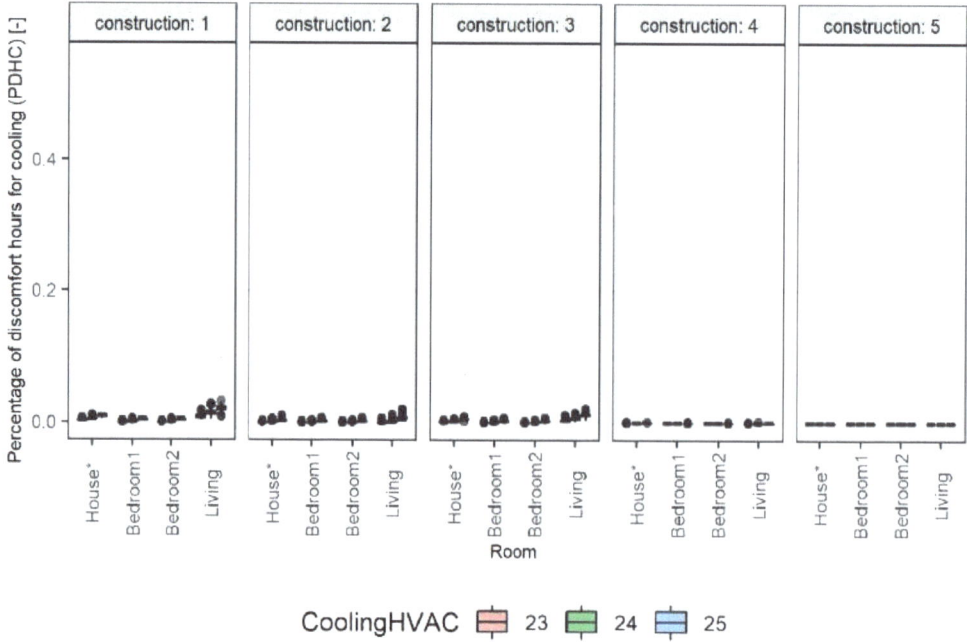

Fig. (30). Boxplot of the Percentage of discomfort hours for cooling (PDHC) grouped by the construction and the setpoint for cooling of the HVAC (CoolingHVAC) for each room and for the house for the hybrid ventilation mode.

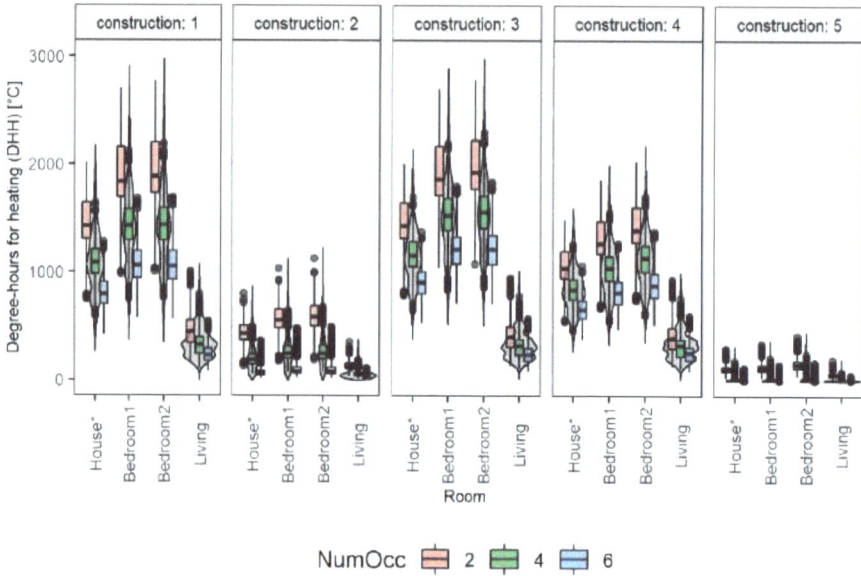

Fig. (31). Boxplot of the Degree-hours for heating (DHH) grouped by the construction and the number of occupants (NumOcc) for each room and for the house for the hybrid ventilation mode.

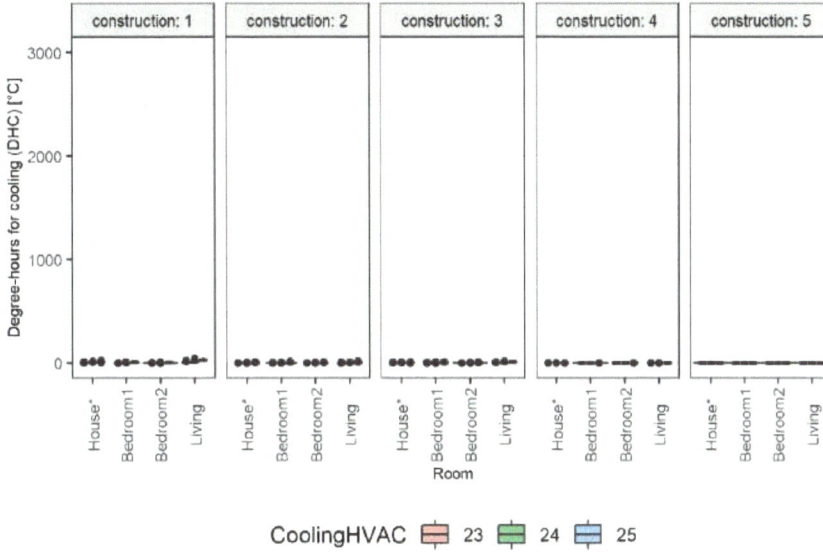

Fig. (32). Boxplot of the Degree-hours for cooling (DHC) grouped by the construction and setpoint for cooling of the HVAC (CoolingHVAC) for each room and for the house for the hybrid ventilation mode.

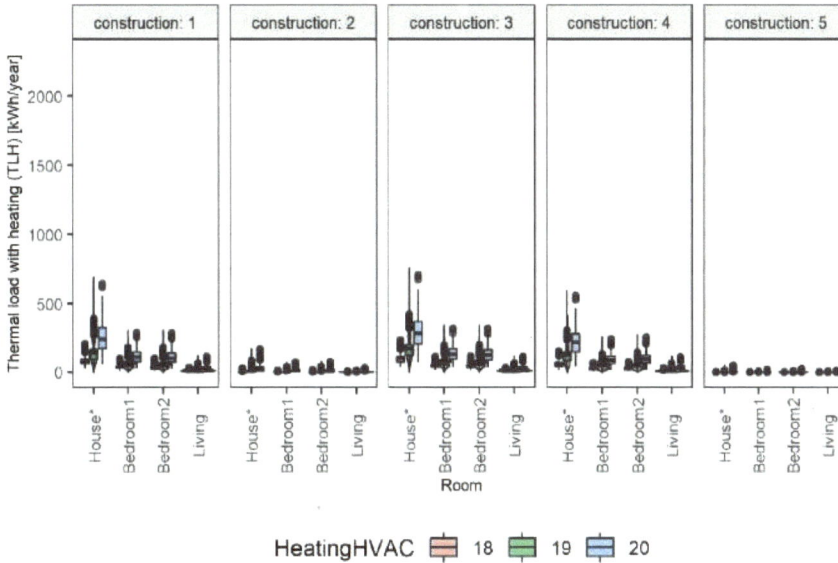

Fig. (33). Boxplot of the Thermal load with heating (TLH) grouped by the construction and the setpoint for heating of the HVAC (HeatingHVAC) for each room and for the house for the hybrid ventilation mode.

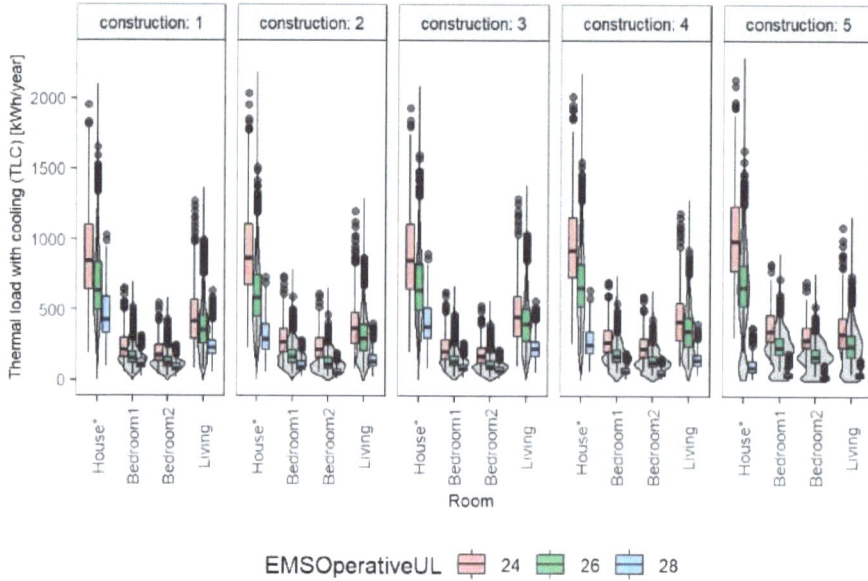

Fig. (34). Boxplot of the Thermal load with cooling (TLC) grouped by the construction and the operative temperature setpoint for natural ventilation (EMSOperativeUL) for each room and for the house for the hybrid ventilation mode.

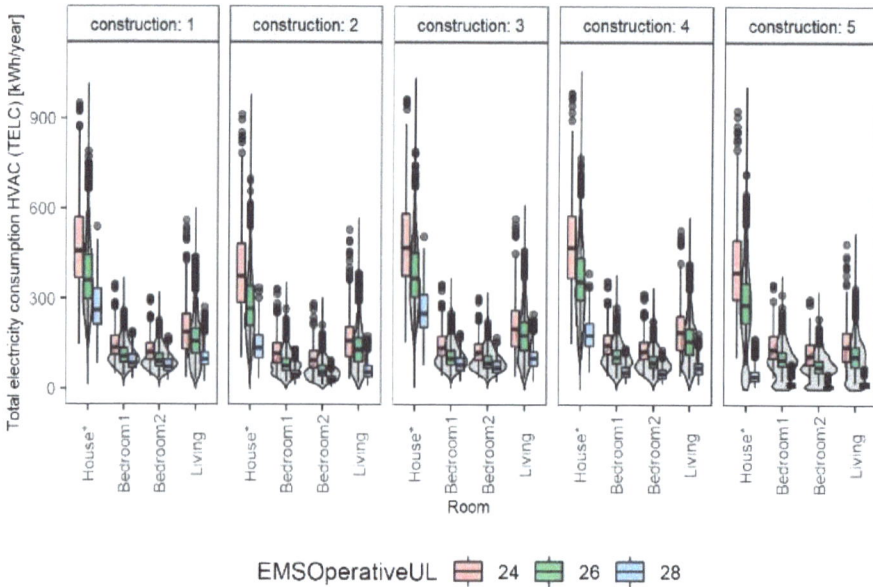

Fig. (35). Boxplot of the Total electricity consumption with HVAC (TELC) grouped by the construction and the operative temperature setpoint for natural ventilation (EMSOperativeUL) for each room and for the house for the hybrid ventilation mode.

Table **9** shows the descriptive statistics of the performance criteria in the hybrid ventilation mode, grouped by the construction system. Some findings:

- For the PDHH, the range was higher for construction 1 (0.2516) and lower for construction 5 (0.1153). The coefficient of variation was higher for construction 5 (99.2%) and construction 2 (48.9%).
- For the DHH, the highest range was found for construction 1 (2017.4 °Ch) and the highest coefficient of variation for construction 5 (134.6%).
- In the case of PDHC and DHC, as the means were very low, it was considered as irrelevant criteria for this hybrid analysis.
- For the TLH, the highest range was found for the construction 3 (668.7 kWh/year). However, the coefficient of variation was higher for construction 5 and 2 due to the low median. Thus, it was not considered statistically significant. In this sense, the highest coefficient of variation was 56.4% for construction 1.
- In the case of TLC, the highest range was found for construction 5 (2123.7 kWh/year), achieving 58.9% of the coefficient of variation.
- For the TELC, the highest range was obtained for construction 4, achieving 930.4 kWh/year and 59.3% of the coefficient of variation for construction 5.

Table 9. Descriptive statistics for the output variables (except for the PDHC and DHC) of the house grouped by the construction for the uncertainty analysis of the hybrid ventilation mode.

Construction	Measure	PDHH [-]	DHH [°Ch]	TLH [kWh/year]	TLC [kWh/year]	TELC [kWh/year]
1	min	0.1605	425.8	24.2	91.8	83.6
	max	0.4121	2017.4	642.9	1952.2	948.5
	range	0.2516	1591.6	618.7	1860.4	864.9
	median	0.2945	1076.6	140.1	633.4	355.0
	mean	0.2947	1091.0	156.3	676.4	374.6
	std.dev	0.0482	294.1	88.1	282.9	128.9
	coef.var	16.4%	27.0%	56.4%	41.8%	34.4%
2	min	0.0098	16.0	0.0	54.0	30.8
	max	0.2413	800.1	163.4	2037.1	911.2
	range	0.2315	784.1	163.4	1983.1	880.4
	median	0.0877	192.1	7.4	568.9	257.3
	mean	0.0966	226.4	13.9	612.9	276.2
	std.dev	0.0473	140.4	19.2	295.5	128.6
	coef.var	48.9%	62.0%	137.9%	48.2%	46.6%

	min	0.1839	527.2	35.0	79.0	78.1
	max	0.3945	2004.3	703.7	1932.8	963.4
	range	0.2106	1477.1	668.7	1853.8	885.3
3	median	0.2944	1144.7	167.6	628.0	360.4
	mean	0.2944	1154.9	183.5	661.5	378.8
	std.dev	0.0394	268.8	96.4	286.3	131.7
	coef.var	13.4%	23.3%	52.5%	43.3%	34.8%
	min	0.1408	382.0	15.4	58.0	53.2
	max	0.3178	1489.1	554.5	2013.2	983.6
	range	0.1770	1107.1	539.1	1955.2	930.4
4	median	0.2397	833.3	119.1	634.8	340.8
	mean	0.2396	841.2	129.3	657.6	354.0
	std.dev	0.0331	199.8	76.4	316.8	143.7
	coef.var	13.8%	23.8%	59.1%	48.2%	40.6%
	min	0.0000	0.0	0.0	11.6	6.3
	max	0.1153	293.2	46.7	2135.3	925.4
	range	0.1153	293.2	46.7	2123.7	919.0
5	median	0.0172	10.5	0.0	645.7	267.9
	mean	0.0293	37.4	0.6	638.9	267.2
	std.dev	0.0290	50.3	3.2	376.4	158.3
	coef.var	99.2%	134.6%	568.6%	58.9%	59.3%

DISCUSSION

The performed analyses were considered robust as many aspects of the sensitivities and uncertainties could be visualized and calculated.

By discussing the natural ventilation *versus* hybrid ventilation modes, one can see that depending on the construction system some good levels of thermal comfort and energy consumption can be achieved. In a general manner, the performance criteria related to the thermal discomfort for heating were much more representative than the cooling criteria. This indicates that a decision-maker could analyze and perform conclusions regarding the performance of the building based only in these indices, rather than in all of them. In respect to the heating, construction 2 and 5 and could be considered good alternatives for the construction of the building for this climate. However, by observing the uncertainties, it is not always clear whether one system is better than the other due to the amplitude ranges and the overlapping of the statistical distribution.

If we opt to consider the coefficient of variation as the uncertainty measure, we can argue that for the natural ventilation mode:

• There is from 13.8% - 49.9% of uncertainty in the estimation of the percentage of discomfort hours for heating (PDHH) (excluding construction 5), and the amplitude range varies from 0.1179 – 0.2622.
• There is from 27.0% - 68.9% of uncertainty in the estimation of the percentage of discomfort hours for cooling (PDHC), and the amplitude range varies from 0.0466 – 0.1229.
• There is from 21.0 – 65.6% of uncertainty in the estimation of the degree-hours for heating (DHH) (excluding construction 5), and the amplitude range varies from 297.1 – 1982.1°Ch.
• There is from 50.0 – 56.2% of uncertainty in the estimation of the degree-hours for cooling (DHC) (excluding construction 5), and the amplitude range varies from 164.0 – 750.1 °Ch.

And, for the hybrid ventilation mode:

• There is from 13.4% - 99.2% of uncertainty in the estimation of the percentage of discomfort hours for heating (PDHH), and the amplitude range varies from 0.1153 – 0.2516.
• There is from 23.3% - 62.0% of uncertainty in the estimation of the degree-hours for heating (DHH) (excluding the construction 5), and the amplitude range varies from 293.2 – 1591.6 °Ch.
• There is from 52.5% - 59.1% of uncertainty in the estimation of the thermal loads with heating (TLH) (excluding construction 2 and 5), and the amplitude range varies from 46.6 – 668.7 kWh/year.
• There is from 41.8% - 58.9% of uncertainty in the estimation of the thermal loads with cooling (TLC), and the amplitude range varies from 1853.8 – 2123.7 kWh/year.
• There is from 34.4% - 59.3% of uncertainty in the estimation of the total electricity consumption with HVAC (TELC), and the amplitude range varies from 864.9 – 930.4 kWh/year.

These values are relatively high and caused uncertainties in the statistical tests of means and individual values, in such a point that, for some comparisons, one cannot conclude which construction system is better than the other (for a specific criterion) due to those operational uncertainties. This was somehow expected, as some authors stated that the uncertainties in occupants profiles for residential buildings tend to be greater than for commercial ones [8], due to a more diverse pattern and use. In the absence of the uncertainty data, one decision-maker could easily fall into Type I error. For example, a null hypothesis could be that the TELC of construction 4 is equal to construction 5, and the alternative hypothesis would be that the TELC of construction 4 is higher than construction 5. The confidence interval and overlapping of the distribution (see Fig. **35** and Table **9**) led a decision-maker to accept the null hypothesis, but it could be rejected in the absence of accurate operational uncertainties information.

To improve the interpretation of the results, new confidence intervals could be created by grouping the found results in the most influent variables to quantify the remaining uncertainty in each group. This grouping was not performed for this study and is a suggestion for further investigation.

In a general manner, the most influent variables for the performance criteria, for the natural ventilation mode, were the number of occupants, the opening factor of the windows and the schedules of occupancy of the bedrooms. And for the hybrid ventilation, the most influent variables were the number of occupants, the heating setpoint of the HVAC, the cooling setpoint of the HVAC and the operative temperature upper limit for natural ventilation of the EMS algorithm.

CONCLUSION

This study performed a series of building performance simulation experiments considering uncertainties related to the operational input variables of a house. Some of the input variables (such as the operational schedules and internal loads) were developed based on field research and statistical data treatment for a specific city in southern Brazil. The simulation experiments considered two modes of natural or hybrid ventilation with different control algorithms, five construction systems with different thermal properties and different performance criteria related to the thermal comfort and energy consumption with HVAC.

The main conclusion is that, for the performed experiments, the uncertainties of the operational input variables were particularly important and should be considered for decision-making purposes or other analyses related to the use of building performance simulation. The literature already discovers that, usually, a small number of input variables represent most of the output variance [53], and the same was identified in this study, regarding the operational input variables.

By considering the most influent variables discovered by the global sensitivity analysis, if one considers a different number of occupants, different values for the heating or cooling setpoint or even different limits for allowing the natural ventilation, it can compromise the comparison between different performance alternatives. In the same sense, the setpoint temperatures and limits of operative for allowing natural ventilation were influent variables for the hybrid ventilation mode. These variables represent real uncertainties of the model, which were not known exactly at the concept or design stage of building development.

For the natural ventilation mode, for instance, up to 49.9% of uncertainty was achieved in the estimation of the percentage of discomfort hours for heating, and up to 65.5% of uncertainty was found for the estimation of the degree-hours for heating. These values were calculated by grouping the data into the different construction systems and observing the coefficient of variation of the distribution. For the hybrid ventilation mode, up to 59.3% of uncertainty was found in the estimation of the total electricity consumption with HVAC, due to the operational uncertainties.

Although the probabilistic approach was based on some of the various methods for modelling the operational characteristics of a building, it has proven to be a solution to make the performance evaluation more reliable, as stated by the study [54].

This study only considered one typology of low-income dwelling for a specific climate of Brazil, and by using specific operational schedules obtained through a field survey. In this sense, case studies are important to investigate certain phenomena but are limited in replicability. Future research could focus on the other strategies for occupancy modelling, such as an agent-based method. Even so, information on occupancy and user behaviour is always needed to produce databases and to help the building simulation efforts to overcome the performance gap issue and improve the accuracy and credibility of the building simulation.

CONSENT FOR PUBLICATION

Not Applicable.

CONFLICT OF INTEREST

The author confirms that this chapter contents have no conflict of interest.

ACKNOWLEDGEMENTS

The author acknowledges the Brazilian agencies FINEP, CNPq, CAPES (Finance code 001), UFSC and UFMS for supporting this research.

REFERENCES

[1]   IEA, *International Energy Agency. Key World Energy Statistics.*, 2020.

[2]   IEA, *International Energy Agency. Energy Efficiency 2019.*, 2019.

[3]   S. Attia, E. Gratia, A. De Herde, and J.L.M. Hensen, "Simulation-based decision support tool for early stages of zero-energy building design", *Energy Build.*, vol. 49, pp. 2-15, 2012.

[http://dx.doi.org/10.1016/j.enbuild.2012.01.028]

[4]   T. Østergård, R.L. Jensen, and S.E. Maagaard, "Early Building Design: Informed decision-making by exploring multidimensional design space using sensitivity analysis", *Energy Build.,* vol. 142, pp. 8-22, 2017.
[http://dx.doi.org/10.1016/j.enbuild.2017.02.059]

[5]   X. Li, and J. Wen, "Review of building energy modeling for control and operation", *Renew. Sustain. Energy Rev.,* vol. 37, pp. 517-537, 2014.
[http://dx.doi.org/10.1016/j.rser.2014.05.056]

[6]   X. Song, J. Zhang, C. Zhan, Y. Xuan, M. Ye, and C. Xu, "Global sensitivity analysis in hydrological modeling: Review of concepts, methods, theoretical framework, and applications", *J. Hydrol. (Amst.),* vol. 523, no. 225, pp. 739-757, 2015.
[http://dx.doi.org/10.1016/j.jhydrol.2015.02.013]

[7]   L.O. Degelman, "A model for simulation of daylighting and occupancy sensors as an energy control strategy for office buildings", *Building Simulation,* pp. 1-9, 1999.

[8]   B.F. Balvedi, E. Ghisi, and R. Lamberts, "A review of occupant behaviour in residential buildings", *Energy Build.,* vol. 174, pp. 495-505, 2018.
[http://dx.doi.org/10.1016/j.enbuild.2018.06.049]

[9]   P. de Wilde, "The gap between predicted and measured energy performance of buildings: A framework for investigation", *Autom. Construct.,* vol. 41, pp. 40-49, 2014.
[http://dx.doi.org/10.1016/j.autcon.2014.02.009]

[10]   T. Hong, J. Langevin, and K. Sun, "Building simulation: Ten challenges", *Build. Simul.,* vol. 11, no. 5, pp. 871-898, 2018.
[http://dx.doi.org/10.1007/s12273-018-0444-x]

[11]   T. Hong, D. Yan, and S. D'Oca, *Build. Environ.,* vol. 114, pp. 518-530, 2017.
[http://dx.doi.org/10.1016/j.buildenv.2016.12.006]

[12]   M. Jia, R. S. Srinivasan, and A. A. Raheem, "From occupancy to occupant behavior: An analytical survey of data acquisition technologies, modeling methodologies and simulation coupling mechanisms for building energy efficiency", *Renew. Sustain. Energy Rev,* vol. 68, pp. 525-540, 2017.
[http://dx.doi.org/10.1016/j.rser.2016.10.011]

[13]   *"International Energy Agency. Annex 66 Report: Definition and simulation of occupant behavior in buildings," Energy in Buildings and Communities Programme.* EBC, 2018.

[14]   D. Yan, "Occupant behavior modeling for building performance simulation: Current state and future challenges", *Energy Build.,* vol. 107, pp. 264-278, 2015.
[http://dx.doi.org/10.1016/j.enbuild.2015.08.032]

[15]   J. Pfafferott, and S. Herkel, "Statistical simulation of user behaviour in low-energy office buildings", *Sol. Energy,* vol. 81, no. 5, pp. 676-682, 2007.
[http://dx.doi.org/10.1016/j.solener.2006.08.011]

[16]   E. Azar, and C.C. Menassa, "A comprehensive analysis of the impact of occupancy parameters in energy simulation of office buildings", *Energy Build.,* vol. 55, pp. 841-853, 2012.
[http://dx.doi.org/10.1016/j.enbuild.2012.10.002]

[17]   K.B. Lindberg, S.J. Bakker, and I. Sartori, "Modelling electric and heat load profiles of non-residential buildings for use in long-term aggregate load forecasts", *Util. Policy,* vol. 58, no. April, pp. 63-88, 2019.
[http://dx.doi.org/10.1016/j.jup.2019.03.004]

[18]   B. Gucyeter, "Evaluating diverse patterns of occupant behavior regarding control-based activities in energy performance simulation", *Front. Archit. Res.,* vol. 7, no. 2, pp. 167-179, 2018.
[http://dx.doi.org/10.1016/j.foar.2018.03.002]

[19]   X. Feng, D. Yan, and C. Wang, "On the simulation repetition and temporal discretization of stochastic occupant behaviour models in building performance simulation", *J. Build. Perform. Simul.,* vol. 10, no. 5–6, pp. 612-624, 2017.
[http://dx.doi.org/10.1080/19401493.2016.1236838]

[20]   S. D'Oca, and T. Hong, "Occupancy schedules learning process through a data mining framework", *Energy Build.,* vol. 88, pp. 395-408, 2015.
[http://dx.doi.org/10.1016/j.enbuild.2014.11.065]

[21]   A. Marszal-Pomianowska, P. Heiselberg, and O. Kalyanova Larsen, "Household electricity demand profiles – A high-resolution load model to facilitate modelling of energy flexible buildings", *Energy,* vol. 103, pp. 487-501, 2016.
[http://dx.doi.org/10.1016/j.energy.2016.02.159]

[22]   B. Jeong, J-W. Jeong, and J.S. Park, "Occupant behavior regarding the manual control of windows in residential buildings", *Energy Build.,* vol. 127, pp. 206-216, 2016.
[http://dx.doi.org/10.1016/j.enbuild.2016.05.097]

[23]   S. Firląg, "Control algorithms for dynamic windows for residential buildings", *Energy Build.,* vol. 109, pp. 157-173, 2015.
[http://dx.doi.org/10.1016/j.enbuild.2015.09.069]

[24]   *Standard 90.1 - energy standard for buildings except low-rise residential buildings.," american society of heating.* Refrigerating and air-conditioning engineers: Atlanta, GA, 2013.

[25]   Portaria n° 372, de 17 de setembro de 2012. Regulamento Técnico da Qualidade para a Eficiência Energética de Edificações Comerciais, de Serviços e Públicas (RTQ-C).*Ministério do Desenvolvimento, Indústria e Comércio Exterior - MDIC.* Instituto Nacional de Metrologia, Qualidade e Tecnologia - INMETRO: Rio de Janeiro, RJ, 2012.

[26]   *Portaria n° 18, de 16 de janeiro de 2012. Regulamento Técnico da Qualidade para a Eficiência Energética de Edificações Residenciais (RTQ-R) [Quality Technical Regulation for Energy Efficiency of Residential Buildings (RTQ-R)],* 2012.

[27]   A.S. Silva, F. Luiz, A.C. Mansur, A.S. Vieira, A. Schaefer, and E. Ghisi, "Knowing electricity end-uses to successfully promote energy efficiency in buildings : a case study in low-income houses in Southern Brazil", *International Conference on Energy & Environment,* 2013.

[28]   A. S. Silva, F. Luiz, A. C. Mansur, A. S. Vieira, A. Schaefer, and E. Ghisi, "Knowing electricity end-uses to successfully promote energy efficiency in buildings : a case study in low-income houses in Southern Brazil," Int. J. Sustain. Energy Plan. Manag., vol. 2, no. 2012, pp. 7–18, 2014, doi: 10.5278/ijsepm.2014.2.2.,

[29]   A. Schaefer, and E. Ghisi, "Method for Obtaining Reference Buildings", *Energy Build.,* vol. 128, pp. 660-672, 2016.
[http://dx.doi.org/10.1016/j.enbuild.2016.07.001]

[30]   S. Siegel, *Estatística Não-paramétrica para Ciências do Comportamento.,* 2006.

[31]   D.B. Johnson, and T. Mizoguchi, "Selecting the K th Element in $X + Y$ and $X_1 + X_2 + \cdots + X_m $", *SIAM J. Comput.,* vol. 7, no. 2, pp. 147-153, 1978.
[http://dx.doi.org/10.1137/0207013]

[32]   A. Chalom, C.Y. Mandai, and P.I.K.L. Prado, "R Package 'pse' - Parameter space exploration with Latin Hypercubes", *CRAN,* 20152015.https://cran.r-project.org/web/packages/

[33]   "Department of Energy. EnergyPlus Simulation Software", *Energyplus,* 20202020.https://energyplus.net/

[34]   NREL, "National Renewable Energy Laboratory. Open Studio", *U.S. Department of Energy, Office of Energy Efficiency and Renewable Energy,* 2020. https://www.openstudio.net/

[35]   M.V. Swami, and S. Chandra, "Correlations for pressure distribution on buildings and calculation of

natural-ventilation airflow", *ASHRAE Trans.,* vol. 94, no. 1, pp. 243-266, 1988.

[36]   ABNT, *Associação Brasileira de Normas Técnicas. NBR 15220-2 - Desempenho térmico de edificações. Parte 2: Método de cálculo da transmitância térmica, da capacidade térmica, do atraso térmico e do fator solar de elementos e componentes de edificações,* 2005.

[37]   M. Kottek, J. Grieser, C. Beck, B. Rudolf, and F. Rubel, "World Map of the Köppen-Geiger climate classification updated", *Meteorol. Z. (Berl.),* vol. 15, no. 3, pp. 259-263, 2006. [http://dx.doi.org/10.1127/0941-2948/2006/0130]

[38]   "EnergyPlus weather data", *EnergyPlus,* 2020. https://energyplus.net/weather

[39]   ASHRAE, *Standard 55 - Thermal Environmental Conditions for Human Occupancy,* 2017.

[40]   W. Tian, "A review of sensitivity analysis methods in building energy analysis", *Renew. Sustain. Energy Rev.,* vol. 20, pp. 411-419, 2013. [http://dx.doi.org/10.1016/j.rser.2012.12.014]

[41]   W. Tian, "A review of uncertainty analysis in building energy assessment", *Renew. Sustain. Energy Rev.,* vol. 93, no. May, pp. 285-301, 2018. [http://dx.doi.org/10.1016/j.rser.2018.05.029]

[42]   CB3E, "Centro Brasileiro de Eficiência Energética em Edificações", *Proposta de Instrução Normativa Inmetro para a Classe de Eficiência Energética de Edificações Residenciais,* Florianópolis, SC, Brasil, 2018. http://cb3e.ufsc.br/

[43]   Handbook of Fundamentals.*American society of heating* Refrigerating and air-conditioning engineers: Atlanta, GA, 2017.

[44]   ISO society of heating refrigerating and air-conditioning engineers,

[45]   M. Hughes, J. Palmer, V. Cheng, and D. Shipworth, "Global sensitivity analysis of England's housing energy model", *J. Build. Perform. Simul.,* vol. 8, no. 5, pp. 283-294, 2015. [http://dx.doi.org/10.1080/19401493.2014.925505]

[46]   A. Saltelli, P. Annoni, I. Azzini, F. Campolongo, M. Ratto, and S. Tarantola, "Variance based sensitivity analysis of model output. Design and estimator for the total sensitivity index", *Comput. Phys. Commun.,* vol. 181, no. 2, pp. 259-270, 2010. [http://dx.doi.org/10.1016/j.cpc.2009.09.018]

[47]   M.J.W. Jansen, "Analysis of variance designs for model output", *Comput. Phys. Commun.,* vol. 117, no. 1–2, pp. 35-43, 1999. [http://dx.doi.org/10.1016/S0010-4655(98)00154-4]

[48]   G. Pujol, B. Iooss, and A. Janon, "R Package 'sensitivity' - A collection of functions for factor screening, global sensitivity analysis and reliability sensitivity analysis of model output", *CRAN,* 2015. https://cran.r-project.org/web/packages/

[49]   N. Delgarm, B. Sajadi, K. Azarbad, and S. Delgarm, "Sensitivity analysis of building energy performance: A simulation-based approach using OFAT and variance-based sensitivity analysis methods", *J. Build. Eng,* vol. 15, pp. 181-193, 2018. [http://dx.doi.org/10.1016/j.jobe.2017.11.020]

[50]   A.S. Silva, and E. Ghisi, "Estimating the sensitivity of design variables in the thermal and energy performance of buildings through a systematic procedure", *J. Clean. Prod.,* vol. 244, p. 118753, 2020. [http://dx.doi.org/10.1016/j.jclepro.2019.118753]

[51]   S. Yang, W. Tian, E. Cubi, Q. Meng, Y. Liu, and L. Wei, "Comparison of Sensitivity Analysis Methods in Building Energy Assessment", *Procedia Eng.,* vol. 146, pp. 174-181, 2016. [http://dx.doi.org/10.1016/j.proeng.2016.06.369]

[52]   P. Bratley, and B.L. Fox, "ALGORITHM 659: implementing Sobol's quasirandom sequence generator", *ACM Trans. Math. Softw.,* vol. 14, no. 1, pp. 88-100, 1988. [http://dx.doi.org/10.1145/42288.214372]

[53]   P. Aude, L. Tabary, and P. Depecker, "Sensitivity analysis and validation of buildings ' thermal models using adjoint-code method", In: *Energy Build* vol. 31. , 2000.

[54]   N. Heijmans, P. Wouters, and X. Loncour, "Assessment of innovative ventilation in the framework of the EPBD—A probabilistic approach", *Build. Environ.,* vol. 43, no. 8, pp. 1354-1360, 2008. [http://dx.doi.org/10.1016/j.buildenv.2007.01.044]

CHAPTER 8

Indoor Climate Management of Museums: the Impact of Ventilation on Conservation, Human Health and Comfort

Hugo Entradas Silva[1,*] and **Fernando M. A. Henriques**[1]

[1] *Departamento de Engenharia Civil, Faculdade de Ciências e Tecnologia, FCT, Universidade NOVA de Lisboa, 2829-516 Caparica, Portugal*

Abstract: Cultural heritage plays an important role in society, not only in cultural terms but also due to its touristic interest. From a purely economic point-of-view, the increasing number of visitors can be a way to achieve financial sustainability. However, it is necessary to ensure that conservation and comfort conditions are not affected, since the human body releases heat, moisture, CO_2 and odours.

A suitable relation between ventilation and occupancy may be used to minimize some of these effects, but it is not easy to reach because there is no unanimity in the literature on comfort and health issues. Besides, the information about ventilation and occupancy that is used in cultural heritage buildings is scarce, even after the recent publication of the EN 15759-2.

In this chapter, a sensitivity simulation study using a hygrothermal simulation model of a generic museum is developed. This chapter aims to analyse the impact of the binomial ventilation *vs.* occupancy, simulating various combinations of ventilation and air recirculation on the indoor air quality, conservation and energy consumption in museums. Since the visits to major national museums take usually long periods, the concept of adaptation was analysed to reduce the airflow of fresh air per visitor. The study was carried out using the software *WUFI® Plus* for the hygrothermal and energy simulation.

Keywords: Air recirculation, Cultural heritage, Conservation, Computational simulation, Indoor air quality, Museums, Occupancy, Preventive conservation, Ventilation.

---

* **Corresponding author Hugo Entradas Silva:** Departamento de Engenharia Civil, Faculdade de Ciências e Tecnologia, FCT, Universidade NOVA de Lisboa, 2829-516 Caparica, Portugal; Tel: +351 964356293; E-mail: h.silva@campus.fct.unl.pt

Enedir Ghisi, Ricardo Forgiarini Rupp and Pedro Fernandes Pereira (Eds.)

INTRODUCTION

Museums play an important role in storing collections, allowing different forms of art to be presented to the world. Museums must ensure the conservation of collections and at the same time a pleasant experience for visitors, providing adequate conditions of comfort and health [1].

Despite the unquestionable importance of visitors, it is important to note that they release heat, moisture, carbon dioxide (CO_2), odours and act as an open door for exterior pollutants, which can affect the equilibrium of the indoor microclimate.

Although historic buildings are characterized by high thermal inertia, they usually present a poor hygrothermal response, which can contribute to an unstable microclimate and render it difficult to obtain a serious compromise between conservation, comfort and sustainability [2]. Sometimes the use of powerful climate control systems is unavoidable and the impossibility of changing the building façades to avoid identity losses [3] means that one of the adopted strategies for energy reduction is linked to ventilation [4]. However, it cannot be dissociated from the human occupancy since it influences the air renewal and consequently the moisture, pollutants and odours.

An excessive occupancy can constitute a serious risk to the microclimate stability and a challenge to heritage management due to degradation of the indoor air quality (IAQ) and the increase of moisture. Some articles attest the visitors' effect on the indoor climate [5, 6] and the risks of undue occupancies for conservation [7 - 9]. There are some cases of common sense in which it was necessary to limit the number of visitors or their impact to mitigate conservation risks, such as the Scrovegni Chapel in Padova [10, 11]. With the growing interest in cultural tourism, limitations to the visits may become a necessity all over Europe.

The choice of a suitable relationship between occupancy and ventilation is crucial for heritage sustainability. However, this management should be based on sound fundamentals, since ventilation has a major impact on climate stability and energy consumption, and an unnecessary limitation of the number of visitors induces revenue cuts on buildings.

There are some standards and guidelines focusing on IAQ based on comfort with wide international acceptance, but they vary in assumptions and proposed values [12 - 16]. Concerning ventilation requirements and occupancy for cultural heritage, the issue is even more ambiguous. It is assumed that excessive ventilation disrupts the microclimate stability [4], while reduced ventilation associated with a high occupancy contribute to high humidity levels, mould risk and surface condensation [10, 11, 17]. Despite this evidence, it was not possible to find methods that specifically help in ventilation and occupancy management in museums. The European standard EN 15759-2 [18] was recently published on ventilation in cultural heritage to ensure optimum preservation of buildings and collections while ensuring comfort. This document is a useful tool for the ventilation management, namely by presenting a step-by-step approach to identify factors and areas of risk, but it is more descriptive than prescriptive and does not clarify what ventilation to use.

It is necessary to develop more research in this area so that museum managers and climate designers have tools to support decision making on a subject that can have a catastrophic impact on conservation, comfort and health. In this work, the authors sought to gather information on the topic, seeking to satisfy the needs of conservation, comfort and health. A climate simulation study was developed for three European cities with different climates to test the impact of various ventilation strategies.

STATE-OF-THE-ART

General Considerations

Ventilation concerns date back to the late 18th century when humans were considered to be the main cause of pollution inside non-industrial buildings. Until the mid-1800s, the air expelled by humans was believed to be toxic mainly due to carbon dioxide, until it was demonstrated that this gas was harmless for the concentrations usually found inside buildings.

At the beginning of the 20th century, there was a paradigm shift when it was proven that the air expelled in the breathing process was non-toxic, however during the first third of the century the main concerns were related to the health and the reduction of risk contagion of endemic diseases. This approach would once again be changed when around 1930 it was possible to conclude that the spread of diseases through the air was not one of the main forms of contagion. From here, ventilation began to be seen as a comfort factor that aims to ensure that the occupants of a given space perceive the air quality as acceptable. Although humans continue to be seen as the main sources of pollution, from this period onwards the main concerns were linked to released odours [19]. Despite some adaptations, this approach remains valid until now.

IAQ Based on Comfort Perception

Traditional Approach

According to comfort issues, the indoor air quality (IAQ) can be defined as the occupants' perception of the IAQ of a certain space. This perception is not linear and can vary according to various factors [13, 20], such as the emission of CO_2 and other gases, odours emitted by the occupants and by the building components [12] and subjective factors, as the expectancy and cultural adaptation. Yaglou, in 1936, carried out the first internationally accepted tests to evaluate and predict the comfort sensation in humans, using climate-controlled chambers and occupied buildings with different ventilation rate and adopting various scales, including a scale of odour intensity [21]. These results were used for more than 50 years in standards and guidelines of around the world.

In the 1980s and 1990s, new experimental campaigns were carried out with a higher number of people (mainly office workers and university students), better test conditions and modern hygiene habits. Results obtained in Europe [22, 23], USA [24] and Japan [25] achieved a great acceptance in the international community, presenting a good correlation between them, that support the robustness of the conclusions. These levels of acceptance were obtained for people not adapted to the environment. For adapted people, the rate of ventilation per occupant to achieve the same level of acceptance can be estimated as one-third of the value for non-adapted people [26]. The European results [22, 23] (see Fig. **1**) have been the basis of several international documents, as the ASHRAE 62.1 [15] or the EN 16798-1 [27].

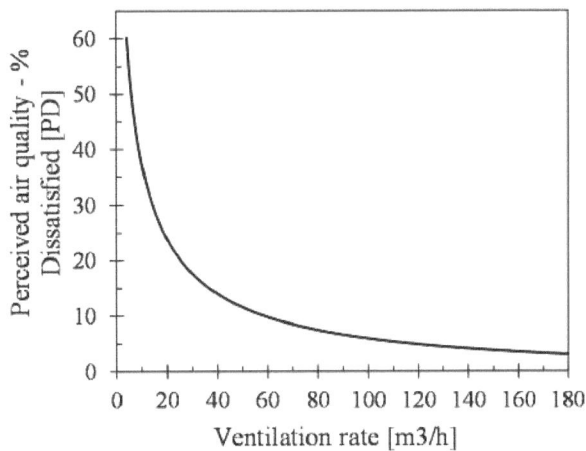

Fig. (1). Relationship between the percentage of occupants dissatisfied with air quality and the ventilation flow for a standard person [28].

IAQ can be evaluated as a percentage of dissatisfied occupants according to their mean vote and approximated with a high level of confidence by equations (1) and (2) [28]:

$$PD = 395 \cdot e^{(-1.83 \cdot (q/3.6)^{0.25})} \qquad \text{for } q \geq 1.15 \text{ m}^3/\text{h.olf} \qquad (1)$$

$$PD = 100 \qquad \text{for } q \leq 1.15 \text{ m}^3/\text{h.olf} \qquad (2)$$

where PD is the percentage of dissatisfied people (%), q is the airflow (m^3/h.olf) and one olf corresponds to the bio effluents emitted by a standard person.

A relationship between the percentage of dissatisfied people and the concentration of indoor CO_2 was achieved and proved as a good indicator to evaluate the IAQ, since while people are releasing CO_2 are also releasing odours. The work that gave rise to these results allowed to conclude that an airflow of 27 m^3/h per person corresponds to 80% of satisfied people, which corresponds to a CO_2 concentration of about 650 ppm above the exterior concentrations [26]. In this way, the difference between indoor and outdoor CO_2 concentrations can be used as an expeditious indicator of the IAQ, since the CO_2 concentration is usually more easily obtained than the ventilation rate. The estimation of airflow per person can be obtained with the application of a mass balance equation according to the usual CO_2 emission per person [26].

Usually, this analysis can be made in stationary conditions to simplify the method. Analysis in steady-state conditions does not consider the volume, affecting the necessary time to reach the equilibrium conditions. This is especially important for the cases where pollutant emissions (the CO_2 in this particular case) occurs in a limited period. In these cases, using a steady-state equation can overestimate the necessary ventilation to keep the pollutant within the desired threshold. For a continuous and constant occupancy, steady-state mass balance equations may be used with satisfactory results [16]:

$$q_{tot} = \frac{n \cdot G_{CO2} \cdot 10^6}{\varepsilon_v \cdot (C_{CO2,i} - C_{CO2,e})} \qquad (3)$$

where q_{tot} is the total airflow (m^3/h), n the number of visitors (-), G_{CO2} the emission rate of CO_2 (m^3/h), $C_{CO2,i}$ the allowed indoor concentration of CO_2 (ppm), $C_{CO2,e}$ the concentration of CO_2 in the exterior (ppm) and ε_v the ventilation efficiency (-). Usually, the assumption of totally efficient ventilation is considered admissible.

Despite the utility of the described results, it is evident that there are other sources of pollution inside buildings other than human beings, as building components, furniture, carpets and air conditioning systems that emit a wide variety of volatile organic compounds (VOC's). Although each VOC typically occurs at concentrations below the odour and irritation threshold, the set of several VOC's can significantly influence the occupants' perception of IAQ. This question was firstly considered by Fanger [28] and is fully described in the study [13].

Several polluting loads for some types of buildings can be found in the study [13]. To consider these sources of pollution, Fanger [28] suggested that the pollution emitted by each source should be expressed in terms of the pollution emitted by a standard person, introducing the concept of "olf" (from the Latin term *olfactus*) which is nothing more than the odour emitted by an average European adult engaging in sedentary activity. This method allows considering the totality of sensory pollution loads (olf's) by the sum of the components associated with the occupants and the building.

Besides, Fanger [28] proposed to refer to indoor air quality decipol (from the Latin *pollutio*) to classify the IAQ in a space with a pollution source generated by a standard person and ventilated by 36 m^3/h of fresh air, *i.e.* 1 db = 0.36 olf (m^3.h). However, these concepts have fallen into disuse and today they are practically not mentioned in international standards.

The Adaptation Concept

From the initial studies of Yaglou in climatic chambers [21] the results of the people perceived IAQ were based on their first impressions, neglecting the adaptation in time.

Although some adaptations have occurred over time, such as a significant decrease in the perception of odours, the standards and guidelines have been based on the results obtained with the first perception because it is considered that this impression is important and because it is considered unrealistic to ask people to wait a few minutes before they adapt. However, when the people's presence in a space is prolonged, the consideration of adaptation may make sense.

To clarify this point, Gunnarsen and Fanger [29] studied the change in the perception of air quality during the first 15 minutes of exposure in controlled environment chambers. They used 16 women and 16 men aged from 18 to 30 to evaluate the IAQ in 42 trials. Occupants were exposed to various concentrations of bio effluents (from 500 to 4 000 ppm of CO_2), tobacco smoke and emissions from buildings.

The participants of the study assessed IAQ every 2 minutes by using scales of odour intensity and IAQ acceptability while they were exposed to constant levels of pollution. Odour tolerance was found to increase after a few minutes independently of the CO_2 concentration. It was also noted that sensorial irritation is usually constant and may increase over time, concluding that although there is a strong adaptation to odours, adaptation to possible pollutants from buildings is reduced [29]. The authors verified that 95% of the people changed their perception of IAQ, concluding that a good adaptation occurs after the first 6 minutes. They further verified that after adaptation, the levels of pollution tested did not influence the level of satisfaction and that adaptation to one pollutant makes exposure to others more acceptable. Overall, this study has concluded that ventilation for comfort can be greatly reduced if a few minutes of discomfort are accepted or if exposure to increasing levels of pollution is performed gradually within the first 10 minutes.

This approach seems to apply to cultural heritage, with a special focus on museums, since there tends to be a defined tour circuit where visitors enter and buy the ticket in a space near the exterior - well ventilated – and gradually walk through the building. The duration of the visits is usually longer on major national museums, so it seems plausible to question the comfort of the initial minutes. This approach can contribute significantly to increasing indoor air stability and reduce energy consumption, clearly contributing to environmental and economic sustainability.

Calculations of the required airflow per adapted person may be obtained by multiplying the required airflow for an un-adapted person by a factor of 0.33 or by multiplying the CO_2 concentration by a factor of 3 to achieve the same comfort level. The airflow due to the pollution from the building components should be maintained [26].

International Guidelines and Standards

As described in the introduction, there is no agreement around this theme, so it is considered pertinent to present the values and concepts defined in the international standards/guidelines, such as EN 16798-1 [27], ANSI/ASHRAE 62.1 [15], CIBSE Guide A [16] and EN 15759-2 [18].

As mentioned earlier, these documents were mainly based on tests carried out on office workers and university students. However, its use has been extrapolated to other types of buildings, such as museums. It is important to understand its bases and evaluate its applicability to cultural heritage.

EN 16798-1

EN 16798-1 [27] that supersedes EN 15251 [12], has been prepared by Technical Committee CEN/TC 156 "Ventilation for buildings" to supply constraints or issues that were not included in documents such as EN ISO 7730 [30] and CR 1752 [13] and to provide parameters for energy calculation and long-term assessment of the indoor climate [12]. Regarding IAQ, EN 16798 [27] is based on the results described previously for non-adapted people, considering both the emissions generated by humans and buildings and their components.

Four categories are presented depending on the level of occupant's expectations, and the buildings use. Category I (the most demanding) is designed for high levels of expectation and is recommended for spaces occupied by very sensitive and fragile people. Category II is intended for a normal expectation level and should be used in new buildings and renovations. Category III focus a moderate level of expectation and can be used in existing buildings. Category IV covers all remaining cases. As regards the building components, the document presents three pollution classes, from very-low polluting buildings to non-low polluting. The respective class should be chosen by the designer according to the materials and type of building. According to this standard, total ventilation requirements (q_{tot} in m^3/h) are defined as the sum of the airflow per person ($q_{p\,in}$ m^3/h) according to the bio effluents and airflow per floor area (q_b in m^3/h) to consider the pollution released by the building components, as referred by eq. 4. The respective values are shown in Table 1. The possibility of only using the component associated with human odours is not considered.

Table 1. Ventilation rate\s according to the human bio effluents (q_p) and building pollution level (q_b) [27].

Category	Expected Percentage of Dissatisfied Non-Adapted People [%]	q_p – Airflow Per Person [m^3/(H.Person)]	q_b – Airflow Per Floor Area [m^3/(h.m^2)]		
			Very Low Polluting Building	Low Polluting Building	Non-Low Polluting Building
I	15	36	1.8	3.6	7.2
II	20	25.2	1.26	2.52	5.04
III	30	14.4	0.72	1.44	2.88
IV	40	9	0.54	1.08	2.16

$$q_{tot} = n \cdot q_p + A \cdot q_b \tag{4}$$

Where n is the number of people in the analysed room (-) and A is the floor area (m^2).

It is possible to simplify the analysis and present the required ventilation according to the floor area by multiplying the component associated with human emissions by the occupancy density of the space as shown in Fig. (**2**). Despite not recommended, also the hypothesis of non-polluted buildings was considered for this analysis.

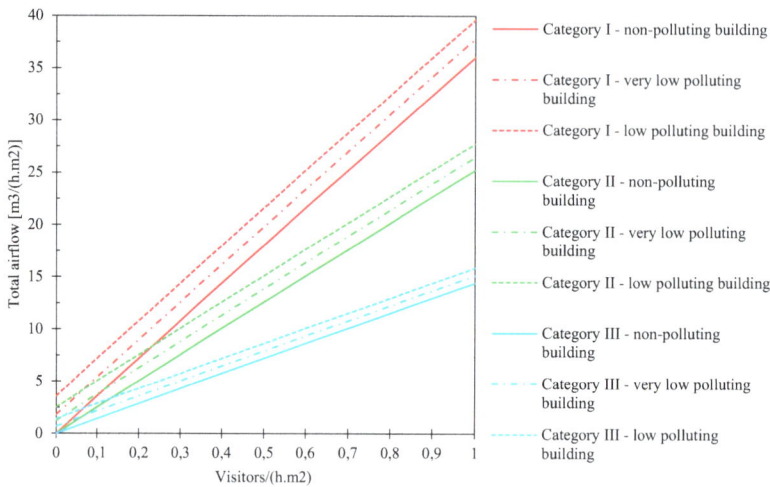

Fig. (2). Ventilation according to comfort category, occupancy and building pollution.

This standard also addresses the possibility of calculating the ventilation rates based on a steady-state mass balance equation, considering the typical differences between the indoor and outdoor CO_2 concentrations presented in Table **2** for a standard CO_2 emission of 0.02 m^3/h.person.

Table 2. Typical CO_2 concentration above outdoors and corresponding airflow per person [27].

Category	Typical CO_2 Concentration Above Outdoors [ppm]	Minimum Airflow Per Person [m^3/(H.Person)]
I	550	36
II	800	25.2
III	1350	14.4
IV	>1350	<14.4

ASHRAE 62.1

The ASHRAE 62.1 [15] specifies the minimum airflow and other measures to ensure indoor air quality to an acceptable level for humans and to minimize health risks. This standard considers the IAQ as acceptable when the air contains no known contaminants at harmful concentrations and where a substantial majority of exposed people classify the air as comfortable (80% or more). This document, although based on results presented in the sub-chapter *"Traditional approach"* as the other international guidelines or standards here presented, has the particularity of considering the adaptation concept.

Although the ASHRAE 62.1 presents values for museums, churches and libraries, the consideration of ventilation flows for adapted people can be considered too low by certain authors since the target public of some of this type of buildings remains inside them for a short period. The standard indicates that ventilation must be obtained for peak occupation, for typical use or in cases where it fluctuates by an average value, defining standard occupation densities for use whenever it is not possible to estimate the actual occupation.

The standard presents three different procedures to reach the required IAQ level, that may return different values according to the adopted assumptions. The first method, named *Ventilation Rate Procedure*, which is more prominent in this work because it eases the relationship between occupancy and ventilation, is a prescriptive method that determines ventilation rates based on the type of use, occupancy and floor area, considering the emission of bio effluents and pollutants by the building components.

As mentioned for EN 16798-1, the minimum airflow must be calculated using equation 4, using data presented in Table **3** for several cultural heritage buildings. This table also shows the acceptance levels for people adapted and not adapted to the proposed ventilation flows.

According to the *Ventilation Rate* Procedure, ASHRAE 62.1 recommends that the total ventilation for museums should consider an airflow of 13.68 m^3/h per person and 1.08 m^3/h per square meter of floor area; for churches, the standard recommends the use of airflow of 9 m^3/h per person and 1.08 m^3/h per square meter of floor area. The standard has a default occupancy density for design purposes of 0.4 people/m^2 for museums and 1.2 people/m^2 for churches.

Table 3. Ventilation flows based on emissions from occupants (q_p) and the building itself (q_b) according to ASHRAE 62.1 [15].

Occupancy Category	Expected Percentage of Dissatisfied People [%]		q_p - Airflow Per Person [M³ /(H.Person)]	q_b - Airflow Per Floor Area [M³/(H.M²)]	Occupant Density [Person/M2]
	Adapted People	Non-Adapted People			
Places of religious worship	19	40	9	1.08	1.2
Libraries	19	40	9	2.16	0.1
Museums (children's)	14	31	13.68	2.16	0.4
Museums/galleries	14	31	13.68	1.08	0.4

The second method, called *IAQ Procedure*, is based on the analysis of contamination sources, concentration limits and perceived level of air quality acceptance. The use of this method presupposes the identification of possible sources of contamination, the use of the limit value and tabulated periods of exposure to avoid health risks. The definition of the design airflow shall be the highest obtained by the mass balance equation for the maximum concentration of a given contaminant, the level of perceived air quality by the occupants or by comparison with the defined airflow for similar areas.

As regards the CO_2 concentration, the ASHRAE 62.1 cite some external references, as the OSHA (5,000 ppm), MAK (5,000 ppm), Canadian guidelines (3,500 ppm), NOSH (5,000 ppm) and ACGH (5,000 ppm).

The third method, called *Natural Ventilation Procedure*, is based on a set of geometric constraints that allow indoor air quality to be guaranteed with the use of natural ventilation while considering the possibility of incorporating mechanical systems.

CIBSE Guide A

CIBSE Guide A [16] is one of the most widely used sources for designers and researchers in the field of building climate behaviour. This document presents three methods to determine the ventilation rates. The first one, described as a prescriptive method, provides an airflow per person based on the data presented previously for non-adapted people, in which all sources other than body odour are not considered. For buildings usually considered as cultural heritage, such as museums and churches, an airflow of 36 m^3/h per occupant is recommended.

The second method should be used in situations where the pollutants are released into the room at a known rate, and local ventilation is not practicable. To apply this method, it is necessary to use a mass balance equation and the concentration limits for each pollutant. The 2006 version of this handbook further advances the existence of a third method that considers odours emitted both by humans and the building itself, although not describing it because it has not gained worldwide acceptance and referring to its application CR 1752 [13].

EN 15759-2

EN 15759-2 [18] was recently published to present procedures for the sustainable management of ventilation in cultural heritage to ensure its preservation and human comfort. This standard presents ventilation management guidelines with a special focus on a step-by-step approach to identify factors and risk-areas and forms of mitigation and a set of procedures to measure and control indoor climate parameters.

Despite the expectations around EN 15759-2, the standard is more descriptive than prescriptive, except the step-by-step approach, and, although a useful tool for heritage management, it does not provide all the necessary management and design answers for stockholders, namely on the choice of design ventilation rates, the relation between ventilation and occupancy, and on how to define the maximum occupancy. In its first appendix, the standard presents two examples of application: a museum located in a cultural heritage building with a collection composed of different objects and a modern storage room. In the first case, the application of EN 15757 [31] was considered to limit mechanical degradation, assuming 35% and 75% of RH to limit mould risk and the abrupt fall of RH due to winter heating. Concerning temperature, a comfort interval between 18 and 26°C was considered. In the second case, the standard goes further and prescribes a maximum ACH of 0.04 h^{-1}, limiting RH between 30 and 50% and T between 7 and 22 °C.

IAQ Based on Human Health

In addition to comfort issues, IAQ should ensure a low risk for the occupant's health. The effects of IAQ on humans may be acute and of short duration (as ocular irritations) or developed over a longer period (as cancer) [20]. To reduce health risks, maximum concentrations and exposure times for each pollutant should be defined according to their specificities. In this chapter, only CO_2 was considered.

Several standards and guidelines use CO_2 as an indicator of comfort, relating it to the perceived air quality [12 - 16], however, their limits are based on comfort and not on health. The CO_2 concentration can be associated with some symptoms as fatigue, headaches, increased perception of heat and unpleasant odours for concentrations between 500 and 3 200 ppm [32]. An exposure of several weeks to concentrations above 7 000 ppm may increase the blood acidity [26, 32], but severe effects are not expected for concentrations below 10 000 ppm [33]. Changes in the respiratory, cardiovascular and nervous systems may be only evidenced for concentrations above 10 000 ppm, but these risks are usually evidenced only for values above 30 000 or 50 000 ppm [32, 33].

There are several publications, mainly for industrial environments, that present the maximum concentration of several pollutants in a way that does not cause damages to health, where it is possible to highlight the limits defined by ACGIH's Threshold Limit Values (TLVs) [34] that defines a maximum concentration of 5,000 ppm for exposures up to 8 h and 30,000 ppm for exposures up to 10 minutes. These limits are widely used and corroborated by other documents [35 - 37], although they are not defined to protect the most sensitive individuals [15].

In addition to the severe impact that high CO_2 concentrations have on health, the discussion about the effect of moderate concentrations on the well-being of individuals has not been worthy of a huge discussion in the scientific community [38 - 40], and several authors have argued the use of more demanding threshold values.

For residential buildings, for example, a document requested by the Government of Canada [32] setts a maximum level of 3 500 ppm. This document assumes that the adopted limit provides a sufficient margin to protect against undesirable changes in the acid-base balance of blood and release of calcium from bones.

Kim *et al.* [41] concluded that the homeostasis of a group of individuals when exposed to an environment with a CO_2 concentration above 2 000 ppm and work-stress showed changes. According to Satish *et al.* [42], exposures of 2.5 h up to 2500 ppm affected the decision making of a group of individuals when compared to concentrations of 600 ppm. For exposures up to 4 h to concentrations up to 1,400 ppm, Vehviläinen *et al.* [43] concluded that individuals did not show physiological changes, however, for concentrations between 2 000 and 4 000 ppm significant symptoms were reported, which led the authors to recommend a maximum concentration of 1500 ppm. Maula *et al.* [44] concluded that an exposure of 4 h to 2 260 ppm in open-space offices with minimalist furniture increased perceived fatigue slightly, as in the perception of air quality, however, there were no symptoms of health problems.

Zhang *et al*. [33, 45] compared the response of a set of individuals when exposed for 4.25 h at 500 ppm and 3 000 ppm. Authors tested two different approaches that made interesting conclusions possible: a) they added CO_2 in the room until it reached 3 000 ppm; b) reducing ventilation until reaching 3 000 ppm. In the first case, they concluded that there were no significant changes in the perception of air quality, health symptoms or cognitive performance. In the second case, although the CO_2 concentration is identical, a significant reduction in the perception of air quality has been reported, increasing the intensity of acute general health symptoms (neuro-behavioural) without increasing respiratory or mucous membrane symptoms and decreasing the cognitive response. According to this study, CO_2 individually is not responsible for health damage at concentrations up to 3 000 ppm, however, when associated with the production of bio-effluents, it significantly affects individuals. Despite this conclusion, it should be noted that in non-industrial buildings, the main source of CO_2 is usually related to human metabolism. Thus, according to these results, concentrations above 3 000 ppm should be avoided.

IAQ in Museums

In common buildings, the indoor climate and air quality are controlled according to human health and comfort needs, but when the analysis is concerned with cultural heritage, particularly in buildings used to exhibit or store valuable artefacts, it is not possible to ignore the needs of the collections, avoiding degradation phenomena that can be triggered by poor climates or high relative humidity [16].

Several risks to cultural heritage are directly or indirectly attributed to the deficient relationship between ventilation and occupancy, namely excessive RH fluctuations, too high or too low RH levels, surface condensation, water evaporation and salt damage, and the transport and deposition of atmospheric pollutants [18]. High ventilation rates contribute to high energy consumptions and high RH fluctuations [4], which increase the risk of mechanical degradation of organic hygroscopic materials [3, 31]; on the other hand, low ventilation rates associated with a large number of occupants contribute to high indoor RH, which may lead to condensation, mould germination, salt deliquescence or metal oxidation [3, 4]. Adopting indoor air recirculation can improve preventive conservation [46].

There are numerous studies published in the international bibliography addressing the theme under analysis. However, in many cases, the analysis focuses on energy efficiency, comfort and health, leaving out the conservation. Although it is not intended to make an exhaustive state-of-the-art review, some works considered relevant for the study of ventilation in museums will be presented.

In cultural heritage buildings, ventilation must avoid air stagnation and the creation of microclimates near the collections and contribute to the removal of pollutants. According to [47], a minimum recirculation of 6 air changes per hour should be considered. In measurements made at the National Archives Building (Washington, D.C.), air changes of fresh air from 0.9 to 1.2 h^{-1} were obtained.

In a study comparing the behaviour of two museums in London (one naturally ventilated and the other with mechanical ventilation), Cassar *et al.* [48] obtained a fresh air change rate of 1.3 h^{-1} for the air-conditioned museum. Concerning to air recirculation, a complete recirculation of approximately 6.5 h^{-1} was estimated.

In 1999, Camuffo *et al.* [49] published a study approaching the indoor air quality at the Correr museum in Venice. The authors reported that the museum did not have a central ventilation system for the intake of fresh air, considering, however, that this was not a problem, since the rooms have high dimensions, with high ceilings and a moderate number of visitors. The building's ventilation occurs mainly by opening the windows twice a day to operate the shutters. Inside, there is a mechanical system that allows the air recirculation, with speeds varying between 2 and 3 m/s 20 cm below the fans. The fact that the outdoor climate usually has high humidity values is a problem for exchanges between indoor and outdoor. It was also found that ventilation through the opening of windows is a serious risk due to the entry of pollutants.

In 2000, Druzik *et al.* [50] published a paper reporting the results of a monitoring campaign carried out in 1988 in five museums and galleries in Southern California: the Norton Simon Museum, J. Paul Getty Museum, Scott Gallery of the Huntington Art Gallery and Library, the Southwest Museum, and the Sepulveda House.

The authors found buildings naturally ventilated by opening the doors and windows and others with complete air conditioning systems. In the first case, outdoor air exchanges from 1.6 to 3.6 h^{-1} were obtained, while in air-conditioned buildings they obtained outdoor air exchanges below 1 h^{-1} and recirculation rates ranging from 5 and 8 h^{-1}. The authors recommend the use of carbon recirculation filters to avoid high levels of indoor pollution.

Mazzei *et al.* [51] developed a simulation model to evaluate the effect of air conditioning systems in museums, adopting an infiltration rate of 0.25 h^{-1} and an airflow of 21.6 m^3/h of fresh air per person. The authors state that the recirculation rate should vary between 6 and 8 renewals per hour [47,52].

Martens assessed the impact of ventilation on the indoor climate and energy consumption using a generic computer model of a non-controlled museum simulated for the weather conditions of De Bilt, The Netherlands [4]. The author concluded that an increase in the air change rate (ACH) from 0.1 to 1 h^{-1} has a reduced influence on hourly RH fluctuations, but significant influence on the weekly fluctuations with an increase of around 15% RH, hence the recommendation of a maximum ACH of 0.1 h^{-1} if the objective is to maintain the climate stability. To evaluate the impact of ventilation on energy consumption, the author considered the climatization of the model to 18-22 °C and 48-52% RH, concluding that ACH higher than 0.1 h^{-1} led to a considerable increase in energy consumption. However, Martens emphasizes that the use of low ventilation rates can be dangerous if associated with high occupancy rates.

In 2016, Zorpas and Skouropatis [53] studied the indoor air quality in two museums located in Cyprus. They presented the country reality, noting that many museums are characterized by natural ventilation and that few works focused on air quality in cultural heritage in that country were carried out. Concerning CO_2, they found concentrations between 631.55 and 698.77 ppm in the Cypriot-Archaeological Museum and values up to 748.22 ppm in the Byzantine Museum. According to the Cyprus legislation, the usual CO_2 values in offices must be between 600 and 800 ppm, with a maximum of 1 000 ppm being allowed. About ventilation, there is a rate of 14.4 to 21.6 m^3/h of fresh air per visitor in museums and galleries.

Human presence also plays a major role in the quality of the indoor environment. On some studies of the Scrovegni chapel [10, 11], the authors advocate the use of an airflow of 60 m^3/h per person to remove the generated water vapour without causing excessive temperature gradients and defend the need of an air treatment system for the case-study to avoid the entry of pollutants through openings, ensuring the required air quality for visitors. Authors evidence that air velocity should not influence heat and mass changes on painted surfaces and recommend a velocity of less than 1 m/s in the intake zone and less than 0.1 m/s near the frescos.

To guarantee these conditions, an ACH less than 1 h^{-1} during visitation periods and a daily average of 0.5 h^{-1} were required.

Kramer *et al.* [54], studying the Hermitage Museum Amsterdam, develop a simulation model, evaluating the ventilation, composed by infiltration and mechanical ventilation activated only when the interior concentration of CO_2 exceeds the limit of 1 000 ppm. Note that in Amsterdam, the external concentration of CO_2 is around 600 ppm. The mechanical ventilation based on

CO_2 typically occurs between 2 pm and 5 pm, with recirculation between 7 am and 7 pm and air infiltration at an average rate of 0.11 h^{-1} throughout the day. The museum applies a high recirculation rate of 7.5 h^{-1}.

METHODOLOGY

General Considerations

As referred previously, there is no consensus on cultural heritage literature about IAQ. In addition to issues related to conservation, comfort and health, it is important to remember the influence that ventilation can have on energy consumption in museums. To find a compromise between ventilation, conservation, comfort, health and energy, it was considered relevant to develop a simulation campaign testing various scenarios and analysing their impacts on the building's response using the hygrothermal simulation software *WUFI®Plus* [55]. This software is based on the numerical model presented in [56], and it was widely tested and validated in several cases, including historical buildings and museums [57 - 61].

The control of the indoor climate in museums is inevitable, due to the comfort of the occupants and, above all, to guarantee the conservation of the collections. This theme is far from generating consensus in the international scientific community. Historically, very demanding temperature and relative humidity set-points have been used, such as 20 °C and 50% RH with small fluctuations [62]. Since the last decade of the 20th century, there has been an attempt to use fewer demanding approaches based on new knowledge about the conservation of materials and the need to reduce energy consumption, both for economic and environmental reasons.

In this chapter, a less demanding approach was used: 16-25 °C and 40-60% RH. This set-point is following the recommendations published by the Group of Organizers of Large-scale Exhibitions (Bizot Group, that comprises the directors of the world's leading museums and galleries), the Australian Institute for the Conservation of Cultural Materials (AICCM) and the Association of Art Museum Directors (AIC) [63 - 66].

Case Study – Geometry, Envelope and Internal Gains

Building Geometry

To study the effect of ventilation on cultural heritage, a typical building was used, with geometry inspired by the National Museum of Ancient Art (Lisbon) and to represent museums installed in old palaces. The model presents a regular plan of quadrangular shape, with facades exposed in the four main orientations. It is composed by two floors and a total of 16 rooms. It was considered the presence of a basement floor for the reserve, conservation and restoration workshops. It was decided not to model this floor, considering that there is no exchange of heat and mass between the basement and the ground floor. The horizontal plan can be seen in Fig. (**3**) and the facades in Fig. (**4**). The building dimensions are in Table **4**.

Table 4. Building dimensions.

Floor	Room	Area [m²]	Height [m]	Volume [m³]	Ratio Openings/Floor [-]
Ground Floor	Hall	308	3.5	1078	0.08
	1.1	506.25	3.5	1771.875	-
	1.2	180	3.5	630	0.1
	1.3	64	3.5	224	0.1
	1.4	180	3.5	630	0.1
	1.5	64	3.5	224	0.1
	1.6	180	3.5	630	0.1
1ˢᵗ Floor	2.1	506.25	4.66	2358	0.08
	2.2	180	4	720	0.1
	2.3	64	4	256	0.1
	2.4	180	4	720	0.1
	2.5	64	4	256	0.1
	2.6	180	4	720	0.1
	2.7	64	4	256	0.1
	2.8	180	4	720	0.1
	2.9	64	4	256	0.1

Ground floor   1st floor

Roof

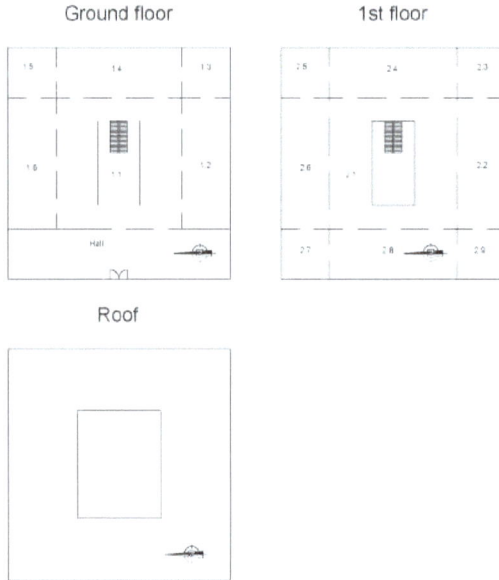

Fig. (3). Building geometry – horizontal plan.

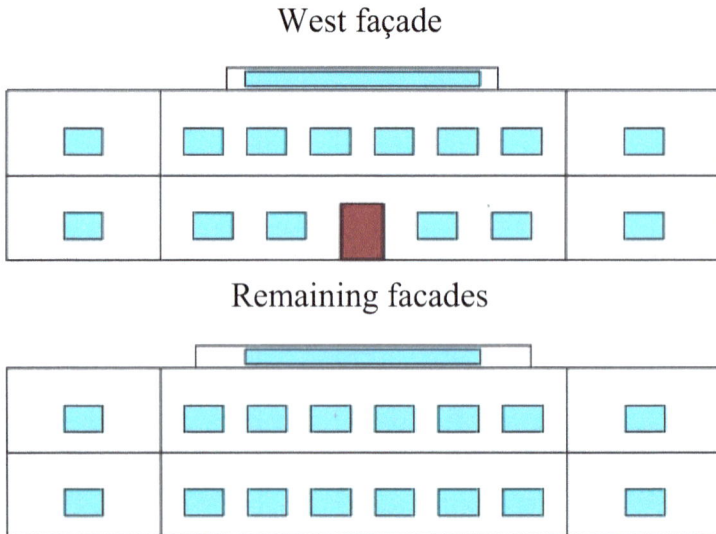

West façade

Remaining facades

Fig. (4). Building geometry – facades.

Envelope

Constructive solutions representative of the old buildings in southern Europe, more specifically in Portugal, were adopted with the choice falling to the use of simple walls composed by mortared limestone and lined on both sides with lime-based mortar [67]. For the horizontal envelope, a more modern solution composed by reinforced concrete was adopted. For windows, double glazed aluminium frames were considered, including interior shading elements to limit the risk of chemical degradation due to solar radiation. The building's elements assemblies and U-value, as properties of the materials, can be seen in Table **5**.

Table 5. Simulated building element assemblies and U-value, materials' thicknesses and basic hygrothermal proprieties [55, 67, 68].

Element	Assembly	Thickness (e, m)	Bulk Dnsity (ρ, kg/m^3)	Specific Heat Capacity, Dry (c_p, J/kg.K)	Thermal Conductivity, Dry (λ_0, W/m.K)	U-Value (W/m^{2}K)
External Walls Outside → Inside	Lime mortar	0.03	1785	850	0.70	
	Mortared-limestone	0.54	2122	850	1.76	1.78
	Lime mortar	0.03	1785	850	0.70	
Internal walls Outside → Inside	Lime mortar	0.03	1785	850	0.70	
	Mortared-limestone	0.24	2122	850	1.76	2.07
	Lime mortar	0.03	1785	850	0.70	
Ceilings Outside → Inside	Old oak	0.02	740	1400	0.15	
	Concrete screed – top layer	0.005	1890	850	1.6	
	Concrete screed – middle layer	0.04	1970	850	1.6	
	Concrete screed – bottom layer	0.005	1990	850	1.6	0.56
	Aerated concrete	0.20	460	840	0.12	
	Reinforced concrete	0.25	2350	850	2.3	
	Lime mortar	0.03	1600	850	0.70	

(Table 5) cont.....

	Generic gravel	0.1	1400	1000	0.7	
	Geomembrane	-	-	-	-	
	XPS	0.04	40	1500	0.03	
	PVC Roof membrane	10E-4	1000	1500	0.16	
	Concrete screed – top layer	0.005	1890	850	1.6	
Roof Outside → Inside	Concrete screed – middle layer	0.04	1970	850	1.6	0.34
	Concrete screed – bottom layer	0.005	1990	850	1.6	
	Aerated concrete	0.10	460	840	0.12	
	Reinforced concrete	0.25	2350	850	2.3	
	Lime mortar	0.03	1600	850	0.70	
Doors	Old oak	0.05	740	1400	0.15	2.01
Windows	Double gazed aluminium frames	$U_w = 2.45$ W/m^{2}K	$F_f = 0.70$	SHGC=0.15	$\varepsilon = 0.15$	

Internal Loads

Visitors Flux

When trying to study the relationship between ventilation and occupancy, one of the first problems starts with the occupancy to be used. Ideally, typical values recorded in the museum under analysis or similar museums should be used. Usually, it is easy to obtain the number of visitors per month or even the daily occupation, however, knowing the occupation profile of each room at each moment is practically impossible.

When there is no real data, it is usual to assume a typical occupation of 0.4 people per square meter, as shown in ASHRAE 62.1 [15]. However, it is easy to understand that it will be difficult to obtain a constant occupation rate throughout the day and year. Even so, it is necessary to adopt a typical occupation profile that allows studying the impact of other factors, namely the ventilation, on which more attention falls in this work.

In the framework of the present study variable rates throughout the day were considered adequate. Based on a study developed at the Monastery of Jerónimos (Portugal) [69], a daily average profile was estimated, adopting a maximum occupant density of 0.4 visitors/m^2, as can be seen in Fig. (5).

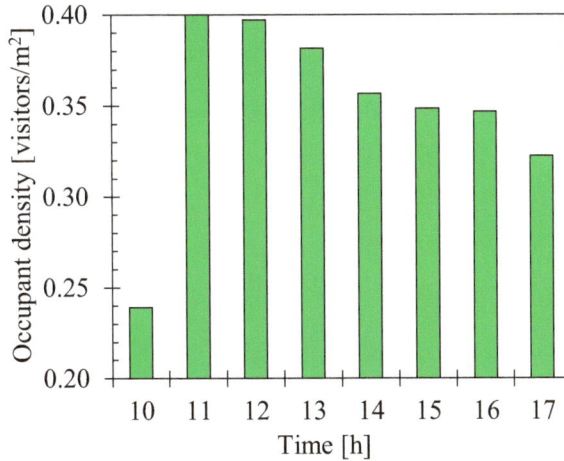

Fig. (5). Occupant density according to the study [69].

Sensible and Latent Heat

Internal loads are defined as the amount of sensible and latent heat that is emitted by a source in a given space (*e.g.* the human body, lighting and other electric equipment). Additionally, also the CO_2 generation should be considered as an important factor.

While the fraction of latent heat is responsible for an increase of water-vapour concentration, the sensible heat fraction is absorbed and stored by the surrounding surfaces. It is estimated that between 20–60% of the emitted sensible heat assumes the radiation form, which depends on several aspects as clothing, mean radiant temperature and air velocity [16].

The heat generated by the human body occurs due to oxidation and it depends on the person, namely its gender, age and activity level. This type of heat is named as metabolic rate and usually presented in *met*. A unit metabolic rate (1 met) corresponds to the amount of energy generated by a standard European adult male sitting at rest, with a body area of $1.8 \ m^2$, which corresponds to an energy flow of $58.2 \ W/m^2$ [70, 71]. The metabolic rate for each activity is based on the amount of heat that is emitted by an average person resting. In other words, 1.6 metabolic rate means that the person is emitting 1.6 more heat than the amount of heat produced by an average person while resting.

The amount of energy emitted by a person can be determined through the metabolic rate multiplied by the surface skin area, proposed by DuBois [71]:

$$A_D = 0.202 \cdot m^{0.425} \cdot l^{0.725} \tag{5}$$

Where A_D is the surface skin area (usually named Dubois area) (m²), m is the person's mass (kg) and l is the person's height (m).

It is commonly accepted to assume that an average male weighs 70 kg and is 1.73 m tall [70, 71] adopting a reduction factor of 0.85 for women and 0.75 for children [72].

This makes it easy to estimate the average metabolic rate for a group of people that are in the same space and performing the same activity:

$$met \cdot 58.2 \cdot A_D \cdot (\%M \cdot Rf_M + \%F \cdot Rf_F + \%C \cdot Rf_C)/100 \tag{6}$$

Where *%M, %F and %C* means the percentage of male adults, female adults and children's, respectively, and Rf_M, Rf_F e Rf_C are the respective reduction factors.

However, the metabolic rate is not always easily determined and authors typically use tabulated values corresponding to each activity, such those present in ASHRAE Fundamentals [71], EN ISO 8996 [73], ASHRAE 55 [70] or EN ISO 7730 [30], or through physiological measurements. According to [16], for most engineering-related purposes it is possible to use tabulated values with reasonable accuracy (*i.e.* ± 20%) for metabolic rates below 1.5 met.

Concerning the metabolic rate used in cultural heritage, there is no consensus in the scientific community, and values ranging from 1.2 to 1.7 met were found. From the observation of a group of 15 people at the Van Abben Museum in Eindhoven, Kramer *et al.* [74] found that, on average, visitors spent 8.5 minutes standing (1.2 met) and 0.5 minutes walking calmly (2 met), resulting in a value of 1.24 met. Kramer *et al.* [75] obtained an average value of 1.5 met for the Hermitage Amsterdam museum by solving the PMV equation in order of the metabolic rate according to the thermal sensations and clothing recorded by users in 1121 surveys and using more than a year of monitoring of the ambient conditions. Maekawa *et al.* [76] used a value of 1.5 met in the study of the Museu Casa de Rui Barbosa in Rio de Janeiro. Yau *et al.* [77] through a field study at the National Museum of Malaysia reached an average value of 1.7 met. La Gennusa *et al.* [78] analysed the conflict between conservation and comfort in museums using the PMV model with a metabolic rate of 1.2 met, as usually used for offices or educational establishments. Silva *et al.* [79] assumed that the activity in a museum is equivalent to a light activity such as that developed when shopping in a shopping centre and used a value of 1.6 met [30]. Bellia *et al.* [80] assumed a total heat gain per person of 147 W (85 W for sensible heat and 62 for latent heat). Ascione *et al.* [81] adopted a metabolic rate of 1.5 met. Ferdyn-Grygierek [82] assumed that each visitor would be responsible for 75 W of sensible heat

(50% was emitted by convection and the other 50% by radiation) and 55 W of latent heat, which means a total heat of 130 W. Kramer *et al.* [83] assumed that each visitor was responsible for 100 W in the exhibition room. Karyono *et al.* [84] assumed that in the cathedral the visitors would have a metabolic rate of 1.0 met (which corresponds to seated, relaxed) and in the museum, it would vary between 1.0 (which corresponds to seated, relaxed) and 1.2 (which corresponds to standing, at rest). Balocco *et al.* [85], who modelled an 11th century church in Florence, assumed that each visitor emitted 75 W of sensible heat and 55 W of latent heat, which means a total heat of 130 W. Camuffo *et al.* [86], who tested a heating system in two Italian churches, considered that metabolic rate of a person varied between 1.0 met (seated, quiet person), 1.2 met (standing, relaxed person), 1.5 met (singing) and 2.0 met (walking). Aste *et al.* [87], who studied several heating strategies for churches, assumed a total heat load of 108 W per person.

At this point, it was decided to estimate the average metabolic rate based on the type of audience and assuming as a reasonable behaviour that visitors spend 40% of their time walking calmly through the museum (1.7 met - walking about [70]) and 60% of the time stopped to observe the works (1.2 met - standing, relaxed [70]), achieving an average metabolic rate of 1.4 met. To define the profile of the average visitor, a study developed by the Directorate General for Cultural Heritage (DGPC) of the visitors of the national museums in Portugal was used [88]. This study was carried out between 3rd December 2014 and 2nd December 2015 and covered 14 museums. A slight prevalence of the female public was observed, reaching 56% of the visits. It was decided to consider the presence of men/women without incorporating the influence of age, obtaining an average value of 134 W:

$$(0.4 \cdot 1.7 + 0.6 \cdot 1.2) \cdot 58.2 \cdot (0.56 \cdot 0.85 + 0.44 \cdot 1) \approx 134(W) \tag{7}$$

The heat generated by the human body is dissipated from the body surface in a combination of radiation, convection, evaporation and breathing. The sensible heat loss rate (losses by convection and radiation) and latent heat loss rate (losses by evaporation and breathing) are not constant and depend on the metabolic rate and ambient conditions and its quantification is a lengthy and difficult process.

The Department of Energy of the United States of America presented and applied to the EnergyPlus software a polynomial equation (8) [89] to estimate the amount of sensitive heat according to the metabolic rate and air temperature:

$$\begin{aligned} = {}& 6.461927 + 0.946892 \cdot M + 0.0000255737 \cdot M^2 + 7.139322 \cdot T - 0.0627909 \\ & \cdot T \cdot M + 0.0000589172 \cdot T \cdot M^2 - 0.198550 \cdot T^2 + 0.000940018 \\ & \cdot T^2 \cdot M - 0.00000149532 \cdot T^2 \cdot M^2 \end{aligned} \tag{8}$$

where S is the amount of sensible heat (W); M is the metabolic rate (W) and T is the air temperature (°C).

According to Chapter 9 of the ASHRAE Handbook – Fundamentals [52], the sensible heat can be divided into radiative and convective fractions as 0.6 and 0.4, respectively, for typical office conditions. Here, this ratio was considered admissible.

Knowing the total amount of heat and the sensible fraction, the amount of latent heat can be easily achieved:

$$L = M - S \tag{9}$$

Where L means the amount of latent heat (W), M means the amount of total heat (W) and S is the sensible heat (W).

However, the latent loads are usually presented through water vapour production rates in g/h. This conversion can be obtained through the quotient between the latent heat and the water evaporation enthalpy [54]:

$$G_{w,vap} = \frac{L}{\Delta h_{vap}} \cdot 3600 = \frac{M - S}{\Delta h_{vap}} \cdot 3600 \tag{10}$$

where $G_{w,vap}$ is the water vapour generation rate per person (g/h), L is the amount of latent heat (W), H is the total heat (W), S is the amount of sensible heat (W) and Δh_{vap} means the water evaporation enthalpy (2257 J/g [54]).

Taking a total heat generation of 134 W and assuming an average indoor temperature of 20°C, a sensible heat of 90 W (60% emitted by radiation and 40% by convection) and latent heat of 44 W were obtained. A water vapour generation rate of 70 g/h was achieved.

CO_2 Generation Rate

The human metabolism consumes oxygen and releases carbon dioxide depending on the physical activity intensity, the person's height and type of diet. The rate of oxygen consumption per person (VO_2) is determined through the following equation [26]:

$$V_{O_2} = \left(\frac{0.00276 \cdot A_D \cdot met}{(0.23 \cdot RQ + 0.77)}\right) \cdot 3.6 \tag{11}$$

Where V_{O_2} is the rate of oxygen consumption [m^3/h], met is the metabolic rate (met), A_D is the body surface area in accordance to Dubois equation (1.8 m^2 for the average European male), 3.6 is the coefficient to convert the rate of oxygen consumption from l/s to m^3/h and RQ is the respiratory quotient (ratio between the CO_2 volumetric production rate for a specific rate of oxygen consumption).

The respiratory quotient depends on a person's diet, activity intensity and physical condition. For an average adult conducting a sedentary activity, the respiratory quotient is 0.83. The multiplication of this quotient by the rate of oxygen consumption allows to determine the CO_2 production rate:

$$V_{CO_2} = \left(\frac{0.00276 \cdot A_D \cdot met}{(0.23 \cdot RQ + 0.77)}\right) \cdot 3.6 \cdot RQ \tag{12}$$

Assuming a metabolic rate of 1.4 met and an average body surface area of 1.67 m^2, a production rate of 0.020 m^3/h is achieved. Considering the CO_2 density in the gaseous state of 1.83 kg/m^3 at 20 °C and 1 atm, the CO_2 production rate is 37 g/h.

Lighting

Concerning to lighting, all the electrical energy used by a lamp is transformed into heat and dissipated by convection and radiation. In addition to the influence of the lamp's properties, the luminaire itself has an important influence on the division of energy by affecting the ratio between radiant and convective heat.

The lighting level is either presented in terms of lighting power per area (W/m^2) or by the level of lighting required for space (lux). However, only the latter shows the lighting level of the space, since the former corresponds to the amount of consumed/released heat. Some documents, such as ASHRAE Standard 90.1 [90] and ASHRAE Fundamentals [71, 91], present values of lighting for spaces in W/m^2, as shown in Table 6 for cultural heritage buildings. Despite this variable is quite useful for thermal calculations and as an input for thermal and hygrothermal simulation, it does not present any type of information about the luminous performance of the solution.

Table 6. Examples of lighting power found in the literature [71,91].

Type of Building	Lighting Power [W/m²]
Museums	
Exhibition area	11.3
Restoration/conservation area	11.0
Storage area	8.6
Libraries	
Card file and Cataloguing	7.8
Reading area	10.0
Stacks	18.4
Religious buildings	
Main nave	16.5
Hall	6.9
Pulpit, choir	16.5

On the other hand, CIBSE Guide A [16] presents the recommended lighting depending on the type of building according to the required lighting level, as shown in Table 7.

Table 7. Examples of illumination levels found in the literature [16].

Type of Building	Maintained Illuminance [lux]
Museums and art aalleries	
Display area	200
Storage area	50
Libraries	
Lending area	200
Reading area	500
Store area	200
Religious buildings	100-200

The power necessary to meet the required lighting level depends on the type of lamp to be used. The comparison between incandescent, fluorescent and LED lamps allow a simplified understanding of the performance of each type of lighting: incandescent lamps have high consumption for reduced lighting levels since a good part of the energy is transformed into heat; fluorescent lamps appear at an intermediate level, with higher efficiency than that presented by incandescent lamps, but they continue to spend much of the energy consumed in the generation of heat; finally, for LED, all the energy consumed is transformed into light, which translates into much lower energy consumption for the same lighting level.

It is possible to relate the lighting power with the illuminance and the floor area through the equation:

$$P = \frac{E_v \cdot A}{\eta} \qquad (13)$$

Where P is the electric power (W), E_v is the illuminance (lux), A is the surface area (m^2) and η is the luminous efficacy for the adopted solution (lm/W). Typical lighting efficiencies for various lighting solutions are presented in Table **8**.

Table 8. Luminous efficiency according to the lamp type [92, 93].

Lamp Type	Luminous Efficiency [Lumens/Watt]
Tungsten incandescent light bulb	10-15
Halogen lamp	20
Fluorescent lamp	50-90
LED lamp	70-95
Metal halide lamp	75-90

The proportion of heat that enters the compartment depends on the type and location of the lighting fixtures. When the luminaires or lamps are suspended from the ceiling or mounted on the walls, all the heat will appear as an internal gain. When the luminaires are embedded in the ceiling or in the false ceiling, part of the emitted power results in a gain of heat for the structure that surrounds it.

The form of heat dissipation for various luminaire configurations with fluorescent lamps is presented in Table **9** according to the values present in the *EnergyPlus* software user manual [94].

Table 9. Ways of dissipating heat from various configurations of light sources (adapted from [94]).

Type of Dissipation	Lighting Sources Configurations			
	Suspended	Surface-Mounted	Recessed	Luminous and Recessed Ceiling
Radiant	0.6	0.9	0.55	0.55
Convective	0.40	0.10	0.45	0.45

Since the LED technology is increasingly widespread, it was considered that all the model's lighting is LED. Considering an illumination of 200 lux during the opening hours – from 9.00 am to 7:00 pm, and 20 lux at night and average lighting efficiency of 82.5 lumens/watt, LPD of 2.42 W/m^2 and 0.24 W/m^2, respectively, were achieved. Suspended lighting was considered, assuming 0.6 of radiant emission and 0.40 of convective.

Simulation Study

Concerning to comfort, the airflow per visitor varies according to the document and with (or without) the assumption of the concept of adaptation. According to standard EN 16798-1 (for non-adapted people), category II seems admissible for new museums and category III for museums in historic buildings. Category II (for an acceptance level of 80%) provides for an airflow of 25.2 m^3/(h.person), while Category III (for an acceptance level of 70%) provides for an airflow of 14.4 m^3/(h.person). Other documents, such as CIBSE Guide A, recommend highest values, namely an airflow 36 m^3/(h.person). For large national museums, where visits can take several hours, it seems plausible to consider the concept of adaptation, allowing air flows three times greater for the same level of comfort. The ASHRAE 62.1 standard is an example of this approach, recommending an airflow rate of 13.68 m^3/(h.person) for an acceptance level of 85%.

As regards the health hazard, the previous analysis concluded that for concentrations up to 3 000 ppm, no significant effects were found. It was found that these CO_2 concentrations cause complaints, especially when associated with metabolic activity, something that is covered by the approach focused on comfort.

Considering the adaptation concept, and adopting a maximum CO_2 threshold of 3 000 ppm, the data published by EN 1679-1 were adapted and assumed as reference for the current study, as can be seen in Table **10**.

Table 10. Typical CO_2 concentration above outdoors according to EN1679-1 [27] for adapted and non-adapted people and considering health issues.

Category/PD (%)	CO_2 Above Outdoors (ppm)		q_b – Airflow Per Floor Area [m^3/(h.m^2)]		
	Adapted People[1)]	Non-Adapted People[2)]	Very Low Polluting Building	Low Polluting Building	Non-Low-polluting Building
II (20 %)	Min (3 000-$C_{CO2,o}$[3)];2 400)	800	1.26	2.52	5.04
III (30 %)	3 000-$C_{CO2,o}$[3)]	1 350	0.72	1.44	2.88

(Table 10) cont.....

> [1] for visitors that remain more than 18 minutes inside the building;
> [2] for visitors that remain less than 18 minutes inside the building;
> [3] for health reasons the absolute concentration of CO_2 should not exceed 3 000 ppm.
> $C_{CO2,o}$ means the outdoor CO_2 concentration

The airflow was estimated using the mass balance equation for stationary conditions presented in (3). Concerning the exterior CO_2 concentration to be used, it is common to find values from 300 to 450 ppm [14], although there is no consensus around an exact number since it depends on the location and level of pollution. For this case, an outdoor concentration of 400 ppm was adopted according to the recommendation of EN 13779 [14] for polluted city centres. A CO_2 generation rate of 0.02 m3/h per person was considered according to EN 16798-1 [27], as can be seen in Table **11**.

Table 11. Ventilation rates according to the human bio effluents (q_p) and building pollution level (q_b) for an outdoor CO_2 concentration of 400 ppm and a CO_2 generation rate equal to 0.02 m³/h.

Category/ PD (%)	Adapted People[1]		Non-Adapted People[2]		q_b – Airflow Per Floor Area [m³/(h.m²)]		
	q_p [m³ /(h.person)]	CO_2 Above Outdoors (ppm)	q_p [m³ /(h.person)]	CO_2 Above Outdoors (ppm)	Very Low Polluting Building	Low Polluting Building	Non-Low-polluting Building
II (20 %)	8.3	2400	25.2	800	1.26	2.52	5.04
III (30 %)	7.7	2600	14.4	1350	0.72	1.44	2.88

[1] for visitors that remain more than 18 minutes inside the building;
[2] for visitors that remain less than 18 minutes inside the building

As regards the studies concerning museums, it was possible to find a pattern, concluding the frequent use of an air change rate of about 1 h^{-1} from the exterior and recirculation rates between 5 and 8 h^{-1}. In some cases, natural ventilation was found. Even when buildings are equipped with mechanical ventilation, it may be only activated when necessary. Despite the importance of ensuring the stability of the interior microclimate in space and time, sometimes the recirculation and mechanical ventilation are turned off during the night.

In this chapter, it was decided to test various ventilation strategies based essentially on the comfort and health requirements. It was defined that mechanical ventilation would only work during the museum's opening hours. We also tested the possibility of the ventilation system being equipped with a heat recovery system with an efficiency of 0.85. Based on the usual practice in museums, several solutions were tested with internal air recirculation, activating mechanical ventilation only when necessary and limiting total ventilation to 1.5 ach. All simulated combinations can be seen in Table **12**.

Table 12. Simulation campaign.

#	Infiltration (0.2 ach)	Constant Mechanical Ventilation (9 am – 7 pm)			Mechanical Ventilation on CO$_2$ Demand (max 1.5 ach)			Recirculation (6 to 8 ach)		Heat Recovery
		7.7 (m³/h. visitor)	14.4 (m³/h. visitor)	25.2 (m³/h. visitor)	800	1350	2600	All day	9 am – 7 pm	
1	X									
2	X	X								
3	X		X							
4	X			X						
5		X								X
6			X							X
7				X						X
8	X				X				X	
9	X					X			X	
10	X						X		X	
11	X				X				X	X
12	X					X			X	X
13	X						X		X	X
14	X				X			X		
15	X					X		X		
16	X						X	X		
17	X				X			X		X
18	X					X		X		X
19	X						X	X	X	X

Weather File

To understand the impact of the climate, it was decided to simulate three cities: Lisbon, Amsterdam and Krakow. These cities were chosen because they have different climates and because they are associated with several reference articles in the area of museum climate [3, 54, 61, 75].

The climatic classification according to Köppen [95], the annual averages, maximum and minimum values of temperature and relative humidity and the annual global solar radiation can be seen in Table **13**. For Lisbon a test reference year (TRY) develop by Silva [62] was adopted. The remaining weather files were taken from the *EnergyPlus weather database* [96].

Table 13. Climate characterization.

#	City	Köppen Class	Temperature [°C]			Relative Humidity [%]			Global Solar Radiation [kW/m²]
			Mean	Min	Max	Mean	Min	Max	
1	Lisbon	Csa	16.9	4.0	36.2	70	13	100	1863
2	Amsterdam	Cfb	10.0	-8.4	32.7	84	30	100	982
3	Krakow	Dfb	8.3	20.2	31.9	79	27	100	1045

Data Analysis

Fulfilling the objective of this chapter, the impact of each ventilation strategy on energy consumption, compliance with the comfort and health limits related to CO_2 and the stability of the indoor climate according to conservation was analysed. The analysis of energy consumption was carried out for the entire building. In turn, the characterization of climate stability overtime was made only for room 1.2, while spatial stability was assessed through the maximum time differences between room 1.2 and 2.6.

The indoor climate was characterized by seasonal fluctuations, short-term fluctuations and daily cycles. The seasonal cycle is calculated as a centred 30-day moving average. In turn, seasonal fluctuations are obtained as the biggest differences above and below the annual average. Short-term fluctuations are calculated as the hourly differences between the data obtained and the seasonal cycle. According to standard EN 15757 [31], typical fluctuations are obtained through the 7th and 93rd percentiles of these differences, thus allowing 14% of the largest fluctuations to be excluded. Daily fluctuations were obtained by the difference between the maximum and minimum values obtained in 24 consecutive hours. To obtain typical daily fluctuations, 10% of the largest fluctuations were excluded.

To relate seasonal cycles and short-term fluctuations to the risk for conservation, it was decided to use an expedited method published in the handbook of the American Society of Heating and Air-Conditioning Engineers, Inc – ASHRAE [52] and entitled "Museums, galleries, archives and libraries. According to this document, ideal short-term fluctuations of ± 5% RH and ± 2°C have been defined, although fluctuations of up to ± 10% RH are allowed if no seasonal fluctuations were found. The same conditions have been adopted for space fluctuations. For the seasonal cycle, it was decided not to impose limits since normally the materials are more capable of adapting to long fluctuations.

RESULTS

To analyse the impact of various ventilation strategies in museums, a simulation campaign was carried out for three European cities in a total of 57 simulations.

The ventilation is usually defined according to the indoor air quality science, based on a certain airflow per occupant to guarantees an acceptable level of the perceived air quality. This approach is used in several buildings; however, it is important to clarify whether it is valid for museums. The impact of ventilation on energy consumption, comfort, health and conservation was analysed.

In many museums located in historic buildings, natural ventilation is common. Thus, a strategy with constant ventilation was tested throughout the museum with 0.2 ach. Then, three levels of comfort were tested, with different fresh airflows per visitor through constant mechanical ventilation during the opening hours. The hypothesis of a mechanical ventilation system with and without heat recovery was tested. Based on published studies on museums, several cases were also tested with air recirculation systems and the intake of fresh air by mechanical ventilation based on CO_2 concentrations, limiting the total ventilation to 1.5 ach.

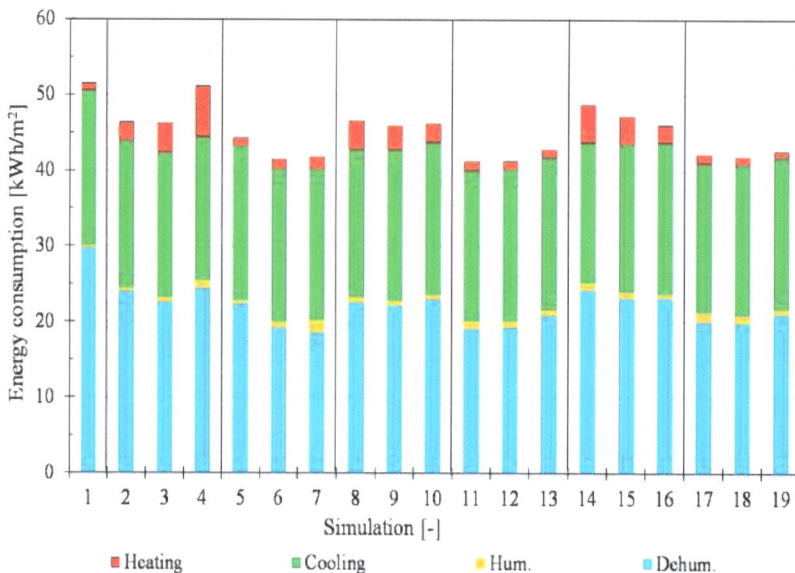

(Fig. 6) contd.....

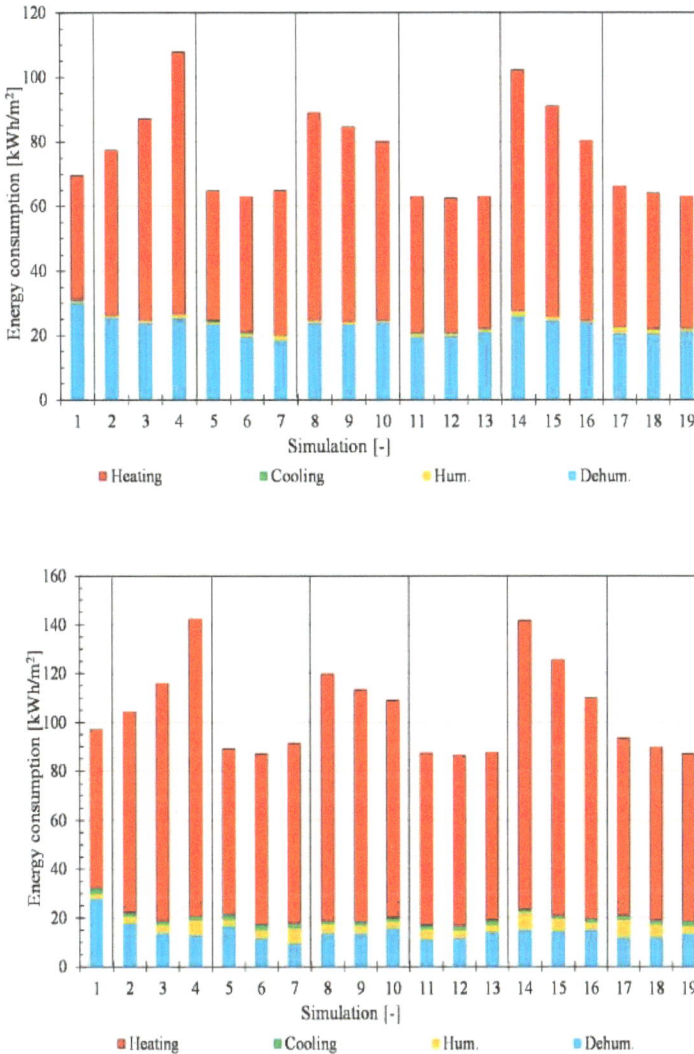

Fig. (6). Energy consumption for the 19 ventilation strategies for Lisbon, Amsterdam and Krakow.

It is important to note that occupancy rates of around 0.4 visitors/m² were used, as argued in ASHRAE 62.1. This occupation can be considered too high for most museums and returns high ventilation rates. In this study, daily ventilation rates between 0.8 a 2.5 ach were obtained, which translates into a difference of 2.3 ach between open hours and night in the most severe case, which can cause an important imbalance in climatic stability. The impact of the ventilation strategies on energy consumption can be seen in Fig. (**6**). As expected, it is possible to distinguish a different behaviour for the three cities under analysis. For the case of

Lisbon, the total energy consumption varies between 41.3 and 51.5 kWh/m². The consumption associated with heating and humidification is practically insignificant when compared to the energy needs for dehumidification and cooling. The highest value is obtained for constant ventilation of 0.2 ach, resulting in great needs for dehumidifying and cooling. The fact that Lisbon is characterized by a more temperate climate than other cities and the production of heat and water vapour in the interior for low ventilation rates are responsible for these data. The strategy incorporating mechanical ventilation with heat recovery on demand for CO_2 concentrations above 1200 ppm (800 + 400 ppm) with recirculation only during the open hours and heat recovery is the most efficient, allowing a 20% reduction compared to the strategy with the highest consumption. In general, it is possible to verify that the best results are obtained when adopting a mechanical ventilation system with heat recovery. The data obtained seem to suggest that the use of night ventilation for cooling can be an important strategy for climates in southern Europe.

For Amsterdam and Krakow, the consumption panorama is different. In Amsterdam, total consumptions between 62.5 and 108 kWh/m² were achieved, where heating needs take the main role, followed by the dehumidification needs. For Krakow, total energy consumptions between 142.3 and 86.3 kWh/m² were found.

If in the case of Lisbon, the increase in ventilation to certain levels made it possible to balance the consumptions, the same is not true for Amsterdam and Krakow. Increased ventilation always causes an increase in energy consumption. The use of mechanical ventilation with heat recovery allows the best results, with the various strategies reaching very close values. Thus, it is concluded that the ventilation strategy should not be defined essentially based on energy, since the use of heat recovery systems allows to balance consumptions.

As regards to the concentration of CO_2, the analysis of Fig. (**7**) allows verifying the ineffectiveness of natural ventilation for buildings with high occupancy, reaching a maximum concentration of about 8 000 ppm, which in addition to the comfort problems, exceeds the limits considered safe for humans. Any of the ventilation strategies used in simulations 2 to 7 is valid. The differences obtained are based on the adopted approaches, which are valid concerning to indoor air quality and health.

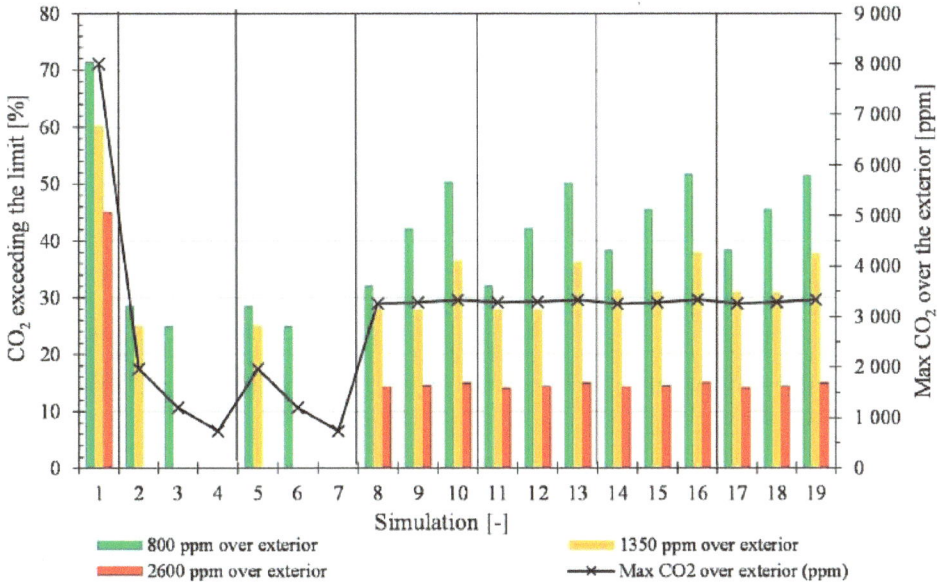

Fig. (7). Indoor air quality and maximum CO_2 concentrations according to the ventilation strategies.

For the remaining ventilation strategies, despite being based on studies developed in museums, they were found to be ineffective in controlling CO_2 concentration, since the intake of fresh air was limited to 1.5 ach. The use of these approaches seems appropriate for service buildings in general and museums in particular, where recirculation contributes to a redistribution of indoor air, however, the number of visitors should be reduced.

By rearranging the mass balance equation presented in equation 3, the occupancy rate per unit area can be estimated:

$$\frac{n}{A} = \frac{h \cdot ach \cdot \varepsilon_v \cdot (C_{CO2,i} - C_{CO2,e})}{G_{CO2} \cdot 10^6} \qquad (14)$$

Where n the number of visitors (-), G_{CO2} the emission rate of CO_2 (m³/h), $C_{CO2,i}$ the allowed indoor concentration of CO_2 (ppm), $C_{CO2,e}$ the concentration of CO_2 in the exterior (ppm), ε_v the ventilation efficiency (-), A means the area in m² and h is the room height (m).

In addition to comfort, health and energy consumption, museums are also interested in ensuring the stability of the indoor climate for the conservation of collections.

To limit the risk of degradation, setpoints of temperature and relative humidity suitable for this type of building were defined, namely 16-25 °C and 40-60% RH, respectively.

Despite it is considered that within these intervals the risk should be low, it is mandatory to characterize the fluctuations of the indoor climate in time and space. Thus, seasonal fluctuations, short-term fluctuations and typical daily variation for room 1.2 were calculated and the spatial difference concerning hourly and seasonal differences between room 1.2 and 2.6 was analysed. This data can be seen from Table **14** to Table **19** for temperature and relative humidity in the three cities.

Table 14. Climate characterization according to temperature for Lisbon.

Simulation	Annual Mean [°C]	Seasonal Cycle [°C]	Short-Term Fluctuations [°C]	Daily Span [°C]	Spatial Drift [°C]	
					Hourly	Seasonal
1	22.6	-4.4/+2.4	-1.0/+0.9	1.9	1.6	1.5
2	22.0	-4.9/+2.9	-0.9/+0.9	2.0	1.2	1.2
3	21.7	-5.0/+3.3	-1.0/+1.0	2.0	1.1	1.0
4	21.3	-5.0/+3.6	-1.0/+1.0	2.3	0.9	0.8
5	22.5	-4.5/+2.5	-1.0/+0.9	1.9	1.5	1.5
6	22.4	-4.6/+2.6	-0.9/+0.9	1.9	1.5	1.4
7	22.3	-4.7/+2.7	-0.9/+0.9	1.9	1.4	1.4
8	21.6	-5.1/+3.4	-1.0/+1.0	1.5	1.9	0.9
9	21.7	-5.0/+3.3	-0.9/+1.0	1.5	1.0	1.0
10	21.9	-5.0/+3.1	-0.9/+1.0	1.5	1.0	1.0
11	22.2	-4.8/+2.8	-0.9/+0.9	1.6	1.1	1.1
12	22.3	-4.8/2.7	-0.9/+0.9	1.6	1.2	1.2
13	22.3	-4.7/+2.7	-0.9/+0.9	1.6	1.2	1.2
14	21.4	-5.0/+3.6	-1.1/+1.0	1.7	0.7	0.6
15	21.5	-5.0/+3.6	-1.0/+1.0	1.6	0.7	0.7
16	21.7	-5.0/+3.2	-1.0/+1.0	1.5	0.8	0.8
17	22.1	-4.9/+2.9	-0.9/+0.9	1.7	0.9	0.9
18	22.1	-4.9/+2.9	-0.9/+0.9	1.7	0.9	0.9
19	22.2	-4.8/+2.8	-0.9/+0.9	1.7	0.9	0.9

Temperature plays a less prominent role concerning conservation, which raises the possibility of using less demanding setpoints than those historically used. For Lisbon, the seasonal cycle varies around the annual average with from –5.5 to 3.6 °C. Concerning short-term fluctuations, they reach maximum values of ± 1°C around the seasonal cycle. Typical daily cycles vary between 1.5 and 2.3 °C, and the spatial variation also shows low fluctuations. Despite the fact that the recirculation reduces the shortest fluctuations, these differences are not enough to influence decision making. These data can be found in Table **14**. For Amsterdam (Table **16**) and Krakow (Table **18**), the annual averages are lower than those obtained for Lisbon, but the magnitude of fluctuations are similar.

Relative humidity cycles have a more severe impact on the conservation of collections and should be observed more carefully. It is possible to verify that there are seasonal fluctuations in all cases, which implies the use of a short-term fluctuation interval of ± 5% RH around the seasonal cycle according to the ASHRAE classification. For Lisbon (Table **15**), it appears that short-term fluctuations always exceed this limit. Fluctuations above the seasonal cycle, for example, are always greater than 11% RH. As regard to daily cycles, it is possible to find several cases in which the relative humidity fluctuates between the minimum and maximum limits imposed by the air conditioning system. In turn, the spatial dispersion presents values within the limits. For Amsterdam (Table **17**) and Krakow (Table **19**), the same trend was observed.

Table 15. Climate characterization according to relative humidity for Lisbon.

Simulation	Annual Mean [RH]	Seasonal Cycle [%RH]	Short-term Fluctuations [%RH]	Daily Span [%RH]	Spatial Drift [%RH]	
					Hourly	Seasonal
1	54.4	-4.4/+3.3	-6.6/+12.1	20.0	3.1	1.7
2	53.9	-5.2/+3.8	-6.8/+11.7	20.0	3.1	1.6
3	53.7	-6.5/+4.0	-7.2/+11.5	19.1	2.5	1.5
4	53.4	-6.8/+4.1	-7.4/+11.4	16.9	1.9	1.3
5	52.4	-5.7/+3.9	-7.4/+11.7	20.0	3.8	2.1
6	52.6	-7.1/+4.4	-8.1/+11.4	18.8	3.8	2.3
7	51.8	-7.8/+5.0	-8.8/+11.0	16.9	3.8	2.5
8	53.5	-6.6/+4.2	-7.3/+11.5	18.1	2.5	1.4
9	53.5	-6.5/+4.2	-7.3/+11.5	18.2	2.5	1.5
10	53.7	-6.4/+4.0	-7.1/+11.6	20.0	2.5	1.6
11	52.4	-6.9/+4.6	-8.1/+11.3	17.6	3.1	2.0
12	52.7	-6.7/+4.4	-7.9/+11.4	18.0	3.1	1.9

(Table 15) cont.....

13	53.0	-6.8/+4.2	-7.6/+11.5	20.0	3.1	1.9
14	53.7	-6.3/+3.9	-7.1/+11.5	18.4	1.7	1.1
15	53.6	-6.3/+4.0	-7.2/+11.5	18.8	1.9	1.2
16	53.8	-6.4/+3.8	-7.0/+11.6	20.0	2.1	1.4
17	52.5	-6.7/+4.5	-7.9/+11.4	18.3	2.4	1.7
18	52.7	-6.5/+4.4	-7.8/+11.5	18.3	2.4	1.7
19	53.1	-6.5/+4.1	-7.4/+11.5	20.0	2.5	1.8

Table 16. Climate characterization according to temperature for Amsterdam.

Simulation	Annual Mean [°C]	Seasonal Cycle [°C]	Short-Term Fluctuations [°C]	Daily Span [°C]	Spatial Drift [°C]	
					Hourly	Seasonal
1	19.1	-2.9/+5.2	-1.2/+1.0	1.9	1.5	1.4
2	18.5	-2.5/+5.3	-1.1/+1.1	1.8	1.2	1.1
3	18.2	-2.2/+5.1	-1.1/+1.1	1.8	1.0	1.0
4	18.0	-2.0/+4.8	-1.2/+1.1	1.9	0.8	0.8
5	19.0	-2.8/+5.2	-1.2/+1.0	1.8	1.4	1.4
6	18.9	-2.8/+5.3	-1.1/+1.0	1.8	1.4	1.3
7	18.8	-2.7/+5.3	-1.1/+1.0	1.8	1.3	1.2
8	18.1	-2.1/+5.1	-1.1/+1.0	1.6	0.9	0.8
9	18.2	-2.2/+5.1	-1.1/+1.0	1.6	0.9	0.9
10	18.3	-2.3/+5.2	-1.1/+1.0	1.6	1.0	0.9
11	18.7	-2.6/+5.3	-1.1/+1.0	1.7	1.1	1.0
12	18.7	-2.6/+5.3	-1.1/+1.0	1.7	1.1	1.0
13	18.8	-2.6/+5.3	-1.1/+1.0	1.7	1.1	1.0
14	17.9	-1.9/+4.8	-1.1/+1.0	1.8	0.7	0.6
15	18.1	-2.1/+5.0	-1.1/+1.0	1.6	0.7	0.6
16	18.3	-2.2/+5.1	-1.1/+1.0	1.6	0.8	0.7
17	18.6	-2.5/+5.3	-1.2/+1.0	1.7	0.9	0.8
18	18.6	-2.5/+5.3	-1.1/+1.0	1.7	0.9	0.8
19	18.7	-2.5/+5.3	-1.1/+1.0	1.7	0.9	0.8

Table 17. Climate characterization according to relative humidity for Amsterdam.

Simulation	Annual Mean [RH]	Seasonal Cycle [%RH]	Short-Term Fluctuations [%RH]	Daily Span [%RH]	Spatial Drift [%RH]	
					Hourly	Seasonal
1	53.8	-4.0/+4.0	-8.6/+11.5	20.0	2.5	1.4
2	53.2	-4.9/+5.1	-10.0/+10.3	20.0	2.5	1.3

(Table 17) cont.....

3	53.0	-5.6/+5.6	-10.1/+10.1	20.0	2.2	1.2
4	52.5	-6.6/+6.3	-8.9/+9.8	20.0	1.9	1.0
5	52.8	-4.4/+4.7	-9.9/+10.5	20.0	3.1	1.6
6	52.2	-4.8/+5.3	-9.7/+10.7	20.0	3.1	1.8
7	51.3	-5.5/+6.1	-9.1/+10.3	19.4	3.1	2.0
8	52.8	-5.7/+5.8	-9.9/+10.2	20.0	1.9	1.1
9	52.9	-5.4/+5.6	-10.1/+10.4	20.0	2.1	1.2
10	53.1	-5.3/+5.4	-9.9/+10.4	20.0	2.5	1.2
11	52.0	-5.1/+5.6	-9.6/+10.7	20.0	2.5	1.5
12	52.2	-4.9/+5.4	-9.7/+10.7	20.0	2.5	1.5
13	52.5	-4.9/+5.2	-9.6/+10.7	20.0	2.6	1.6
14	52.9	-6.1/+5.9	-10.0/+9.9	20.0	1.7	0.9
15	52.9	-5.7/+5.6	-10.2/+10.3	20.0	1.9	1.1
16	53.1	-5.3/+5.2	-9.9/+10.3	20.0	2.0	1.2
17	52.0	-5.3/+5.7	-9.6/+105	20.0	2.3	1.4
18	52.2	-5.0/+5.4	-9.8/+10.8	20.0	2.3	1.4
19	52.6	-4.9/+5.0	-9.6/+10.5	20.0	2.3	1.4

Table 18. Climate characterization according to temperature for Krakow.

Simulation	Annual Mean [°C]	Seasonal Cycle [°C]	Short-Term Fluctuations [°C]	Daily Span [°C]	Spatial Drift [°C]	
					Hourly	Seasonal
1	19.1	-3.1/+5.3	-1.0/+0.9	1.9	1.4	1.2
2	18.8	-2.8/+5.5	-1.2/+0.9	1.8	1.1	1.0
3	18.6	-2.6/+5.4	-1.3/+1.0	1.9	1.0	0.8
4	18.4	-2.4/+5.3	-1.5/+1.1	2.1	0.8	0.6
5	19.1	-3.0/+5.3	-1.1/+0.9	1.9	1.4	1.2
6	19.0	-3.0/+5.4	-1.1/+0.9	1.8	1.3	1.2
7	18.9	-2.9/+5.4	-1.1/+0.9	1.8	1.3	1.1
8	18.5	-2.5/+5.4	-1.3/+1.0	1.7	0.9	0.7
9	18.6	-2.6/+5.4	-1.2/+1.0	1.7	0.9	0.7
10	18.6	-2.6/+5.5	-1.2/+1.0	1.7	0.9	0.8
11	18.9	-2.9/+5.5	-1.1/+0.9	1.7	1.1	0.9
12	18.9	-2.9/+5.5	-1.1/+0.9	1.7	1.1	0.9
13	18.9	-2.9/+5.4	-1.1/+0.9	1.7	1.1	1.0
14	18.3	-2.3/+5.2	-1.4/+1.1	1.9	0.7	0.5

(Table 18) cont.....

15	18.4	-2.4/+5.3	-1.4/+1.0	1.8	0.7	0.6
16	18.6	-2.6/+5.5	-1.2/+1.0	1.7	0.8	0.6
17	18.8	-2.8/+5.5	-1.1/+0.9	1.7	0.8	0.7
18	18.8	-2.8/+5.5	-1.1/+0.9	1.7	0.8	0.7
19	18.9	-2.8/+5.5	-1.1/+0.9	1.7	0.9	0.7

Table 19. Climate characterization according to relative humidity for Krakow.

Simulation	Annual Mean [RH]	Seasonal Cycle [%RH]	Short-Term Fluctuations [%RH]	Daily Span [%RH]	Spatial Drift [%RH]	
					Hourly	Seasonal
1	51.8	-5.1/4.5	-12.2/+10.3	20.0	2.5	1.5
2	50.9	-5.7/5.2	-13.7/+10.0	20.0	2.5	1.4
3	50.2	-7.4/6.2	-10.0/+9.7	20.0	2.2	1.3
4	49.2	-8.4/7.7	-7.7/+8.8	17.5	1.9	1.2
5	50.7	-5.4/4.9	-13.6/+10.0	20.0	2.5	1.8
6	49.5	-6.7/5.5	-10.1/+9.6	20.0	3.1	1.9
7	48.1	-7.3/6.5	-8.7/+9.1	16.6	3.1	2.1
8	49.9	-7.5/6.5	-9.6/+9.7	19.8	1.9	1.3
9	50.2	-7.0/6.1	-10.5/+9.8	20.0	1.9	1.3
10	50.3	-6.7/5.8	-10.7/+9.8	20.0	2.2	1.3
11	49.2	-6.8/5.8	-9.3/+9.5	19.4	2.5	1.6
12	49.6	-6.4/5.4	-10.5/+9.6	19.5	2.5	1.6
13	49.9	-6.3/5.3	-10.6/+9.8	20.0	2.5	1.6
14	50.0	-7.7/6.6	-9.7/+9.6	19.9	1.7	1.1
15	50.1	-7.5/6.3	-10.0/+9.8	20.0	1.8	1.2
16	50.3	-6.9/5.7	-10.4/+9.7	20.0	1.9	1.3
17	49.2	-6.8/5.8	-9.4/+9.6	19.5	2.2	1.4
18	49.4	-6.8/5.6	-10.0/+9.8	19.6	2.2	1.4
19	49.8	-6.5/5.4	-10.5/+9.8	20.0	2.2	1.5

This analysis allowed us to conclude that none of the strategies serves the interests of museums concerning conservation. The high fluctuations in relative humidity can be related to several factors. Namely: the high variation of ventilation between the daily and night time; too high ventilation causes high fluctuations; too high occupancy, preventing the ventilation used to be able to remove all excess of water vapour.

This chapter showed the weakness of the science of indoor air quality when applied to museums, concluding that the ventilation rates usually used for the occupancy rate of 0.4 visitors/m^2 are not adequate.

Further research on the topic should be developed to establish an adequate ratio between occupation and ventilation in museums, testing various types of climates.

CONCLUSION

Throughout this chapter, we sought to study the origins of ventilation rates proposed in various international standards, to analyse the impacts of CO_2 on health and comfort and the impact of ventilation on conservation.

Various ventilation approaches have been found. Concerning comfort and health, the concept of adaptation was considered pertinent, however, its effect on the stability of the indoor climate must be analysed.

It was concluded that natural ventilation usually plays an important role and prevails over any other type of ventilation in several museums around the world. It is common to allow the use of intermittent mechanical ventilation activated by certain levels of CO_2, such as the air recirculation to prevent the deposition of pollutants and improve the spatial stability of the climate. In these cases, ventilation rates of up to 1.5 ach and recirculation rates ranging from 5 to 8 ach are usually found.

Based on the simulation study carried out in this chapter, it was concluded that the three cities have considerably different energy needs. For Lisbon, for example, the greatest needs are related to cooling and dehumidification, while for the remaining cases, the heating needs are preponderant. These results seem to indicate the opportunity to use night ventilation for cooling in some cities of southern Europe.

Regarding the impact of ventilation on energy consumption, it was concluded that the use of mechanical systems with heat recovery reduces the differences and allows the use of high fresh airflows without dramatically increasing energy consumption.

As regard to comfort and health, it was concluded that the exclusive use of natural ventilation for the frequent renovation rates in historic buildings may be insufficient and cause serious problems of discomfort and even health. It is also concluded that the ventilation strategy based on what is common in museums, limiting total ventilation to 1.5 ach is insufficient for the occupancy rate of 0.4 visitors/m^2 suggested in the standard ASHRAE 62.1, for example.

Focusing the analysis on the stability of the indoor climate, it was found that none of the simulated strategies provides adequate conditions for the conservation of the collections. The results obtained seem to indicate an inadequate relationship between ventilation and occupation. The solution could be to increase the ratio between ventilation and occupancy but at the same time reducing the number of occupants. The need to develop more research on this topic is evident. The ratio between ventilation and occupation that serves museum conservation needs should be studied.

CONSENT FOR PUBLICATION

Not Applicable.

CONFLICT OF INTEREST

The author confirms that this chapter contents have no conflict of interest.

ACKNOWLEDGEMENT

The study received support from the FCT - Foundation for Science and Technology under the PhD scholarship PD/BD/52654/2014.

REFERENCES

[1]　G. Thomson, *The museum environment.* 2nd ed. Butterworths: London, 1986.

[2]　H.E. Silva, and F.M.A. Henriques, "Hygrothermal analysis of historic buildings: Statistical methodologies and their applicability in temperate climates", *Struct. Surv.,* vol. 34, no. 1, pp. 12-23, 2016.
[http://dx.doi.org/10.1108/SS-07-2015-0030]

[3]　H.E. Silva, and F.M.A. Henriques, "Preventive conservation of historic buildings in temperate climates. The importance of a risk-based analysis on the decision-making process", *Energy Build,* vol. 107, pp. 26-36, 2015.
[http://dx.doi.org/10.1016/j.enbuild.2015.07.067]

[4]　M.J. Martens, Climate Risk Assessment in Museums, PhD Diss., Technische Universiteit Eindhoven, 2012.,

[5]　A. Martinez-Molina, P. Boarin, I. Tort-Ausina, and J-L. Vivancos, "Assessing visitors' thermal comfort in historic museum buildings: Results from a Post-Occupancy Evaluation on a case study", *Build. Environ.,* 2018.
[http://dx.doi.org/10.1016/j.buildenv.2018.02.003]

[6]　E. Schito, and D. Testi, "A visitors' presence model for a museum environment: Description and validation", *Build. Simul.,* vol. 10, pp. 977-987, 2017.
[http://dx.doi.org/10.1007/s12273-017-0372-1]

[7]　D. Camuffo, R. Van Grieken, H-J. Busse, G. Sturaro, A. Valentino, A. Bernardi, N. Blades, D. Shooter, K. Gysels, F. Deutsch, M. Wieser, O. Kim, and U. Ulrych, "Environmental monitoring in four European museums", *Atmos. Environ.,* vol. 35, pp. S127-S140, 2001.
[http://dx.doi.org/10.1016/S1352-2310(01)00088-7]

[8]　K. Gysels, F. Delalieux, F. Deutsch, R. Van Grieken, D. Camuffo, A. Bernardi, G. Sturaro, H-J. Busse,

and M. Wieser, "Indoor environment and conservation in the Royal Museum of Fine Arts, Antwerp, Belgium", *J. Cult. Herit.,* vol. 5, pp. 221-230, 2004.
[http://dx.doi.org/10.1016/j.culher.2004.02.002]

[9]   P. Merello, F-J. García-Diego, P. Beltrán, and C. Scatigno, "High Frequency Data Acquisition System for Modelling the Impact of Visitors on the Thermo-Hygrometric Conditions of Archaeological Sites: A Casa di Diana (Ostia Antica, Italy) Case Study", *Sensors (Basel),* vol. 18, no. 2, p. 348, 2018.
[http://dx.doi.org/10.3390/s18020348] [PMID: 29370142]

[10] C. Bonacina, P. Baggio, F. Cappelletti, P. Romagnoni, and A.G. Stevan, "The Scrovegni Chapel: The results of over 20 years of indoor climate monitoring", *Energy Build.,* vol. 95, pp. 144-152, 2015.
[http://dx.doi.org/10.1016/j.enbuild.2014.12.018]

[11] P. Baggio, C. Bonacina, P. Romagnoni, and A.G. Stevan, "Microclimate Analysis of the Scrovegni Chapel in Padua - Measurements and Simulations", *Stud. Conserv.,* vol. 49, pp. 161-176, 2004.
[http://dx.doi.org/10.1179/sic.2004.49.3.161]

[12] *Indoor environmental input parameters for design and assessment of energy performance of buildings addressing indoor air quality, thermal environment, lighting and acoustics. EN Standard 15251.* European Committee for Standardization: Brussels, 2007.

[13] CEN; CR Report 1752–1998, "Ventilation for buildings - design criteria for the indoor environment", *European Committee for Standardization,* Brussels, 1998.

[14] *Ventilation for non-residential buildings – performance requirements for ventilation and room-conditioning systems. EN Standard 13779.* European Committee for Standardization: Brussels, 2007.

[15] *Ventilation for acceptable indoor air quality. ANSI/ASHRAE Standard 62.1–2013.* American Society of Heating Ventilating and Air Conditioning Engineers: Atlanta, 2013.

[16] *Environmental design.* CIBSE Guide A. Chartered Institution of Building Services Engineers: London, 2006.

[17] E. Lucchi, "Review of preventive conservation in museum buildings", *J. Cult. Herit.,* vol. 29, pp. 180-193, 2018.
[http://dx.doi.org/10.1016/j.culher.2017.09.003]

[18] *Conservation of cultural heritage – Indoor climate – Part 2: Ventilation management for the protection of cultural heritage buildings and collections. EN Standard 15759-2.* European Committee for Standardization: Brussels, 2018.

[19] P.O. Fanger, Chapter 22, Perceived Air Quality and Ventilation Requirements.*Indoor Air Quality Handbook* JD Spengler, JM Samet, and JF McCarthy, 2001, pp. 22-1.

[20] *Building environment design - indoor air quality - methods of expressing the quality of indoor air for human occupancy. ISO Standard 16814.* International Organization for Standardization: Geneva, 2008.

[21] J.D. Spengler, J.M. Samet, J.F. McCarthy, Ed., *Indoor Air Quality Handbook.* McGraw-Hill Book Co: New York, NY, 2000.

[22] B. Berg-Munch, G. Clausen, and P.O. Fanger, "Ventilation requirements for the control of body odor in spaces occupied by women", *Environ. Int.,* vol. 12, pp. 195-199, 1986.
[http://dx.doi.org/10.1016/0160-4120(86)90030-9]

[23] P.O. Fanger, and B. Berg-Munch, "Ventilation requirements for the control of body odor", *Proceedings of Engineering Foundation Conference on Management of Atmospheres in Tightly Enclosed Space,* American Society of Heating Refrigerating and Air Conditioning Engineers, pp. 45-60, 1983.

[24] W.S. Cain, B.P. Leaderer, R. Isseroff, L.G. Berglund, R.J. Huey, E.D. Lipsitt, and D. Perlman, "Ventilation requirements in buildings—I", *Control of occupancy odor and tobacco smoke odor, Atmospheric Environment,* vol. 17, no. 1983, pp. 1183-1197, 1967.

[25]   G. Iwashita, K. Kimura, and S. Tanabe, "Indoor air quality assessment based on human olfactory sensation, Journal of Architecture", *Planning and Environmental Engineering.,* vol. 410, pp. 9-19, 1990.

[26]   *Standard guide for using carbon dioxide concentrations to evaluate indoor air quality and ventilation. D 6245.* American Society for Testing and Materials: West Conshohocken, USA, 2012.

[27]   *Energy performance of buildings - Ventilation for buildings - Part 1: Indoor environmental input parameters for design and assessment of energy performance of buildings addressing indoor air quality, thermal environment, lighting and acoustics - Module M1-6. EN Standard 16798-1.* European Committee for Standardization: Brussels, 2019.

[28]   P.O. Fanger, "Introduction of the olf and the decipol units to quantify air pollution perceived by humans indoors and outdoors", *Energy Build.,* vol. 12, pp. 1-6, 1988.
[http://dx.doi.org/10.1016/0378-7788(88)90051-5]

[29]   L. Gunnarsen, and P. Ole Fanger, "Adaptation to indoor air pollution", *Environ. Int.,* vol. 18, pp. 43-54, 1992.
[http://dx.doi.org/10.1016/0160-4120(92)90209-M]

[30]   *Ergonomics of the thermal environment – Analytical determination and interpretation of thermal comfort using calculation of the PMV and PPD indices and local thermal comfort criteria. EN ISO Standard 7730.* European Committee for Standardization: Brussels, 2006.

[31]   *Conservation of Cultural Property – Specifications for temperature and relative humidity to limit climate-induced mechanical damage in organic hygroscopic materials. EN Standard 15757.* European Committee for Standardization: Brussels, 2010.

[32]   Environmental Health Directorate, "Exposure guidelines for residential indoor air quality", *A report of the federal–provincial advisory committee on environmental and occupational health,* Ottawa, Canada, 1989.

[33]   X. Zhang, P. Wargocki, and Z. Lian, "Physiological responses during exposure to carbon dioxide and bioeffluents at levels typically occurring indoors", *Indoor Air,* vol. 27, no. 1, pp. 65-77, 2017.
[http://dx.doi.org/10.1111/ina.12286] [PMID: 26865538]

[34]   ACGIH; Cincinnati OH, "Documentation of the Threshold Limit Values and Biological Exposure Indices", *American Conference of Governmental Industrial Hygienists,* 2011.

[35]   Maximum Concentrations at the Workplace and Biological Tolerance Values for Working Materials, *Commission for the Investigation of Health Hazards of Chemical Compounds in the Work Area,* Federal Republic of Germany, 2000.

[36]   U.S. Department of Labor, "Occupational Safety and Health Administration", *Code of Federal Regulations,* 1910.www.osha.gov

[37]   NIOSH, "NIOSH Pocket Guide to Chemical Hazards (NPG)", *National Institute for Occupational Safety and Health,* 2004. www.cdc.gov/niosh/npg/npg.html

[38]   P.F. Pereira, and N.M.M. Ramos, "The impact of mechanical ventilation operation strategies on indoor CO2 concentration and air exchange rates in residential buildings", *Indoor and Built Environment,* 2020.

[39]   K. Azuma, N. Kagi, U. Yanagi, and H. Osawa, "Effects of low-level inhalation exposure to carbon dioxide in indoor environments: A short review on human health and psychomotor performance", *Environ. Int.,* vol. 121, no. Pt 1, pp. 51-56, 2018.
[http://dx.doi.org/10.1016/j.envint.2018.08.059] [PMID: 30172928]

[40]   P.F. Pereira, N.M.M. Ramos, and A. Ferreira, "Room-scale analysis of spatial and human factors affecting indoor environmental quality in Porto residential flats", *Build. Environ.,* vol. 186, p. 107376, 2020.
[http://dx.doi.org/10.1016/j.buildenv.2020.107376]

[41] J. Kim, M. Kong, T. Hong, K. Jeong, and M. Lee, "Physiological response of building occupants based on their activity and the indoor environmental quality condition changes", *Build. Environ.,* vol. 145, pp. 96-103, 2018.
[http://dx.doi.org/10.1016/j.buildenv.2018.09.018]

[42] U. Satish, M.J. Mendell, K. Shekhar, T. Hotchi, D. Sullivan, S. Streufert, and W.J. Fisk, "Is CO_2 an indoor pollutant? Direct effects of low-to-moderate CO2 concentrations on human decision-making performance", *Environ. Health Perspect.,* vol. 120, no. 12, pp. 1671-1677, 2012.
[http://dx.doi.org/10.1289/ehp.1104789] [PMID: 23008272]

[43] T. Vehviläinen, H. Lindholm, H. Rintamäki, R. Pääkkönen, A. Hirvonen, O. Niemi, and J. Vinha, "High indoor CO_2 concentrations in an office environment increases the transcutaneous CO_2 level and sleepiness during cognitive work", *J. Occup. Environ. Hyg.,* vol. 13, no. 1, pp. 19-29, 2016.
[http://dx.doi.org/10.1080/15459624.2015.1076160] [PMID: 26273786]

[44] H. Maula, V. Hongisto, V. Naatula, A. Haapakangas, and H. Koskela, "The effect of low ventilation rate with elevated bioeffluent concentration on work performance, perceived indoor air quality, and health symptoms", *Indoor Air,* vol. 27, no. 6, pp. 1141-1153, 2017.
[http://dx.doi.org/10.1111/ina.12387] [PMID: 28378908]

[45] X. Zhang, P. Wargocki, Z. Lian, and C. Thyregod, "Effects of exposure to carbon dioxide and bioeffluents on perceived air quality, self-assessed acute health symptoms, and cognitive performance", *Indoor Air,* vol. 27, no. 1, pp. 47-64, 2017.
[http://dx.doi.org/10.1111/ina.12284] [PMID: 26825447]

[46] E. Avrami, *The conservation assessment: a proposed model for evaluating museum environmental management needs.* The Getty ConservationInstitute: Los Angeles, 1999.

[47] *Air quality criteria for storage of paper-based records. NBSIR 83-2795.* National Institute of Standards and Technology: Gaitherburg, MD, 1983.

[48] M. Cassar, N. Blades, and T. Oreszczyn, *Air pollutant levels in air-conditioned and naturally ventilated museums: a pilot study,* 1999.https://discovery.ucl.ac.uk/id/eprint/2274/

[49] D. Camuffo, P. Brimblecombe, R. Van Grieken, H-J. Busse, G. Sturaro, A. Valentino, A. Bernardi, N. Blades, D. Shooter, L. De Bock, K. Gysels, M. Wieser, and O. Kim, "Indoor air quality at the Correr Museum, Venice, Italy", *Sci. Total Environ.,* vol. 236, no. 1-3, pp. 135-152, 1999.
[http://dx.doi.org/10.1016/S0048-9697(99)00262-4] [PMID: 10535149]

[50] J.R. Druzik, and G.R. Cass, A new look at soiling of contemporary paintings by soot in art museums.*The Indoor Air Quality Meeting for Museums Conference Report.* Oxford Brookes University, 2000.

[51] P. Mazzei, "HVAC systems to control microclimate in the museums", *Proc. of the 2nd Mediterranean Congress of Climatization, Climamed,* 2005

[52] American Society of Heating Refrigerating and Air-Conditioning Engineers (ASHRAE), Chapter 9 - Thermal comfort.*ASHRAE Handbook - Fundamentals,* 2013, pp. 173-204.

[53] A.A. Zorpas, and A. Skouroupatis, "Indoor air quality evaluation of two museums in a subtropical climate conditions", *Sustainable Cities and Society.,* vol. 20, pp. 52-60, 2016.
[http://dx.doi.org/10.1016/j.scs.2015.10.002]

[54] R.P. Kramer, A.W.M. van Schijndel, and H.L. Schellen, "The importance of integrally simulating the building, HVAC and control systems, and occupants' impact for energy predictions of buildings including temperature and humidity control: validated case study museum Hermitage Amsterdam", *J. Build. Perform. Simul.,* vol. 10, pp. 272-293, 2017.
[http://dx.doi.org/10.1080/19401493.2016.1221996]

[55] Fraunhofer Institue for Building Physics, WUFI®Plus - Version 3.1.1.0,

[56] H.M. Künzel, "Simultaneous heat and moisture transport in building components", *PhD Diss.,*

University of Stuttgart, 1994.

[57]   A. Holm, H.M. Künzel, and K. Sedlbauer, "The Hygrothermal Behaviour of Rooms : Combining Thermal Building Simulation and Hygrothermal Envelope Calculation", *Eighth International IBPSA Conference,* Eindhoven, The Netherlands, pp. 499-506, 2003.

[58]   F. Antretter, F. Sauer, and T. Schöpfer, *Validation of a hygrothermal whole building simulation software, in: 12th Conference of International Building Performance Simulation Association.*Sydney, Australia, 2011, pp. 1694-1701. http://www.ibpsa.org/proceedings/bs2011/p 1554.pdf

[59]   G.B.A. Coelho, H.E. Silva, and F.M.A. Henriques, "Calibrated hygrothermal simulation models for historical buildings", *Build. Environ.,* vol. 142, pp. 439-450, 2018. [http://dx.doi.org/10.1016/j.buildenv.2018.06.034]

[60]   F. Antretter, T. Schöpfer, and N.M. Kilian, "An approach to assess future climate change effects on indoor climate of a historic stone church", *9ᵗʰ Nordic Symposium on Building Physics,* pp. 600-607, 2011.Tampere, Finland

[61]   J. Radon, F. Antretter, A. Sadlowska, M. Lukimski, and L. Bratasz, "Simulation of energy consumption for dehumidification with cooling in National Museum in Krakow", *3ʳᵈ European Workshop on Cultural Heritage Preservation,* EWCHP, Bolzano, Italy, 2013.

[62]   H. E. Silva, "Indoor climate management on cultural heritage buildings: Climate control strategies, cultural heritage management and hygrothermal rehabilitation", *PhD thesis. Universidade Nova de Lisboa,* Lisboa, Portugal, 2019.

[63]   A.I.C. Environmental Guidelines, *Museum Climate in a Changing World,* 2013. http://www.conservation-wiki.com/wiki/Environmental_Guidelines

[64]   ICOM-CC, *Environmental Guidelines ICOM-CC and IIC Declaration.*.http://www.icom-cc.org/332--icom-cc-documents/declaration-on-environmental-guidelines/#.XKoNLlVKipo

[65]   https://www.nationalmuseums.org.uk/what-we-do/contributing-sector/environmental-conditions/

[66]   J. Bickersteth, "IIC and ICOM-CC 2014 Declaration on environmental guidelines", *Stud. Conserv.,* vol. 61, pp. 12-17, 2016. [http://dx.doi.org/10.1080/00393630.2016.1166018]

[67]   C.A.P. dos Santos, and R. Rodrigues, *Thermal transmittance of building envelope opaque elements.,* 2012.

[68]   C.A.P. dos Santos, and L. Matias, *U-value of building envelope elements (in Portuguese) - ITE 50,* 20ᵗʰ ed. LNEC, Lisbon, Portugal, 2014.

[69]   H.E. Silva, G.B.A. Coelho, and F.M.A. Henriques, "Climate monitoring in World Heritage List buildings with low-cost data loggers: The case of the Jerónimos Monastery in Lisbon (Portugal)", *J. Build. Eng.,* vol. 28, p. 101029, 2020. [http://dx.doi.org/10.1016/j.jobe.2019.101029]

[70]   *ASHRAE 55–2013: Thermal Environmental Conditions for Human Occupancy.* ASHRAE: Atlanta, 2013.

[71]   Refrigeration and air-conditioning engineers, Fundamentals.*ASHRAE Handbook.,* M.S. Owen, Ed., ASHRAE Inc.: Atlanta, 2013.

[72]   *Carrier Air Conditioning Company; Handbook of Air Conditioning System Design.* 2nd ed. McGraw-Hill: New York, 1965.

[73]   ISO 8996, "Ergonomics of the Thermal Environment – Determination of Metabolic Rate", *International Organization for Standardisation,* Geneva, 2004.

[74]   R.P. Kramer, H.L. Schellen, and J.W. van Schijndel, *Towards temperature limits for museums: a building simulation study for four museum zones with different quality of envelopes.* Proc. Healthy Buildings Europe, Kulve M. te: Eindhoven, 2015.

[75]   R. Kramer, L. Schellen, H. Schellen, and B. Kingma, "Improving rational thermal comfort prediction by using subpopulation characteristics: A case study at Hermitage Amsterdam", *Temperature (Austin),* vol. 4, no. 2, pp. 187-197, 2017.
[http://dx.doi.org/10.1080/23328940.2017.1301851] [PMID: 28680934]

[76]   S. Maekawa, C. Carvalho, F. Toledo, and V. Beltran, "Climate controls in a historic house museum in the tropics: a case study of collection care and human comfort", *PLEA2009 – 26ᵗʰ Conference on Passive and Low Energy Architecture,* pp. 22-24, 2009.*Quebec City, Canada,* pp. 22-24, 2009.

[77]   Y.H. Yau, B.T. Chew, and A.Z.A. Saifullah, "A Field Study on Thermal Comfort of Occupants and Acceptable Neutral Temperature at the National Museum in Malaysia", *Indoor Built Environ.,* 2011.
[http://dx.doi.org/10.1177/1420326X11429976]

[78]   M. La Gennusa, G. Lascari, G. Rizzo, and G. Scaccianoce, "Conflicting needs of the thermal indoor environment of museums: In search of a practical compromise", *J. Cult. Herit.,* vol. 9, pp. 125-134, 2008.
[http://dx.doi.org/10.1016/j.culher.2007.08.003]

[79]   H.E. Silva, F.M.A. Henriques, T.A.S. Henriques, and G. Coelho, "A sequential process to assess and optimize the indoor climate in museums", *Build. Environ.,* vol. 104, pp. 21-34, 2016.
[http://dx.doi.org/10.1016/j.buildenv.2016.04.023]

[80]   L. Bellia, A. Capozzoli, P. Mazzei, and F. Minichiello, "A comparison of HVAC systems for artwork conservation", *Int. J. Refrig.,* vol. 30, pp. 1439-1451, 2007.
[http://dx.doi.org/10.1016/j.ijrefrig.2007.03.005]

[81]   F. Ascione, and F. Minichiello, "Microclimatic control in the museum environment: Air diffusion performance", *Int. J. Refrig.,* vol. 33, pp. 806-814, 2010.
[http://dx.doi.org/10.1016/j.ijrefrig.2009.12.017]

[82]   J. Ferdyn-Grygierek, "Indoor environment quality in the museum building and its effect on heating and cooling demand", *Energy Build.,* vol. 85, pp. 32-44, 2014.
[http://dx.doi.org/10.1016/j.enbuild.2014.09.014]

[83]   R.P. Kramer, M.P.E. Maas, M.H.J. Martens, A.W.M. van Schijndel, and H.L. Schellen, "Energy conservation in museums using different setpoint strategies: A case study for a state-of-the-art museum using building simulations", *Appl. Energy,* vol. 158, pp. 446-458, 2015.
[http://dx.doi.org/10.1016/j.apenergy.2015.08.044]

[84]   T. Karyono, E. Sri, J. Sulistiawan, and Y. Triswanti, "Thermal Comfort Studies in Naturally Ventilated Buildings in Jakarta, Indonesia", *Buildings.,* vol. 5, pp. 917-932, 2015.
[http://dx.doi.org/10.3390/buildings5030917]

[85]   C. Balocco, and R. Calzolari, "Natural light design for an ancient building: A case study", *J. Cult. Herit.,* vol. 9, pp. 172-178, 2008.
[http://dx.doi.org/10.1016/j.culher.2007.07.007]

[86]   D. Camuffo, E. Pagan, S. Rissanen, Ł. Bratasz, R. Kozłowski, and M. Camuffo, "An advanced church heating system favourable to artworks: A contribution to European standardisation", *J. Cult. Herit.,* vol. 11, pp. 205-219, 2010.
[http://dx.doi.org/10.1016/j.culher.2009.02.008]

[87]   N. Aste, S.D. Torre, R.S. Adhikari, M. Buzzetti, C. Del Pero, F. Leonforte, and M. Manfren, "Sustainable church heating: The Basilica di Collemaggio case-study", *Energy Build.,* vol. 116, pp. 218-231, 2016.
[http://dx.doi.org/10.1016/j.enbuild.2016.01.008]

[88]   DGPC, *Estudo de Públicos de Museus Nacionais,* 2015.

[89]   U.S. Department of Energy, EnergyPlus Version 8.9.0 Documentation - Engineering Reference, 2018.,

[90]   N. Aste, S.D. Torre, R.S. Adhikari, M. Buzzetti, C. Del Pero, F. Leonforte, and M. Manfren,

"Sustainable church heating: The Basilica di Collemaggio case-study", *Energy Build.,* vol. 116, pp. 218-231, 2016.
[http://dx.doi.org/10.1016/j.enbuild.2016.01.008]

[91]   Refrigeration and air-conditioning engineers, Fundamentals.*ASHRAE Handbook.,* M.S. Owen, Ed., ASHRAE Inc.: Atlanta, 2009.

[92]   N. Aste, S.D. Torre, R.S. Adhikari, M. Buzzetti, C. Del Pero, F. Leonforte, and M. Manfren, "Sustainable church heating: The Basilica di Collemaggio case-study", *Energy Build.,* vol. 116, pp. 218-231, 2016.
[http://dx.doi.org/10.1016/j.enbuild.2016.01.008]

[93]   D. Camuffo, *Microclimate for Cultural Heritage: Conservation, Restoration, and Maintenance of Indoor and Outdoor Monuments.* Elsevier, 2014.

[94]   U.S. Department of Energy, *Input Output Reference,* pp. 1-2657, 2017.

[95]   O.M. Essenwanger, "General Climatology 1C: classification of climates", *Landsberg. World Survey of Climatology,* Edited by H.E, 2001.

[96]   Weather Data, *EnergyPlus,* 2018. https://energyplus.net/weather

SUBJECT INDEX

A

Activities 280, 300, 303
  metabolic 303
  sedentary 280, 300
Activity intensity 300
Adaptation 159, 163, 169, 275, 277, 278, 280,
    281, 303, 316
  cable 159
  cultural 278
Air conditioning systems 257, 280, 289, 312
Airflow 275, 279, 281, 282, 283, 284, 285,
    286, 289, 290, 303, 304, 307
  defined 285
Algorithms 3, 5, 7, 9, 10, 12, 126, 149, 151,
    152, 162, 163, 164, 183, 214, 215, 216,
    224, 225, 232, 243
  clustering 126
  generic 147
  heat balance 232
  hybrid ventilation 243
  learning classification 9
  machine learning 3, 10, 12

B

Binary association measures 191
Black globe thermometer 104, 106
Bluetooth low energy (BLE) 9
Building 1, 2, 5, 6, 18, 39, 40, 115, 138, 139,
    140, 141, 142, 144, 153, 154, 156, 169,
    183, 213, 215, 217, 219, 221, 223, 231,
    232, 233, 269
  accounts 115
  automation systems 5, 6
  information modelling (BIM) 144, 169
  management systems (BMS) 138, 140, 141,
    142, 153, 154, 156
  occupancy data 183
  performance simulation (BPS) 1, 2, 18, 39,
    40, 213, 215, 217, 219, 221, 223, 231,
    232, 233, 269

energy efficiency 138, 139
Building design 4, 5, 13, 17, 23, 39, 233
  actual 233
  occupant-centric 5
Building simulation 39, 40, 63, 140, 144, 213,
    214, 215, 217, 218, 219, 222, 223, 231,
    270
  community 217
  dynamic 63
  initiatives 217
  programme 219, 223, 231
  results 40
  tools 217

C

Calculated employing distinct methods 104
Change point analysis (CPA) 147, 148, 160,
    161, 162, 164, 169
Channel state information (CSI) 11
Circuit breakers 115, 116, 118, 119, 122
  monophasic 115, 116
Climates 2, 4, 62, 208, 215, 217, 236, 237,
    267, 270, 289, 305, 309, 316
  museum 305
  outdoor 289
Coefficients 103, 104, 155, 184, 191, 192,
    200, 239, 300
  air mass flow 239
  convective heat transfer 103
  correlations 155
  discharge 239
  globe emissivity 104
Comparison of qualitative methods 20
Complete 187
  connection method 187
  link method 187
Computational learning methods 105, 131
Condution transfer function (CTF) 232
Constant mechanical ventilation 305
Consumers 73, 74, 75, 76, 79, 80, 83, 92, 93,
    95, 96, 97

building users and energy 74, 93, 95, 97
Consumption 4, 20, 72, 73, 107, 118, 120,
        124, 141, 179, 180, 181, 222, 299, 300,
        302, 309
    building energy 4, 20, 181
    high energy 288
    influencing energy 179
    lower energy 302
    measuring energy 141
    oxygen 299, 300
    predicting energy 124
    primary energy 73
    real energy 179, 180
    residential energy 72
Contingent valuation method 80, 96
CPA methodology 161
Cultural heritage buildings 275, 284, 286, 289,
        300
Cyber-Physical-Social Systems 25, 26

D

Data acquisition 106, 115, 118, 158
    procedure 106
    strategy 158
    system 115, 118
Data mining (DM) 144, 145, 146, 150
Digital twins (DT) 144, 145, 146, 156, 169
DNAS 5, 140
    framework 5
    ontology 140
Dwellings 44, 45, 47, 48, 49, 50, 51, 52, 139,
        140, 156, 189, 213, 221, 222, 225, 226,
        270
    low-energy 45
    low-income 189, 213, 270

E

Economic growth 75, 214
    global 75
Effects, macroeconomic 76, 77
Efficiency 73, 75, 77, 304
    material 77

Efficiency measures 75, 77, 218, 219
    implementing energy 77, 218
    promoting energy 75
Electric appliances 213, 215, 218, 221, 223,
        225, 226, 230, 231
    possessions 215
Electricity consumption 101, 113, 114, 115,
        119, 225, 230, 268
Electrocardiogram 10
Electronic 18, 38, 46, 55, 56
    data collection 18
    measurement 38
    measuring devices methods 46, 55, 56
Emissions 73, 77, 80, 280, 282, 284, 285
    associated carbon 80
    greenhouse gas 77
Energy 10, 73, 74, 77, 78, 93, 95, 97, 101,
        102, 131, 300
    consumers 74, 77, 93, 95, 97
    electric 101
    electrical 300
    harvest 10
    nuclear 73
    renewable 78, 102, 131
Energy consumption 38, 39, 40, 41, 132, 139,
        156, 178, 179, 180, 182, 218, 221, 290,
        291, 306, 308
    forecasting models 132
    impacts 139
Energy-efficiency 73, 74, 75, 92, 93, 95, 97
    boosting 92
    programs 97
Energy in buildings and communities (EBC) 2
Energy management system 6, 12, 112, 219,
        232, 238
Energy performance analyses 182
Energy performance of buildings 78, 93, 138
    and energy efficiency 138
    directive (EPBD) 78, 93
EnergyPlus 213, 238, 246, 249, 298
    programme 213, 238, 246, 249
    software 298
Ethernet-connected device 108
Evaluation 2, 4, 15, 16, 22, 24, 62, 82, 146,
        194, 215

considering questionnaire-based 24
large-scale longitudinal 15
longitudinal survey-based 24

F

Federal Government 233
Fenestration surfaces 109
Fieldbus network 145
Five-factor model (FFM) 6
Floor 59, 115, 118, 141, 157, 158, 232, 234, 235, 236, 292
  basement 292
Fluorescent lamp 301, 302
Forecasting error 112, 126, 128
Function 106, 124, 143, 154, 155, 169, 225, 248
  activation 106
  monotonic 155
  radial-type 106

G

Gambelas house data 126
Gathering contemporaneous 18
Gaussian 106, 149
  distribution 149
  functions 106
Global 111, 121, 238, 239, 243, 248, 249, 250, 251, 252, 258, 259, 260, 261, 306
  sensitivity analysis (GSA) 238, 239, 243, 248, 249, 250, 251, 252, 258, 259, 260, 261, 306
  solar radiation 111, 121, 306
Google 79, 159
  maps 159
  Scholar 79
Gross domestic product (GDP) 77
Ground 232
  reflectance 232
  temperatures, initial 232

H

HARP consortium 79
Health 9, 76, 77, 80, 155, 276, 277, 284, 285, 286, 287, 288, 291, 303, 307, 309, 310, 316
  care 9
  conditions 80, 155
  effects 76
  hazard 303
  problems 287
  public 77
  risks 284, 285, 286
  symptoms 288
Heat-balance equation 103
Heating 57, 72, 73, 74, 75, 79, 80, 82, 92, 93, 298
  and cooling strategy 73
  equipment 80
  stock 72
  strategies 298
  systems 57, 72, 74, 75, 79, 82, 92, 93, 298
Heating appliances 74, 75, 79, 96
  retrofit planning 75
Hidden Markov models (HMM) 147
Hierarchical analysis 205
Historic buildings 276, 303, 307, 316
Home energy management systems (HEMS) 102, 115, 124, 131, 132
Homeostasis 287
Housing 17, 76, 157, 181, 188, 189, 233
  low-income 181, 188, 189
  social 233
Human health 76, 286, 288
  mental 76
HVAC 102, 111, 245
  cooling thermostat 245
  control and home energy efficiency 102
  energy consumption 111
Hybrid 245, 266
  analysis 266
  ventilation function 245
Hydrodynamic model 145

I

Indices 125, 213, 246, 249, 267
 calculating sensitivity 249
Indoor 2, 3, 5, 6, 7, 11, 20, 39, 40, 41, 58, 59, 60, 61, 139, 140
 evaluating carbon dioxide concentration 3
 hinder occupant wellbeing 20
 occupant behaviours 61
 occupant localisation 11
Indoor air quality (IAQ) 12, 38, 275, 276, 278, 279, 280, 284, 285, 286, 288, 289, 290, 291
 monitoring 12
Indoor environment 1, 4, 6, 7, 10, 11, 12, 15, 20, 38, 39, 40, 41, 42, 57, 58, 61, 102, 139, 141, 142, 143, 153, 290
 building's 39
 factors 57, 58
 quality (IEQ) 5, 6, 7, 11, 12, 15, 20, 40, 41, 139, 140, 141, 142, 143, 153
 monitoring systems 102
Indoor humidity 3
Influence 5, 179
 building performance 179
 socioeconomic factors 5
Intelligent 7, 115
 building systems 7
 weather station (IWS) 115
Interdisciplinary survey 5
Intergovernmental panel on climate change (IPCC) 76, 78
Interior 59, 235, 294, 304
 microclimate 304
 Wall 59, 235, 294
International energy agency (IEA) 2, 75, 76, 179, 214, 215
Internet-of-things (IoT) 1, 5, 11, 12, 23, 131, 144
IoT-based 11, 12, 23, 25
 building automation system 25
 frameworks 23
 solutions 11
 systems 11, 12

IoT 11, 12
 frameworks 12
 systems 11

J

Jansen method 248, 250

K

Köppen class temperature 306

L

Lighting 7, 8, 179, 219, 221, 222, 225, 226, 230, 231, 240, 244, 296, 300, 301, 302
 artificial 7, 8
 natural 179
 power 300, 301, 302
 system 179
Logistic regressions 147, 154, 165, 168, 169, 182
Lower energy expenditure 77
Low power radio solutions 109

M

Management 107, 276, 286
 buildings energy 107
 heritage 276, 286
Market, real estate 77, 81
Measure indoor temperature 58
Measurement equipment 221
Mechanical 157, 289, 290, 304, 307, 309, 316
 extract ventilation (MEV) 157
 systems 289, 316
 ventilation 289, 290, 304, 307, 309
Methods 15, 56, 145, 194
 cross-sectional 15
 forecasting 145
 hierarchical 194
 questionnaire-based 56
Model-based predictive control(MBPC) 111

Modelling 145, 146, 182, 215, 216, 270
  ontology 145
  techniques 182
MOGA 126
  design 126
  employing 126
Monitoring system and surveys 158
Multi-objective genetic algorithm (MOGA) 101, 125, 126
Multi-scale simulation 145

N

National 232, 289
  archives building 289
  renewable energy laboratory (NREL) 232
Natural ventilation 221, 222, 223, 238, 239, 242, 243, 245, 246, 248, 249, 250, 251, 252, 256, 265, 267, 268, 269, 270, 285, 304, 307, 309, 316
  condition 252
  function 242
  mode 239, 243, 246, 248, 249, 250, 251, 252, 256, 268, 269, 270
  procedure 285
  purposes 222
Neural networks (NN) 124
Non-dispersive Infrared (NDIR) 58
Non-intrusive load monitoring (NILM) 115
NumOcc variable 241

O

Occupancy 8, 109, 283
  density 283
  detection 8, 109
  indoors 8
Occupant 3, 23, 57, 60
  centric control (OCC) 23
  related factors 57, 60
  related research 3
Operation 2, 4, 5, 20, 21, 122, 123, 132, 138, 156, 157, 180, 182, 190, 195, 197
  occupant windows 165

occurrence of 190, 197
Outdoor 57, 289
  air exchanges 289
  environmental factors 57

P

Passive infrared (PIR) 8, 109, 141
Pearson correlation coefficient (PCC) 117, 119, 154, 155, 156
Post-occupancy evaluations (POE) 19, 20, 22, 24, 25
Power 107, 116, 122, 123, 156, 218, 229, 231, 301, 302
  consumption 218
  electric 122, 302
  supply 156
Process 17, 19, 75, 82, 93, 147, 151, 184, 185, 188, 194, 215, 220
  building renovation 93
  clustering 194
Programmes 2, 77, 189, 215, 216, 231, 232, 238
  computer 238
  energy efficiency retrofit 77
  public housing 189

Q

Quadratic models 41
Qualitative methods 1, 4, 13, 14, 20, 21, 24, 25, 26

R

Radial-basis function (RBFs) 106, 124
Radiant fraction 239, 240, 243, 244
  of luminaires 240, 244
  of occupants 239, 243
Radiant heat transfer 104
Relative humidity (RH) 109, 111, 141, 158, 159, 236, 286, 290, 291, 306, 307, 311, 312, 313, 315
  cycles 312

fluctuates 312
  sensors 141, 158
Relative scaling methods 78
Responses 15, 16, 17, 18, 84, 85, 86, 88, 89,
    90, 91, 92, 93, 95, 96, 276, 288
  cognitive 288
  hygrothermal 276
Room floor area 247

S

Sectors 73, 101, 115, 214, 217
  engineering 217
  residential 115
Seed points 188, 194
Self-powered wireless sensor (SPWS) 108,
    109, 115
Sensations 13, 25, 246
  real-world 13
Sensible heat 296, 297, 298, 299
  amount of 299
  emitted 296
  fraction 296
Sensitivity analysis methods 248
Sensors 7, 8, 9, 10, 11, 43, 57, 58, 59, 61, 107,
    108, 109, 141, 156, 157, 158, 182
  capacitive 8
  hall effect 8
  laser range 9
  magnetic contact 43
  temperature and relative humidity 141, 158
  ultrasonic 61
  wearable 8, 10, 141
Sequential monitoring instants 150
Services 9, 20, 75, 76, 143, 179, 217, 219
  assistive-living 9
  building energy 179
  electronic office 143
Signals 9, 10, 11, 12, 113, 114, 145, 147, 224
  electroencephalogram 10
  propagation 11
   processing 145, 147
Similarity measure 186, 187, 193
Simple 187

connection method 187
  link method 187
Simulation programmes 4, 195, 214, 215, 216,
    217, 219, 231, 233
  computer 195
  eQuest 217
Simulations 179, 180, 181, 182, 183, 192,
    237, 239, 248, 249, 275, 300, 307, 311,
    312, 313, 314, 315
  computational 192, 275
  hygrothermal 300
Smart Plugs (SP) 115, 116
Social 5, 6, 13
  cognitive theory (SCT) 5
  psychology theories 5, 6
  science approaches 13
Software 157, 161, 164, 291
Solar and physical energy 182
Solar energy systems 217
Solar radiation 57, 121, 158, 169, 236, 237,
    294, 306
  annual global 306
  diffuse 236
Spearman's rank correlation coefficient
    (SRCC) 155, 164, 165, 168, 169
Statistical 184, 223, 248, 268
  techniques 184, 248
  tests 223, 268
Storage 115, 301
  area 301
  capacity 115
Strategies 5, 8, 19, 73, 179, 214, 217, 270,
    276, 307, 309, 315
  adopted 276
  machine-learning-based 8
  native deterministic input 217
Subjects 22, 42, 47, 52, 74, 76, 92, 190, 216,
    220, 222, 225
  energy-related 74
Support building design 40
Sustainability 275, 276, 281
  economic 281
  financial 275
  heritage 276
Sustainable energy security package 73

Systems 4, 5, 6, 7, 8, 25, 26, 60, 63, 74, 102, 132, 138, 139, 140, 141, 143, 144, 180, 182, 215, 287, 290, 304, 309
  air-conditioning 180, 182
  air treatment 290
  automatic window control 63
  building management 138, 141
  energy-efficient 74
  heat recovery 304, 309
  intelligent 7
  nervous 287

T

Techniques 9, 43, 146, 147, 148, 154, 156, 160, 169, 183, 184, 185, 187, 188, 193, 204, 248
  analytical 185
  data analysis 183
  effective 148
  exploratory analysis 184
  hierarchical 187, 188, 193, 204
  non-hierarchical clustering 193
  variance-based 248
Technologies, energy-efficient heating 82
Temperature 59, 105, 107, 109, 127, 128, 129, 132, 133, 138, 141, 158, 159, 218, 219, 311, 312, 313, 314
  ambient 127, 128, 132, 133
  atmospheric 129
Theory of planned behaviour (TPB) 3, 5, 6
Thermal 78, 88, 103, 178, 180, 182, 207, 209, 213, 215, 232, 234, 235, 241, 246, 247, 257, 260, 261, 264, 265, 268, 300
  adaptation 246
  calculations 300
  capacity 234, 235
  heat transfer 241
  load 103, 247, 257, 260, 261, 264, 265
Threshold limit values (TLVs) 287
Thresholds 3, 7, 156, 279, 280
  irritation 280
  typical temperature 3
Tools, statistical 138, 147, 154, 216

Total volatile organic compounds (TVOC) 141
Transverse survey 42

U

Universal serial bus (USB) 108

V

Ventilated facade 157
Ventilating and air conditioning 101
Ventilation 157, 213, 232, 238, 246, 249, 257, 262, 269, 275, 277, 278, 284, 285, 286, 289, 304, 306, 307, 308, 309, 310, 316
  binomial 275
  building's 289
  centralised mechanical extract 157
  flows 278, 284, 285
  hybrid 213, 232, 238, 246, 249, 257, 262, 269
  local 286
  management 277
  rate procedure 284
  strategies 277, 304, 306, 307, 308, 309, 310, 316
Ventilation rates 24, 42, 278, 279, 282, 283, 284, 285, 288, 290, 304, 308, 309, 316
  daily 308
  high 288, 308
  low 288, 290, 309
  natural 42
Ventilation system 289, 304, 307
  central 289
  mechanical 307, 309
Virtual representations 12
Visual analysis 144, 146
Volatile organic compounds (VOC) 141, 280

W

Ward's method 193
Water 118, 299
  evaporation enthalpy 299

heater 118
Water vapour 150, 309, 315
  pressure 150
Weather 206, 217, 219, 236, 290
  conditions 290
  predictive 219
Weather station 57, 58, 158
  local 57, 158
  remote 57
Wet bulb temperatures 237
Wi-Fi signals 12
Wilcoxon test 224, 230
  non-parametric 230
Wind 41, 57, 58, 158, 169, 180, 206, 237
  conditions 206
  direction 41, 158, 169, 237
  intensity 58
Window(s) 3, 40, 41, 42, 43, 44, 45, 46, 55,
    56, 61, 62, 157, 218
  dynamic 218
  glass 157
  office 45
  residential 3
  state 41, 42, 43, 44, 46, 55, 56, 61, 62
  usage 40, 42
Windows opening 39, 40, 41, 43, 45, 46, 54,
    55, 56, 62, 161, 162, 163, 164, 165, 166,
    167, 168, 169, 181, 182, 196, 200, 241,
    242, 244
  actions 56
  angle 41, 43, 46, 55, 56, 62
  closing actions 46, 54
  factor 241, 242, 244
  habits 45
Window opening behaviour 38, 40, 41, 42, 43,
    45, 46, 47, 53, 54, 55, 56, 57, 63
  monitor occupant 38, 40
  regarding occupant 40
Wind pressure coefficients 232
Winter 77, 225, 226, 229, 286
  conditions 225, 226, 229
  heating 286
  mortality 77
Wireless 116, 145
  network, technical 116

  sensor network 145
WTP analysis 80

www.ingramcontent.com/pod-product-compliance
Lightning Source LLC
Chambersburg PA
CBHW050807220326
41598CB00006B/141